AF856728

The Political Economy of the Space Program

Pacific Books Science and Technology Management Series

General editors: Terence G. Jackson, Jr., Albert Shapero, Wesley L. Tennant

No. 1 The Political Economy of the Space Program
by Mary A. Holman

No. 2 Adaptations of Scientists and Organizations
by Howard M. Vollmer

The Political Economy of the Space Program

by Mary A. Holman

Pacific Books, Publishers Palo Alto, California

To my father
JONATHAN LANE HOLMAN

Copyright © 1974 by Mary A. Holman.
All rights reserved. No part of this book may be used or reproduced in any manner whatsoever without written permission from the publisher, except in the case of brief quotations embodied in critical articles and reviews.

International Standard Book Number 0–87015–199–1.
Library of Congress Catalog Card Number 79–180902.
Printed and bound in the United States of America.

PACIFIC BOOKS, PUBLISHERS / P.O. BOX 558, PALO ALTO, CALIFORNIA 94302, U.S.A.

Foreword

In looking back over the past decade of space activities, I continue to be surprised and concerned at the lack of both popular and expert understanding of the economic and social consequences of such a structured research and development program. In her usual precise and thoughtful way, Dr. Holman has illuminated this subject and provided a framework for understanding the political and economic consequences of R & D expenditures.

In looking forward, it has always seemed to me self-evident that a vigorous program of directed research and development is essential to the continuation of the growth of our industrial economy. It seems so clear that the affluence we have achieved is the result of the increased productivity of our industrial establishment, which is based upon the creation of new products and new processes made possible by the employment of new technology, new technology that can be developed only by applying new knowledge derived from basic and applied research. One of the outstanding characteristics of space research and development is its ability to contribute directly to each step in this sequence. For it is in the space program alone, with the possible exception of a major war, that every talent and every discipline of a modern industrial society is challenged and forced to produce if the total program is to succeed. It is hard to believe that the obvious present benefits of space research—communication satellites for economical world-wide communications, meteorological satellites for world-wide and long-term weather prediction, microelectronic circuits which have made possible the present computer industry—are not justification in themselves for increasing the resources allocated to space research in the future. It is unfortunate that the present immaturity of our understanding of our economic and social system, along with our inability to describe it analytically, does not permit an unequivocable demonstration of that which I hold to be self-evident. Dr. Holman's book is a beginning toward development of this understanding.

In the more distant future, we can look to the development of a new and radically different space transportation system, which will have as great an effect on our everyday life and economy as did the invention of the airplane. The key to this system is the space shuttle, a fully reusable two-stage vehicle, capable of taking off vertically like a rocket and landing horizontally like an airplane. The first stage, which is about the size and shape of a C-5A, will fly to 200,000 feet, separate from the "orbiter," then re-enter the earth's atmosphere, turn around and fly under its own power back to the take-off field, where it will land horizontally. The second stage or orbiter, which is about the size of a DC-8, will continue into orbit, discharge its passengers and cargo, pick up a return load, re-enter the atmosphere, and land horizontally at the take-off point. This same configuration, without the stop in orbit, can also be used for point-to-point transportation on earth. The development of the first prototype is now beginning, and I fully expect that by the end of the next decade a number of the readers of this book will have flown into and out of space. Consider with me for a moment the possibility of traveling in the space shuttle from, say, London to Los Angeles sometime in the 1980's. You will board the regularly scheduled vertical take-off jitney in downtown London to fly quickly to Heathrow Airport. There, in the center of the field, the Los Angeles-bound shuttle will be erected, being fueled with liquid oxygen and hydrogen from tank trucks alongside. As the fueling operation is completed, your luggage will be put into the baggage pod, and with your flight companions, you will enter the passenger module. The passenger compartment will be swung into place in the orbiter section of the shuttle. The flight crew will be conducting their final check-out of the vehicle, querying their on-board digital computer about the condition and configuration of each system and subsystem.

Passenger seats will swivel so that you will be sitting erect until power is applied, at which time the seats will recline so that the gravity load, not more than three times that of earth, will be more easily accommodated. Your seat will be cushioned and provided with its own shock absorbers to lessen the load.

The shuttle will fly west southwest on a sub-orbital course as you attain maximum speed of approximately 12,000 miles per hour at 100 miles altitude. Seven or 8 minutes after lift-off from London, the booster element will peel off and return for a regular runway landing.

Although there was a huge explosive sound at the airport as the rockets ignited, it lasted only a few seconds, and you will have heard no sound, because it was far below your vehicle. As a matter of fact, your whole trip will be practically noiseless. You will not hear the sound of entry into the atmosphere, but neither will people on the ground, as that entry will be made about 400,000 feet above the earth.

Although the passenger compartment will have no windows, television cameras mounted at strategic positions on the exterior of the orbiter vehicle will bring unobstructed panoramic views of the earth and of the sky into the cabin. As you leave the pad you will see the great thrusts of fire that propel your craft, and the blue of the Atlantic and the glistening clouds will occupy the television screens as you outdistance the sun on your westward journey.

While you are in flight you will, of course, be weightless. A cabin attendant will assist you to float about the cabin, or you may choose to remain in your seat, which can now stay in the reclined position, or not, just as you like. Probably you will want to experiment with this new sensation. Certainly our astronauts seem to enjoy it. However, as descent begins, you will have to return to your seat and belt yourself in—just as in other aircraft. Your views of the sunlit surface will be memorable as the earth seems to come up to meet you at 11,500 miles per hour. For this flight has been not only noiseless and weightless, but without any vibration or feeling of motion after the gravity pull at the launching. Upon entry into the atmosphere, however, you will again encounter the forces of the earth, and your shock-absorbing seats will return you to a reclining position.

Your next sensation will occur when the subsonic jet engines take over for your landing on the runway at Los Angeles International Airport. Air traffic control will be programmed, using precise fixes on altitude and position verified by communications satellites, so you will not encounter any traffic delays. The sound of your craft coming in for its landing will actually be less than that of current jets.

Your landing at Los Angeles International Airport will be a subsonic jet runway landing, and you will board the scheduled vertical take-off jitney there for its quick flight to downtown Los Angeles. Assuming that you took off from London at noon on Monday, you would arrive in Los Angeles before 5 A.M. Monday, in less than an hour's travel time—in time for a swim before breakfast. And if you were really pressed for time and had to return to London as soon as possible, you could, theoretically, leave the West Coast that afternoon and be back in London before the end of that Monday, after one hour of flight and eight hours of sun time.

I see no technological barriers to this kind of development. One lesson we have learned from the lunar landing, "impossible" as it seemed to most people less than a decade ago, is that no technological development can really be called "impossible" any more.

Finally, let me try to outline what it is possible to accomplish in space by 1979 and 1989, using what we have learned from the Apollo Program as a base.

If we pursued a reasonably bold and aggressive program, by the end of 1979 we would have modular space stations in synchronous, equatorial orbit, in low-altitude polar and in low-inclination earth orbits, as well as having our first station in polar orbit around the moon.

The number of people involved may give some idea of the amount of activity I foresee. By the mid-1980's, there could be as many as 100 people either living in or in transit through the space stations in low earth orbit, perhaps 50 living in the station in synchronous orbit, and perhaps 25 in the space station orbiting the moon. Using basically the same or similar equipment, there may be at least one surface base on the moon with 50 people living there. Nearly daily shuttle flights to the low earth orbit station would be required to support these people and to provide for satellite repair and maintenance.

Nuclear-powered space shuttles would be traveling between lunar orbit and earth orbit and between earth orbit and synchronous orbit. They might be flying once every two weeks to the moon and about once every two weeks to synchronous orbit.

There would, of course, be shuttles operating between lunar orbit and the lunar base and between the lunar orbiting station and exploratory sites on the moon.

By the end of 1989, if we have reasonable space activity, and I think, to some extent independent of how much is done in the seventies, I would expect that men would have established their first colony on the moon, and the first permanent base on Mars. And they would be beginning to fly out to the farther planets.

The kind of resources involved are similar to those applied to the Apollo Program, but the economic and political impact will be incalculably greater.

The key to this program is the space shuttle. It opens a completely new era of transportation on the earth and in space.

It is literally man's key to the solar system. It may also be the key that permits mankind to survive the post-industrial economy that we are now creating.

George E. Mueller
President and Chairman of the Board
System Development Corporation
(formerly Associate Administrator for Manned Space Flight, National Aeronautics and Space Administration)

Preface

This volume grew out of economic research undertaken on behalf of the Office of Manned Space Flight, NASA, between September 1966 and September 1968. As early as the spring of 1966, more than three years before the first manned space flight to the moon, George E. Mueller, then Associate Administrator for Manned Space Flight, began to foresee serious problems and distortions in several sectors of the economy as a result of a probable peak in spending for the Manned Space Flight Program at that time. As many sections of the book indicate, a large number of these economic problems became reality shortly after the peak occurred; others were delayed as a result of the military build-up for the Viet Nam conflict.

For the most part, this study is limited to the multifarious aspects of the space program undertaken by the National Aeronautics and Space Administration rather than the entire national space effort, which included, for example, the activities of the Department of Defense and several other government agencies, which spent between $1.8 and $2.1 billion for space exploration in the mid- to late 1960's.

The main reason for limiting the scope of the work to NASA was the readiness with which NASA officials provided necessary data and information. With so much written and said about secret government files, it gives me great pleasure to say that I was never denied access to any information that I requested from a NASA employee. As a matter of fact, the NASA officials appeared to make a concerted effort to provide information in an attempt to promote improved public information about the objectives and the ramifications of the space program.

After I left the Space Agency, the Office of University Affairs, NASA, arranged to support my research on a part-time basis for one academic year at the Program of Policy Studies in Science and Technology, The George Washington University. This book could not have been written without the help and encouragement, in vastly different ways, of five friends, Louis H. Mayo, George E. Mueller, Theodore Suranyi-Unger, Jr., Wesley L. Tennant, and Donald S. Watson.

For their help, I would also like to thank Eleanor M. Burdette, Clarence H. Danhof, Evelyn H. Davis, William H. Harrison, Ronald M. Konkel, William E. Lilly, Sandra W. Malaga, Patricia Romani, Charles T. Stewart, Francis B. Smith, and James E. Webb.

Contents

List of Tables

1 The Economic, Social, and Political Environment of the Space Program

The space program, one of hundreds of objectives pursued by the federal government for the public interest, is something more than government spending for gigantic rockets that launch monkeys, men, and machines to the moon and to other planets.

As an object of public policy, the space program is unique. It arouses a curious combination of excitement and cynicism in the minds of Americans. Excellent coverage of space missions by the mass media has instilled the spirit of exploration in millions of Americans. In some way, almost every American was with the three astronauts when they made the first manned lunar landing and return. At the same time, the planet Earth is exploding with crisis after crisis. The moon, Mars, and Venus are remote, and, of course, the science and engineering embodied in the vehicles that took man to the moon are beyond the comprehension of all but a few Americans.

The purpose of this chapter is to identify some of the economic and political factors that influenced the priority assigned to space exploration and affected the level of spending for space in the late 1950's and the 1960's. There can be little doubt that national and international problems salient at the time influenced Presidential funding requests, Congressional appropriations, and public opinion about the space program.

PUBLIC POLICY

Public policy in our pluralistic society is shaped and executed by a bewildering multitude of institutions, individuals, and ideals. The explicit year-to-year activities of the federal government pursued in response to requirements set forth by public laws, the budget, and ad hoc policy decisions taken in any of the three branches of the government constitute only a fragment of public policy-making in the United States. Within that fragment of public policy, four traditional objectives may be discerned: defense policy, foreign policy, economic policy, and social policy. The direction of public policies, of course, emerges from the value judgments of millions of people.[1] The process takes place through the complex interactions of the political system, the legal system, and the market place.

With the notable exceptions of the economic orientation of the legislation in response to the New Deal and to the Employment Act of 1946, the most significant and farthest-reaching policy decisions affecting the United States economy have been brought about by other than economic quests. The space program is an eminently good example of this. It cannot be claimed that space policy is economic policy and that space objectives are economic objectives, any more than they are defense, foreign, or social objectives. The space program affects and has been affected by each of the four broad policy objectives. But the economic effects of the space program are important enough to place that program among the foremost economic issues of the present time. In that sense, a study of the space program by the economist is a study in economic policy formulation and implementation.

THE PUBLIC INTEREST SERVED BY THE SPACE PROGRAM

The Space Act of 1958 outlines the public interest to be served by the space program:

Sec. 102. (c) The aeronautical and space activities of the United States shall be conducted so as to contribute materially to one or more of the following objectives:

(1) The expansion of human knowledge of phenomena in atmosphere and space;
(2) The improvement of the usefulness, performance, speed, safety, and efficiency of aeronautical and space vehicles;
(3) The development and operation of vehicles capable of carrying instruments, equipment, supplies and living organisms through space;
(4) The establishment of long-range studies of the potential benefits to be gained from, the opportunities for, and the problems involved in the utilization of aeronautical and space activities for peaceful and scientific purposes;
(5) The preservation of the role of the United States as a leader in the aeronautical and space science and technology and in the application thereof to the conduct of peaceful activities within and outside the atmosphere;
(6) The making available to agencies directly concerned with national defense of discoveries that have military value or significance, and the furnishing by such agencies, to the civilian agency established to direct and control non-military aeronautical and space activities, of information as to discoveries which have value or significance to that agency;
(7) Cooperation by the United States with other nations and groups of nations in work done pursuant to this Act and in the peaceful application of the results thereof; and

(8) The most effective utilization of the scientific and engineering resources of the United States, with close cooperation among all interested agencies of the United States in order to avoid unnecessary duplication of effort, facilities and equipment.

If it were possible to ignore the overlapping objectives of the public interest outlined in the Space Act, the first three objectives could be viewed as contributing to social policy (mainly science, technology, and education). Objectives 4 and 8 serve economic goals, whereas objectives 5 and 7 are servants of foreign policy. The sixth objective clearly relates to defense policy. The Space Act, like almost all other laws that set forth national goals, does not specify the ways in which space activities are to be integrated into the existing complex of public policies. Similarly, the Space Act does not indicate the role of space activities relative to other goals in pursuit of the general welfare and the security of the United States.

PUBLIC OPINION

Through the complex operation of the democratic process, public opinion affects public policy. Public opinion, in turn, is affected by interest groups, policy statements made by prominent persons and organizations, and by the mass media.

Interest Groups

The purpose of organized interest groups is to influence public policy. There is no cohesive group behind our space policy comparable to such organized lobbies as the Farm Bureau, the American Medical Association, and the AFL-CIO. This does not mean that business firms performing work for NASA are not members of such trade associations as the Aerospace Industries Association of America, the Electronics Industries Association, or even the National Association of Manufacturers. NASA contractors do belong to these loosely knit trade associations. Further, individuals working on space projects are members of scores of different scientific and professional associations. Yet, support of, or for that matter, opposition to, the space program is widely diffused. It is a personal thing. Space antagonists, of course, may wish to buttress their position by arguing that NASA is part of that "evil" called the "military-industrial complex." But the "military-industrial complex" itself is an amorphous entity.

Clashes in Presidential Opinion

After leaving office, former President Eisenhower repeatedly made disparaging statements about the space program and its budget. Eisenhower derided the manned lunar landing project as "nuts"; he estimated the cost of the objective at $40 billion; and he continually claimed that enormous sums were being wasted on the space program.

In response to one attack on the space program made by Eisenhower, President John F. Kennedy replied:

The United States Congress almost unanimously made a decision that the United States would not continue to be second in space. We are second in space today because we started late. It requires a large sum of money. I don't think we should look with equanimity upon the prospect that we will be second all through the sixties and possibly the seventies. We have the potential not to be. I think having made the decision last year [1961], that we should make a major effort to be first in space. I think we should continue to do so. This is not a new position for him [President Eisenhower]. He has disagreed with this [the space program] at least a year or year and a half ago when the Congress took a different position. It is the position I think he took from the time of Sputnik on. But it is a matter on which we disagree.

If there is a waste, then I think it ought to be cut out by the Congress, and I am sure it will be. But if we are going to get into the question of whether we should reconcile ourselves to a slow pace, I don't think so.[2]

Jerome B. Wiesner, former Special Assistant for Science and Technology to President Kennedy, believes that both Kennedy and Johnson had a deep appreciation of the value of scientific research and technology in our society. This belief, according to Wiesner, was reflected in their support and expansion of scientific programs created by both political parties during the entire postwar period.[3]

Divergent Political and Economic Opinions about the Space Program

Public opinion about national objectives is also influenced by views expressed by the intellectual elite. Predictably, there are many divergent views on the political and economic factors associated with spending for space. Frequently, those expressing the strongest views, either pro or con, are discussing matters outside their fields of expertise.

In the political realm, for example, there is the view of Edwin Diamond, Senior Editor of *Newsweek,* who contends that the inception of the manned lunar landing program was a political maneuver on the part of President Kennedy to restore the prestige of the new administration after the Cuban Bay of Pigs disaster, as well as after the manned space flight of Yuri Gagarin.[4] A moderate political interpretation comes from Joseph G. Whelan, a specialist in Soviet and Eastern European affairs, Legislative Reference Service, Library of Congress, who believes that "The main Soviet thrust in space politics has been to claim space leadership, using this asserted claim to buttress larger political claims for the Soviet system. The effect of Soviet claims and performance and the powerful impact of both upon international politics created for the United States a space challenge of major proportions."[5]

Amitai Etzioni, Associate Professor of Sociology, Institute of War and

Peace Studies, Columbia University, views the space program as both an economic drag and the biggest public works program on earth.[6] Gilbert Burck, *Fortune* staff, contends that the United States might be reluctant to incur the large cost of the space program because it might eliminate the ability to reduce the size of the federal sector of the economy. In addition, Burck argues that the space program will strain and distort the allocation of resources. Burck also contends that such a distortion might ultimately damage the private enterprise system.[7] In contrast is the opinion of Leonard Silk, Economics Editor, *Business Week,* who believes that the most important element for long-term economic growth is the expansion of scientific and technological knowledge. According to Silk, the space program will significantly stimulate technological progress in such diverse fields as electronics, metals, fuels, the life sciences, ceramics, machinery, plastics, instruments, textiles, thermals, and cryogenics.[8]

The Mass Media

The American public is informed, and misinformed, by the mass media: magazines, newspapers, radio, and television. Public opinion necessarily reflects information conveyed to the public by the mass media. By interpreting events, the mass media tell the public what kinds of public policies are "good" and "bad" and what kinds of policies should be maintained, expanded, or eliminated.[9] Television, with essential support from NASA, accomplished a spectacular feat in presenting the actual landing of men on the moon. The United States Information Agency estimates that about one-half billion people throughout the world either listened to or watched the successful mission of Armstrong, Aldrin, and Collins on Apollo 11. That amounts to almost one-fourth of the entire world population.

Surveys sent to managing editors of daily newspapers by the National Association of Science Writers and New York University's Department of Journalism reveal that the space age has shifted journalistic values in favor of increased coverage of scientific news. One survey sent to news executives in 1965 showed that almost 80 percent of those responding had increased news coverage about space by at least 50 percent since the first Russian Sputnik in 1957.[10] News editors, of course, are not devoting more coverage to science for the sake of science. They are doing so because science and technology have become increasingly newsworthy. For the same reason, the mass media are also devoting more coverage to reports about poverty, violence in urban areas, environmental quality, and student riots across the country.

Public Opinion Polls

Although fraught with difficulties, public opinion polls can indicate public attitudes toward public policies. The results of such polls, of course, are by no means adequate as a measure of the value consensus of society. Survey questions can be misleading and biased to give weight in favor of a specific goal. People may be ignorant about the issues on which they are queried. In addition, respondents can vacillate, giving inconsistent replies over relatively short periods of time.[11] Vacillation and instability in responses reflect, in part, rapid changes that can take place in a given public program or in the social and economic environment.

The results of a six-year continuing survey conducted by Trendex, Incorporated, about government spending on space and on the lunar exploration program show the volatile nature of public opinion.[12] The survey was conducted by means of telephone calls to residents of Boston, Cincinnati, Des Moines, Pasadena, Sandusky (Ohio), and St. Paul. In January 1967, immediately before the Apollo fire, 71 percent of those responding said that they favored the government's program to put a man on the moon by 1970. The Apollo accident, which resulted in the deaths of astronauts Grissom, White, and Chaffee, occurred during a test of the Saturn launch vehicle on pad 34 at the Kennedy Space Center in Florida. In September 1967, the rate of favorable replies dropped to 51 percent. This compares with 68 percent in favor of the program in October 1968, shortly after the success of the first manned Apollo flight after the Apollo accident.[13]

THE SPACE PROGRAM AS A NATIONAL GOAL

On July 21, 1969, Neil A. Armstrong stepped out of an ugly and uncomfortable vehicle called the Eagle onto the surface of another world. A long-standing dream of mankind, as well as an American goal, was realized with the successful mission of Apollo 11.

The day after the epic manned lunar landing seemed an especially appropriate time for public soul-searching. It was the time for many officials and private individuals to urge a revamping of national priorities and policies. Senators with widely differing political views, such as Ellender of Louisiana and Muskie of Maine, stated that resources used for space should be diverted to the solution of more pressing domestic problems. Senator Edward Kennedy had made similar statements two months earlier, after the flight of Apollo 10. In contrast, Vice President Agnew used the occasion to advocate a new space commitment—to put a man on Mars by the year 2000.[14]

It is not rhetorical to ask: What was the structure of national goals in July, 1969? To request a change in priorities implies the existence of a national goals structure. Furthermore, it implies that the national goals structure is something more meaningful than party platform statements and platitudes that cover all the aspirations of a free society.

Goals for Americans

Since the end of World War II, a number of pronouncements have been made about national goals.[15] The report of the President's (Eisenhower) Commission on National Goals probably has received the greatest amount of attention of any of these statements.[16] Leonard A. Lecht, Director of the Center for Priority Analysis, National Planning Association, believes that the Eisenhower Commission on National Goals was an historical landmark in articulating public concern over the objectives of the American people.[17] Administered by the American Assembly of Columbia University, the Commission defined fifteen goals to be pursued by the American people. The goals are a composite of public and private endeavors. Eleven of the goals formulated by the Commission cover domestic matters; the other four are principally international objectives. The fifteen goals concern (1) the individual, (2) equality, (3) the democratic process, (4) education, (5) the arts and sciences, (6) the democratic economy (i.e., the maintenance of competition), (7) economic growth, (8) technological change, (9) agriculture, (10) living conditions, (11) health and welfare, (12) foreign trade policy, (13) the defense of the free world, (14) disarmament, and (15) the United Nations. The most cherished goals, (1) through (3), relate to the rights, the development, and the opportunity of the individual. For the most part, they are intangible in the sense that their values defy measurement. The Eisenhower Commission did not identify the space program as a separate national goal. Although not structured in the context of the national goals outlined by the Eisenhower Commission, subsequent sections and chapters indicate how the space program affects and has been affected by all but the first three goals.

Obstacles to an Agreed-Upon Evaluation of National Goals

Although it is possible to have a general consensus on major national objectives, there are wide differences of opinion about priorities and the rates at which objectives should be accomplished. In his work on social indicators and goals, Albert D. Biderman discusses six difficulties encountered in devising an adequate and agreed-upon measurement of national goals. The first obstacle is the limited validity, or outright invalidity, of available indexes supposed to measure social phenomena. An example of this conceptual problem is the use of a $3,000 annual family income figure to denote the poverty line. At any point in time, use of such a ceiling would include many families not living in poverty. Similarly, many poor families with an income in excess of the ceiling would be excluded by the measure. Use of standardized educational tests to measure skill and knowledge is another well-known example of the limited validity of indexes used to measure social conditions.

A second obstacle to goals measurement is inaccuracy of data. Although errors are inevitable, the matter of accuracy can be crucial in obtaining the correct answers to certain kinds of social questions. Conflicting indicators and the lack of any empirical information are the third and fourth obstacles outlined by Biderman. The fifth difficulty is the possible incompatibility of models constructed to depict the social environment. This is true particularly of conceptual systems developed by researchers in different academic disciplines. For example, subjects excluded as a result of simplifying assumptions used by the economist in describing the intricate inter-relation of causal factors might constitute the most essential subjects of inquiry for the sociologist, the psychologist, or the jurist.

Biderman's sixth obstacle is by far the most perplexing to overcome. That is the way in which the value consensus of this society can be determined. It is clear that there is little, if any, national agreement on such abstract notions as preferences, standards, and ethical values.[18] The ways in which economists have attempted to cope with the matter of value consensus are discussed in Chapter 3.

An Executive Office for Appraising National Goals

On the initiative of an economist, Gerhard Colm, and a political scientist, Luther Gulick, the National Planning Association issued a policy statement in July, 1968, calling for the establishment of an Executive Office for Appraisal of National Goals.[19] Two decades earlier Colm had been instrumental in establishing the Council of Economic Advisers.

A principal concern of the National Planning Association's statement is the definition and pursuit of national goals by the private and public sectors. According to Colm and Gulick, planning of government programs must be formulated in full knowledge and understanding of private activities. Another concern of the NPA statement is the current lack of information about goals and national priorities in the private sector, as well as in the executive and legislative branches of the government.

The National Planning Association believes that a new office for the appraisal of national goals should be established, by statute, in the Executive Office of the President. The Office should not be an appendage of either the Bureau of the Budget or the Council of Economic Advisers. Colm and Gulick contend that the emphasis of the Council of Economic Advisers has, by necessity, been on such short-run economic problems as the level of employment and income, price and wage stability, and international economic conditions. Although such short-range problems must be viewed in the perspective of longer-run economic potential, Colm and Gulick believe that the Council of Economic Advisers has not been able to focus adequately on many long-run social and economic programs.

In addition, the National Planning Association urges that a standing citizens' committee on national goals and priorities be established. Membership on the proposed committee would be drawn, on a nationwide basis, from business, labor, agriculture, rural, and urban interests, as well as from professional and research organizations. Further, the National Planning Association proposes that the dual task of (1) examining the definition of national goals used in the formulation of government programs, and (2) the contribution of programs to the achievement of national goals be assigned to a joint Congressional Committee on Goals.

SPENDING FOR NATIONAL GOALS

Government spending, federal, state, and local, can be used as a means of identifying national goals. Government spending for national objectives, however, cannot be interpreted as indicating the value consensus of society. At best, information about government expenditures shows what is being or what has been spent in the public sector, rather than what "should be" or what "should have been" spent for national objectives.

Controllable and Uncontrollable Federal Spending

A discussion of resource allocation to national goals must begin with defense expenditures because of the dominant role played by national security in recent decades. Since the late 1940's, defense spending has comprised 80 to 85 percent of total federal spending for goods and services. This compares with 25 percent in 1939 and 35 percent as late as 1940. Because no national defense program can insure against all national security risks, the federal budget, and ultimately Congressional appropriations, results from an evaluation of external risks and from an appraisal of the economic and social costs of reducing them.[20] Defense spending can and does fluctuate by billions of dollars from year to year, or perhaps more precisely from international crisis to international crisis.

Among non-defense programs, there are groups of federal expenditures that are classed as "relatively uncontrollable civilian expenditures." These expenditures are, in a way, exempt from the budgetary process because they have been established by law. The largest item in this group is interest on the national debt, which is regarded as a contractual obligation to be paid ahead of other kinds of spending. Other "relatively uncontrollable civilian expenditures" include: veterans' pensions, compensation, and insurance; farm price supports (operation of the Commodity Credit Corporation); and federal contributions to Medicare and other public assistance programs. Because the majority of these federal expenditures are transfer payments, they do not appear in the national income accounts, i.e., they are not part of the GNP. Nonetheless, the existence of these budgetary priorities, as well as a large defense commitment, affects the size of budget requests and appropriations, and, ultimately, spending by the federal sector. This means that the main impact of any budget change is borne, and also enjoyed, by "unprotected" nondefense programs. The space program, of course, is an "unprotected" civilian expenditure.

The National Income Accounts

Our social accounting systems are an imperfect method for quantifying the costs of national goals in a way that is commensurate with resource availability. The most widely used accounting system is the national income and product account, i.e., the GNP. The gross national product measures the value of all final goods and services produced during a given period of time. The GNP does not measure expenditures for previously accumulated wealth or expenditures for intermediate (i.e., unfinished) goods and services used in the production process. Government expenditures are the chief means of providing for the satisfaction of those national goals furthered by the public sector. The federal component in the GNP account understates the magnitude of federal spending for national goals because transfer payments made by the federal government are not part of the GNP.

The economist distinguishes between federal government expenditures that are exhaustive (i.e., require the use of resources) and those that are non-exhaustive (i.e., transfer payments). When the government actually purchases final goods and services, it makes exhaustive expenditures. These expenditures are part of the public component in the GNP accounts. In contrast, non-exhaustive government expenditures do not absorb resources. The government receives no equivalent value in the form of goods and services from non-exhaustive expenditures. The federal government makes four main kinds of transfer payments: (1) transfer of income to persons, principally for welfare purposes and programs (unemployment benefits, etc.); (2) transfers in the form of subsidies to businesses or to individuals (agriculture price support programs, subsidized air and merchant marine transport, and low-cost school lunch programs); (3) grants-in-aid to state and local governments; and (4) the transfer payment for all interest on the government debt. Transfer payments exert an indirect impact on the GNP in the form of consumer expenditures, business investment, and state and local government expenditures. During the last decade, federal transfer payments comprised from 40 to 45 percent of total federal spending.

Table 1–1 shows the relative importance of transfer payments for selected national goals in 1968. Virtually all federal spending for national defense and for space is exhaustive; it is included in total in the GNP accounts. The table clearly shows that most federal spending for education, health, labor and welfare, commerce and transportation, and agriculture and national resources takes the form of transfer payments and appears indirectly in the GNP account. Spending by state and local governments appears as such in the GNP accounts; it is principally exhaustive.

TABLE 1–1: GOVERNMENT SPENDING BY FUNCTION, CALENDAR YEAR 1968 (*BILLIONS OF DOLLARS*)

	FEDERAL			STATE AND LOCAL		
NATIONAL GOAL	TOTAL	PURCHASE OF GOODS AND SERVICES	TRANSFER PAYMENTS[a]	TOTAL	PURCHASE OF GOODS AND SERVICES	TRANSFER PAYMENTS
National defense	80.4	78.0	2.4	.5	.5	[b]
Health, labor and welfare	47.0	3.7	43.3	30.2	24.0	6.2
Education	4.5	.8	3.7	43.4	42.9	.5
Commerce and transportation	6.3	1.6	4.7	14.7	16.8[c]	[b]
Agriculture and natural resources	9.2	4.8	4.4	3.7	3.7	[b]
Veterans' benefits	7.6	1.7	5.9	[b]	[b]	[b]
Space research and technology	4.5	4.5	[b]	[b]	[b]	[b]
International affairs	2.7	.6	2.1[d]	[b]	[b]	[b]
Total[e]	181.5	99.5	82.0	107.6	100.7	10.3[c]

[a] Includes transfer payments to persons, interest, grants-in-aid to state and local governments, and subsidies less current surplus of federal government enterprises.

[b] Less than $100 million.

[c] The sum of purchases of goods and services and transfer payments exceeds total spending because of current surplus of state and local government enterprises.

[d] Mainly foreign economic assistance.

[e] Includes government operations and housing and community development.

Source: U.S. Department of Commerce, *Survey of Current Business,* July, 1969, p. 34.

Because almost all federal spending for national defense is exhaustive, government purchases of goods and services for national defense averaged an overwhelming 80 percent of total federal spending in the GNP accounting system between 1958 and 1967. During the same period, federal spending for space comprised about 5 percent of the federal component in the GNP accounts. The defense and the space shares of federal spending drop by about one-half when federal non-exhaustive expendi-

TABLE 1-2: GOVERNMENT SPENDING FOR SELECTED NATIONAL COMMITMENTS

FEDERAL EXPENDITURES FOR NATIONAL COMMITMENTS 1958–1967

STATE AND LOCAL SPENDING FOR NATIONAL COMMITMENTS 1958–1967

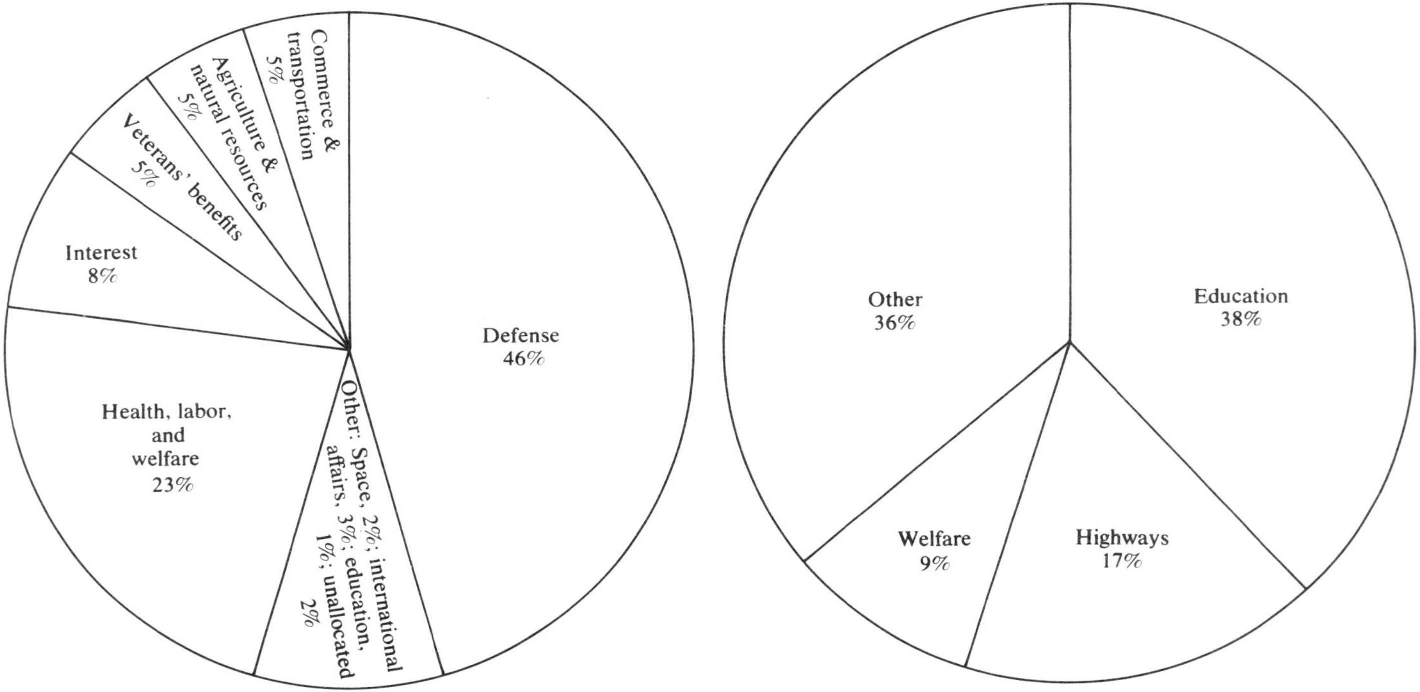

Total: $1,120 billion

Total: $645 billion

Source: Economic Report of the President, February 1968.

tures are considered.[21] Table 1–2 shows the magnitude of federal and state and local government spending for selected national goals between 1958 and 1967.

When state and local government spending is combined with federal, the relative share of government spending for defense drops still further. After adjusting for federal grants-in-aid to state and local governments, it is possible to estimate the relative importance of several national goals in terms of total government spending. Between 1958 and 1967, about 8 percent of total federal spending took the form of grants-in-aid. Programs most affected by grants-in-aid are those for public assistance and relief and those for highways. Considering all three levels of public spending between 1958 and 1967, defense comprises about 30 percent of total spending; education, 15 percent; health, labor, and welfare, 15 percent; commerce and transportation, 8 percent; international affairs, 2 percent; and space, slightly more than 1 percent.

ECONOMIC ISSUES AND FISCAL PHILOSOPHY

Although the extent to which prevailing national and international factors actually affected Presidential requests and Congressional appropriations for spending on the space program in the late 1950's and in the 1960's is uncertain, the development and expansion of the space program harmonized with the domestic and foreign environment existing during the late 1950's and early 1960's.

Table 1–3 outlines the salient economic issues that Presidents Eisenhower, Kennedy, and Johnson believed warranted special public and Congressional attention. The periods listed in the table correspond with either a pronounced change in the economic climate or a change in administration. The major national economic issues listed in the table come from *The Economic Report of the President* for the years 1956 to 1969.[22] The fiscal philosophies listed in the last column of the table are policy recommendations or objectives, many of which either were not accomplished or were effectuated a year or so after the Presidential recommendation.

Revival of Prosperity—1956 and 1957

Continued economic expansion, after the slowdown in economic activity that followed the Korean War, was the dominant economic concern of the Eisenhower Administration in 1956 and 1957. With the return of prosperity, President Eisenhower hoped to balance the federal budget and begin to reduce the national debt. The most troublesome problems facing the Administration were the balance of payments deficit and the accumulation of a large agricultural surplus as a result of the price support

TABLE 1-3: ECONOMIC AND POLITICAL FACTORS AFFECTING FEDERAL SPENDING, 1956-1967

PERIOD	PRESIDENT	MAJOR NATIONAL ECONOMIC ISSUES	FISCAL PHILOSOPHY
1956 and 1957	Eisenhower	Sustain economic expansion; agricultural adjustment; maintain competitive environment; international cooperation; school construction; and area redevelopment assistance	Balanced budget; reduce size of national debt; and postpone personal and corporate tax reductions
1958	Eisenhower	Resumption of high level of economic activity; match Soviet scientific challenge; agricultural adjustment; and area redevelopment assistance	Remove debt limit; harmonize federal credit policy with general fiscal policy; and strengthen unemployment insurance program
1959 and 1960	Eisenhower	Amend Employment Act of 1946 to make price stability an explicit national goal; maintain competition and promote small business; balance of payments equilibrium; area redevelopment assistance; promote research and development; and agricultural adjustment	Balanced budget; debt retirement; tax revision; and raise interest ceiling to enhance debt refinancing
1961, 1962, and 1963	Kennedy[a]	Restore economic prosperity; strengthen and develop manpower base; establish price-wage guideposts; close gap between actual production and productive capacity; and support civilian technology	Request standby for Presidential action to reduce income taxes; enact investment credit system; establish standby program for federal capital improvements; strengthen unemployment insurance system; and tax reduction and revision
1964 and 1965	Johnson	Sustain economic expansion; eliminate gap between actual production and productive capacity; eliminate persistent unemployment; stimulate rapid technological change; balance of payments equilibrium; and area redevelopment	Reduce personal and corporate income taxes in 1964; reduce federal spending; reduce excise taxes; and increase social security benefits
1966, 1967, and 1968	Johnson	Vietnam first priority; combat hard-core unemployment; war on poverty; restore price stability (increasing importance of price-wage guideposts); balance of payments equilibrium; and lift burden on housing industry	Reduce low-priority civilian spending; postpone excise tax reductions; graduate withholding system; accelerate payment of corporate tax liabilities; and surtax on personal and corporate tax liabilities

[a] The Economic Report of the President is usually made in January for economic developments of the preceding year. The 1961 report was made by President Eisenhower.

Source: Economic Report of the President, 1956 through 1969.

program. Rates of unemployment in 1956 and 1957 (between 4.1 and 4.3 percent) were as low as they were to be for the next decade.

In October 1957, when the first Russian Sputnik was launched, the federal government was spending at an annual rate of $76 million for space research and technology—about 15 percent less than the amount spent in 1954.

Economic Decline—1958

During the third quarter of 1957, national production began to decline. In 1958 the rate of unemployment rose to a high 6.8 percent. Despite the recession, the increase in federal spending for goods and services between 1957 and 1958 was less than the increase a year earlier. With a sharp decline in federal revenue, the federal deficit amounted to $2.8 billion in fiscal year 1958 and to $12.4 billion by fiscal year 1959.

In January 1958, President Eisenhower requested an increase in federal spending for science and education to meet the Soviet challenge. The National Aeronautics and Space Administration was created by the Space Act of 1958. In the atmosphere of anxiety created by Soviet scientific successes of that year, Congress also passed the National Defense Education Act, which offered the hope that no talented student would be denied access to higher education because of insufficient financial means.

A Short Period of Prosperity—1959 and 1960

By early 1959 the economy had regained most of the ground lost during the recession. The Eisenhower Administration placed great importance on the contributing role of federal spending to inflation. Because of this concern, President Eisenhower submitted to the Congress a balanced budget for fiscal 1960.[23] During the 1957–1958 recession, wages, costs, and prices rose by about 2 percent despite the declines in production and in employment.

Despite general prosperity the rate of unemployment was 5.5 percent in 1959. During this period unemployment was viewed as frictional, i.e., the result of dislocations attributable to technological change, the transition from agriculture to industry, long-term shifts in product demand and industry location, and to international competition. The President recognized that special remedial measures would be required to match more effectively labor supply and demand in specific geographic areas, industries, and occupations. However, the Administration's view was that the government could not assume major responsibility for these remedial measures. Rather, the federal government should participate in joint programs with the private sector. President Eisenhower believed that local community action could develop new industries in areas of high and persistent unemployment.[24]

The New Frontier—1961 to 1963

The Kennedy Administration inherited an economy that was in the grip of a recession, and the President's most immediate concern became the restoration of reasonably full employment. By the end of 1960 the rate of unemployment had risen to almost 7 percent. Kennedy's Council of Economic Advisers believed that the probable cause of the incomplete and short-lived economic expansion in 1959 and early 1960 was the restrictive fiscal policy pursued in 1960.[25] Yet the Council of Economic Advisers reported in 1962 that the large increase in federal spending for defense and space programs was to insure national security rather than economic stability. The Berlin crisis, the Bay of Pigs crisis, and the first Russian launch of a manned spacecraft all took place in 1961. For economic stabilization, President Kennedy recommended that taxes not be increased to compensate for increased federal spending for defense and space. The Kennedy Administration's largest deficit, about $6.4 billion in fiscal year 1961, was about half the amount incurred by Eisenhower in 1959.

From his economic reports and his budget messages, it appears that President Kennedy strongly favored tax measures, over federal spending, as the means to liven a sluggish economy. In January 1963, President Kennedy asked Congress to enact legislation for tax reform. The request was made in full anticipation of a budget deficit for that fiscal year. The tax legislation was actually enacted a year later, several months after President Kennedy was assassinated.

The GNP Gap

The Kennedy Administration was also concerned about the gap between the productive potential of the economy and the actual size of the gross national product. During the late 1950's and the early 1960's the productive potential or capacity of the economy rose by substantially more than actual production. The level of unemployment of labor, natural resources, and capital is a measure of unused economic capacity. The GNP gap represents output that is permanently lost to society because resources, at that point in time, were not used. The data displayed in Table 1–4 show that the gap between what the nation could have produced and what the nation was actually producing amounted to $42 billion in 1961—the widest gap during a ten-year period. The rate of unemployment amounted to more than 6.5 percent of the labor force in 1961.

The productive activity of our economic system grows over time as a result of technological change, improved management, and growth in the size and quality of the labor force, as well as growth in the size and quality of the nation's stock of capital. Chapter 9 discusses the relation

TABLE 1-4: THE GROSS NATIONAL PRODUCT, POTENTIAL AND ACTUAL, 1956–1967

CALENDAR YEAR	POTENTIAL GNP BILLIONS 1958 DOLLARS	ACTUAL GNP BILLIONS 1958 DOLLARS	DIFFERENCE BILLIONS	RATE OF UNEMPLOYMENT PERCENT
1956	454.2	446.1	− 8.1	4.1
1957	470.0	452.5	−17.5	4.3
1958	486.4	447.3	−39.1	6.8
1959	503.5	475.9	−27.6	5.5
1960	521.1	487.7	−33.4	5.5
1961	539.3	497.2	−42.1	6.7
1962	558.2	529.8	−28.4	5.5
1963	578.4	551.0	−27.4	5.7
1964	600.3	581.1	−19.2	5.2
1965	622.8	616.7	− 6.1	4.5
1966	646.2	652.6	+ 6.4[a]	3.8
1967	672.1	669.3	− 2.8	3.8[b]

[a] A positive difference between the potential GNP, i.e., full-employment GNP, and the actual GNP is reflected in an increase in the price level.

[b] Expansion of the armed forces and antipoverty programs for training the unskilled have a stronger effect on unemployment than output; thus there can be a gap at a rate of unemployment 4.0 percent or less.

Source: Economic Report of the President, January, 1968, pp. 60–61.

of the space program to economic growth. The Council of Economic Advisers estimates that the growth rate for potential output was about 3.5 percent during the 1950's and early 1960's, and about 3.75 percent in the mid-1960's.[26] This means that in recent years the productive capacity of the U.S. economy has been rising by about $30 billion each year.

Unrealized output resulting from the gap between the actual and potential GNP amounted to about $250 billion between 1956 and 1967. Assuming, among other things, perfect substitutability and mobility of resources, unrealized output is equivalent in value to all federal purchases of goods and services made between 1964 and 1967; to all federal purchases made for national defense between 1963 and 1967; to about 10 times the actual amount spent for space between 1957 and 1967; to about 15 times the actual federal natural resources programs between 1957 and 1967; and to about 20 times the amount of federal spending for educational programs during the decade.[27]

To foster economic growth, the Administration proposed measures to encourage civilian research and development and to make the results of federally sponsored R & D more readily accessible to American industry. The measures included the development of a federal-state engineering extension service, adjustment of the corporate income tax laws to give business additional stimulus to invest in research equipment, and support for university training of industrial personnel. As indicated, Kennedy believed that a sizable R & D effort would improve the quality of life because the return from advancing modern technology would bring forth better goods and services at lower prices.[28]

Closing the GNP Gap—1964 and 1965

By the end of 1963, the gap between actual and potential output had been reduced by about 50 percent, but the rate of unemployment was still almost 6 percent. President Johnson viewed the main unfinished task of economic policy to be the war against poverty.

In his first budget message, President Johnson called for a substantial reduction in federal spending and an increase in government efficiency. President Johnson succeeded, although only temporarily, in reducing defense expenditures by about $4 billion between fiscal 1964 and fiscal 1965. His budget message stated that the reduction was possible because of the large and rapid increase (20 percent) in defense outlays between 1961 and 1964. The federal budget deficit dropped from $8.2 billion in fiscal 1964 to about $3.4 billion in fiscal 1965.

Vietnam—1966 to 1968

The rapid military build-up in Vietnam was about six months old when President Johnson made his economic report in January 1966. At that time, he made two statements about Vietnam: First, that he believed that the economic cost of the conflict imposed no unbearable burden on our resources; and second, that prosperity did not depend on the increased military effort. The context of the report, however, was that of a changed economic environment—facing problems associated with an economic system approaching full employment. For the first time in more than a decade, the rate of unemployment dropped below 4 percent of the labor force in 1966.

By January 1967, President Johnson had to tell the American public that increased spending for Vietnam was putting a severe burden on the production capacity of the economy. To shift the gears of the economic system from excessive speed to moderate advance, given the Vietnam commitment, President Johnson asked Congress in January 1967 to enact a surcharge on the tax liabilities of individuals and of corporations. The surcharge was to have provided extra revenue to offset the anticipated increase in defense spending for fiscal 1968.

The anticipated continuation of the boom did not materialize during the first half of 1967. In the spring, the gross national product, in real terms, dropped below that of late 1966. Throughout 1967 the House Ways and Means Committee remained unconvinced that the proposed tax surcharge on personal and corporate incomes was the fiscal remedy that would provide economic growth, full employment, and relative price stability. After the slight decline in economic activity during the first half of 1967, the economy once again began to advance too rapidly. By mid-1968 Congress was ready to pass the Revenue and Expenditure Control Act.

Fiscal management for economic stability becomes difficult when our economic system is operating at full employment, and the federal budget becomes less flexible. Between 1966 and 1968, defense spending conspicuously increased its dominance in the federal budget, and outlays for defense amounted to about 45 percent of the budget during that period. "Uncontrollable" civilian programs, mainly social security, Medicare, veterans' benefits, and interest on the national debt, comprised another 35 percent of total commitments. The margin for adjustment in the allocation of resources to other national goals, i.e., to civilian programs that are relatively "controllable," narrowed sharply between 1966 and 1968. "Discretionary" government spending amounted to only about 20 percent of total federal spending during that period. Almost the entire burden of fiscal restraint for "controllable" civilian programs was on the space program. Between fiscal 1966 and 1968 spending for space dropped by $1.2 billion. This contrasts with increases in spending of $1.0 billion for commerce and transportation, of $1.5 billion for community development and housing, of $2.5 billion for education and manpower development, and an increase of $1.9 billion for agriculture and natural resources.[29]

SOVIET ACCOMPLISHMENTS IN SPACE

The period of dampened economic activity in the United States in the late 1950's and early 1960's was the very time when the Soviet Union accomplished a rapid succession of space feats.

It is not the purpose of this section to analyze whether or not there is a space race, and, if there is one, what the relative position of each of the contestants is in terms of technological advance, military preparedness, or international prestige. Both the Soviet Union and the United States have accomplished triumphs in space; both have also incurred tragic space-related accidents that resulted in setbacks and the death of astronauts.[30] Nevertheless, a look at the timetables on the development of space capabilities strongly suggests that the space program of the United States has been reactive to Soviet accomplishments in space.

Chronology of Space Spectaculars

Over the last decade, the record is quite clear that the Soviet Union has been first to accomplish most space spectaculars. The big exception, of course, is the first successful manned lunar landing made by the United States in July, 1969. As the term is used here, a space spectacular is simply a space-related event that captures the imagination and sympathy of the American public, including the President and members of Congress.

Table 1–5 presents a chronology of space spectaculars accomplished by the USSR and the United States. The list is selective, showing only 13 space payloads between 1957 and July 1969. During the ten years between October 1957 and October 1967, the total number of Soviet space payloads was 269; that for the United States was 529.[31] During that period, the largest number of Soviet space payloads launched in any single year (73) occurred in 1965, whereas the largest number of United States space payload missions was in 1966 (100).

TABLE 1–5: CHRONOLOGY OF SELECTED SPACE SPECTACULARS, 1957–1969[a]

ACHIEVEMENT	UNITED STATES	USSR
First artificial satellite	Explorer I: 2/1/58	Sputnik I: 10/4/57
Experiment with animals	Mercury II with monkey: 1/31/61	Sputnik II with dog: 11/3/57
Spaceship prototype with reliable re-entry system	Mercury IV: 9/13/61	Satellite Ship II: 8/19/60
Manned satellite	Glenn aboard Mercury VI: 2/20/62	Gagarin aboard Vostok I: 4/12/61
Multipassenger satellite	Grissom and Young aboard Gemini III: 3/23/65	Feoktistov, Komarov, and Yegorov aboard Voskhod I: 10/12/64
Extra-vehicular activity	White on tetherline from Gemini IV: 6/3/65	Lenov and Belyayev on tetherline from Voskhod II: 3/18/65
Manned lunar landing	Armstrong, Aldrin, and Collins aboard Apollo XI: 7/20/69	Not yet accomplished

[a]October 1957 to July 1969.

Sources: Adapted from Charles S. Sheldon, *Review of the Soviet Space Program* (Washington, D.C.: U.S. Government Printing Office, 1967), pp. 47–78; and Leonard N. Beck, "A Survey of Space Activities and Space Science," Chapter III, *Soviet Space Programs, 1962–1965, Goals and Purposes, Achievements, Plans, and International Implications*, Staff Report prepared for the Senate Committee on Aeronautical and Space Sciences, December 30, 1966, p. 160.

The Space Age began with the launching of Sputnik I on October 4, 1957. The American public was impressed beyond words. Congress was alarmed and aroused. During the succeeding months, the United States intensified its space research so much that in February 1958 the Missile Command of the Army Ordnance Corps put Explorer I into orbit—only four months after the launch of the first Russian Sputnik. Just a month after the launch of its first artificial satellite, Russia achieved another space first. Sputnik II completed its mission with a live dog aboard. As the table shows, the USSR was also in the lead for: the first man in space, the first multi-manned crew, and the first extra-vehicular activity (i.e., walk in space on a tetherline).

Manned Space Flights

A manned lunar landing became the principal United States space objective, as well as a national goal, when President Kennedy asked Congress on May 21, 1961, to accomplish a manned lunar landing within the decade of the 1960's. Many believe that manned space flight also is the primary objective of the Soviet space program.[32] On April 12, 1961, Yuri Gagarin, aboard Vostok I, became the first man to orbit the earth, logging more than 25 hours in space. The United States had only two sub-orbital flights in 1961. With seven known launches, the USSR had logged a total of 406 hours and 51 minutes of manned spacecraft flying hours by the end of 1964. By the same time, United States manned spacecraft time had totaled 53 hours and 54 minutes. Table 1–6 presents a chronology of manned space flights between 1961 and July 1969.

In terms of manned spacecraft hours, the United States exceeded the number logged by the USSR on December 4, 1965, with the 330-hour flight duration of Gemini VII. By the end of 1966, the United States had more than twice the total number of manned spacecraft hours of the USSR. As the table shows, the Soviet space program had no known manned launches between the spring of 1965 and the spring of 1967. Soviet cosmonaut Komarov was killed on the April 1967 manned space mission. Between January 1, 1968 and July 31, 1969, the Soviet Union launched three manned spacecraft, and the United States launched five.

Measures of Soviet Resource Allocation to Science and to Space

Because of the critical role played by science and technology in the attainment of military power and strength, detailed information about Soviet spending for scientific research, including the exploration of space, is shrouded in secrecy. Experts on Soviet affairs indicate that there are at least two separate budget outlays for research and development in the USSR, one being a so-called "secret budget."[33]

In discussing the secrecy surrounding resources allocated by the USSR to its space program, an authority on Soviet affairs cites the fact that it was only after his death that Sergei P. Korolev received public acclaim for his contribution to Soviet space research. Until his death in January 1966, the official Soviet press referred to Korolev infrequently, and then not by name but as the "chief designer of space vehicles."[34] To cite another example on the web of secrecy surrounding the Soviet space program: in June of 1961 some 7,000 persons were awarded medals and other honors for their participation in the successful launching of the first

TABLE 1-6: CHRONOLOGY OF MANNED SPACE FLIGHTS, 1957–1969[a]

United States Crew and Space Vehicle	Duration Hours	Duration Minutes	USSR Crew and Space Vehicle	Duration Hours	Duration Minutes
			1961		
Shepard aboard Mercury-Redstone 3: May 5	0	15	Gagarin aboard Vostok I: April 12	1	48
Grissom aboard Mercury-Redstone 4: July 31	0	15	Titov aboard Vostok II: August 6	25	18
			1962		
Glenn aboard Mercury-Atlas 6: February 20	4	55	Nikolayev aboard Vostok III: August 11	94	35
Carpenter aboard Mercury-Atlas 7: May 24	4	56	Popovich aboard Vostok IV: August 12	70	57
Schirra aboard Mercury-Atlas 8: October 3	9	13			
			1963		
Cooper aboard Mercury-Atlas 9: May 15	34	20	Bykovsky aboard Vostok V: June 14	119	6
			Tereshkova (female) aboard Vostok VI: June 16	70	50
			1964		
No manned launches			Feoktistov, Komarov, and Yegorov aboard Voskhod I: October 12	24	17
			1965		
Grissom and Young aboard Gemini III: March 23	4	50	Belyayev and Leonov aboard Voskhod II: March 18	27	2
McDivitt and White aboard Gemini IV: June 3	97	56			
Conrad and Cooper aboard Gemini V: August 21	190	55			
Borman and Lovell aboard Gemini VII: December 4	330	35			
Schirra and Stafford aboard Gemini VI: December 15	25	51			

			1966		
Armstrong and Scott aboard Gemini VIII: March 16	10	42	No manned launches		
Cernan and Stafford aboard Gemini IX: June 3	72	21			
Collins and Young aboard Gemini X: July 18	70	47			
Conrad and Gordon aboard Gemini XI: September 15	71	17			
Aldrin and Lovell aboard Gemini XII: November 11	94	35			
			1967		
			Komarov aboard Soyuz I: April 22	23	37
			1968		
Cunningham, Eisele, and Schirra aboard Apollo VII: October 11	259	12	Beregovoy aboard Soyuz III: October 26	94	51
Anders, Borman, and Lovell aboard Apollo VIII: December 21	147	1			
			1969		
McDivitt, Schweickart, Scott aboard Apollo IX: March 3	241	1	Shatalov aboard Soyuz IV: January 14	71	21
Cernan, Young, and Stafford aboard Apollo X: May 18	192	3	Volynov, Yovgenty, and Khrunov aboard Soyuz V with crew transfer to Soyuz IV: January 15	72	41
Armstrong, Aldrin, and Collins aboard Apollo XI: July 16	195	18			

[a] April 1961 to July 1969.

Source: Office of Manned Space Flight, National Aeronautics and Space Administration.

manned space flight. All but one of the recipients remained unidentified. The only honored one named was Premier Khrushchev, lauded as patron of the sciences.[35] Despite the measurement difficulties, including the matter of secrecy, specialists on Soviet affairs have made estimates of the magnitude of resource allocation into the scientific sector of the Soviet economy. Furthermore, within a few percentage points, there is general agreement among the several different estimates.

Incommensurables in Making Inter-Country Comparisons of Resource Allocation

The two most common ways of measuring resource allocation to various sectors of an economic system are by the diversion of labor and by the division of the GNP. Apart from the difficulties in determining the extent of the USSR's scientific effort, which are caused by secrecy, there are many incommensurables that should be recognized when inter-country comparisons of scientific effort are made.

The first problem encountered in obtaining yardsticks to measure R & D effort is that of concepts and definitions. More simply, what is being measured? There are many qualitative and quantitative differences in defining a "scientist," a "scientific activity," and "research and development." On the basis of definitions used by the Organization for Economic Co-operation and Development, for example, the R & D expenditure of the United States amounted to 3.4 percent of the GNP in fiscal 1963.[36] This compares with 2.9 percent, using definitions developed by the National Science Foundation for the same year.

Another reason for proceeding with caution in making international comparisons of resource allocation is the fact that different countries have different scales of preferences, priorities, and standards. Such differences affect the relative levels of welfare that can be attained with equivalent amounts of funds (i.e., the actual rate of exchange that exists between convertible currencies).

In comparisons between the United States and the USSR, this problem is augmented by the fact that the Russian ruble is not a freely convertible currency. The official dollar-ruble rate of exchange is determined by the Soviet government rather than on the open market. In addition, more than one "official" dollar-ruble rate of exchange has existed simultaneously. For example, before 1960, one official rate of exchange was 4 rubles for 1 dollar. Another "official" rate, offered at the same time to American tourists, was 10 rubles for 1 dollar. This means that the ruble was worth both $.25 and $.10 at the official rate of exchange.[37]

Soviet Expenditures for Scientific Research

For perspective, it is useful to discuss estimates of Soviet expenditures for space after a discussion of resource allocation to science and technology in general. According to estimates made by Leon M. Herman, the number of engineers employed in the Soviet economy grew at almost twice the average annual rate at which this number grew in the United States between 1958 and 1964. During the decade of the 1940's the growth pattern had been the reverse. Table 1–7 displays data that show the comparative growth in the number of engineers employed in the United States and in the USSR.[38] The rapid increase in the number of Soviet technical personnel resulted from the Soviet government's systematic effort to improve the quality of its industrial labor force. The rapid increase was made possible by a correspondingly rapid increase in the network of secondary- and higher-level technical schools in the Soviet Union. Other data collected by the Organization for Economic Co-operation and Development for total personnel employed in science in the USSR show an even more rapid increase in the number of scientific personnel employed in the Soviet Union—about 12.5 percent per annum between 1958 and 1963.[39]

TABLE 1–7: COMPARATIVE GROWTH IN THE NUMBER OF ENGINEERS EMPLOYED IN THE UNITED STATES AND IN THE USSR, 1940–1964

	UNITED STATES		USSR	
YEAR	NUMBER (THOUSANDS)	INDEX 1958 = 100	NUMBER (THOUSANDS)	INDEX 1958 = 100
1940	170	32.1	295	32.4
1950	316	59.6	392	43.1
1958	530	100.0	910	100.0
1960	590	111.3	1,135	124.7
1962	650	122.6	1,325	145.6
1964	700	132.1	1,498	164.6
	AVERAGE ANNUAL RATE OF GROWTH			
1940–1950	6.4		2.9	
1958–1964	4.7		8.6	

Source: Leon M. Herman, "Soviet Economic Capabilities for Scientific Research," Chapter IV, Part III, *Soviet Space Programs, 1962–1965: Goals and Purposes, Achievements, Plans, and International Implications,* Staff Report prepared for the Senate Committee on Aeronautical and Space Sciences, December 30, 1966, p. 401.

The structure of higher education in the Soviet Union is markedly different from that in the United States. Emphasis in the Soviet Union is on technical education. There is no institution in the USSR comparable to the American liberal arts college; almost all education in that country is professional.[40] Korol estimates that the number of graduating engineering students in the Soviet Union in 1960 was about three times the number in the United States. The stress on scientific and technical training in the USSR is disclosed in the data displayed in Table 1–8, which show the distribution of college graduates in the civilian labor force by selected academic fields. In 1959, the number of graduate scientific and technical personnel employed in the Soviet economy comprised more than one-half of all college graduates, compared with about one-third of all graduates in the United States. However, it is the belief of many experts on Soviet affairs that graduates of Soviet engineering schools are on a lower professional level than those graduating from engineering schools in the United States.

TABLE 1-8: COMPARATIVE DISTRIBUTION OF COLLEGE GRADUATES IN THE CIVILIAN LABOR FORCE, SELECTED ACADEMIC FIELDS, 1959

	UNITED STATES		USSR	
ACADEMIC FIELD	NUMBER (THOUSANDS)	PERCENT	NUMBER (THOUSANDS)	PERCENT
Engineering	720	11.5	894	29.5
Physical and life sciences	705	11.3	134	4.5
Agriculture and forestry	195	3.1	209	6.9
Health sciences	597	9.5	404	13.3
Scientific and technical	2,217	35.4	1,641	54.2
Education	1,030	16.4	1,009	33.3
Business, economics, and law	1,276	20.4	238	7.9
Other	1,742	27.8	139	4.6
Non-technical	4,048	64.6	1,386	45.8
Total	6,265	100.0	3,027	100.0

Source: Leon M. Herman, "Soviet Economic Capabilities for Scientific Research," Chapter IV, Part III, *Soviet Space Programs, 1962–1965: Goals and Purposes, Achievements, Plans, and International Implications,* Staff Report prepared for the Senate Committee on Aeronautical and Space Sciences, December 30, 1966, p. 407.

Annual Soviet expenditures for scientific research doubled between 1951 and 1957. As Table 1-9 shows, such expenditures doubled again within three years. It will be recalled that by April 1961 the Soviet Union had successfully launched the first manned orbital payload. Leon M. Herman conjectures that as much as 25 percent of the Soviets' scientific expenditures in the mid-1960's was allocated to their space program. The estimate is based on his belief that the boom in Soviet scientific expenditures gathered its greatest momentum after the first Sputnik was launched successfully. This means that the Russian space program became an irresistible tool in the military and political rivalry of the USSR with the West. For comparison, Herman estimates that Soviet appropriations for space were equal to the total Soviet outlay for higher education in 1963 and only slightly smaller than the Soviet outlay for capital to expand all consumer goods industries in 1963.[41]

TABLE 1-9: EXPENDITURES FOR SCIENTIFIC RESEARCH MADE IN THE UNITED STATES AND IN THE USSR, 1950-1966[a]

	UNITED STATES		USSR	
YEAR	BILLIONS OF DOLLARS	INDEX 1958 = 100	BILLIONS OF RUBLES	INDEX 1958 = 100
1950	3.4	31.5	.8	33.3
1952	4.5	41.7	.9	37.5
1954	5.7	52.8	1.0	41.7
1956	8.5	78.7	1.3	54.2
1958	10.8	100.0	2.4	100.0
1960	13.7	126.9	3.3	137.5
1962	15.6	144.4	4.3	179.2
1964	19.2	177.8	5.2	216.7
1966	21.9	202.8	6.5	270.8

[a] Based on the work of Nancy Nimitz, *Soviet Expenditures on Scientific Research* (Santa Monica, Calif.: Rand Corp., 1963), pp. 40–41.

Sources: Leon M. Herman, "Soviet Economic Capabilities for Scientific Research," Chapter IV, Part III, *Soviet Space Programs, 1962–1965: Goals and Purposes, Achievements, Plans, and International Implications,* Staff Report prepared for the Senate Committee on Aeronautical and Space Sciences, December 30, 1966, p. 411. Data for the United States between 1954 and 1966 from the National Science Foundation; United States data for 1950 and 1952 from Dexter M. Keezer, "The Outlook for Expenditures on Research and Development during the Next Decade," *American Economic Review* (May 1960), p. 361.

R & D as a Percentage of the Soviet GNP

After cautioning the reader about the hazards of making inter-country comparisons of resource allocation, especially between the USSR and Western countries, it is not inappropriate to present estimates made by experts for research outlays as a percentage of the Soviet GNP. Alexander Korol believes that Soviet spending for R & D comprised about 2.9 percent of the Soviet GNP in 1960.[42] This compares with 2.7 percent for the United States (National Science Foundation definition) in 1960.

The Organization for Economic Co-operation and Development estimates that R & D was 2.5 percent of the Soviet GNP in 1960, 2.7 percent in 1961, and 3.0 percent in 1962. The OECD estimate uses Soviet GNP figures, adjusted to agree as closely as possible with Western definitions.[43] With Western definitions, the Soviet GNP between 1960 and 1962 was about 10 to 15 percent lower than Soviet published figures.

Space Spending as a Percentage of the Soviet GNP

In an attempt to avoid the almost insurmountable complexities of measuring selected components of the United States GNP and the Soviet GNP, Charles S. Sheldon estimates the Soviet GNP allocation to space in terms of visible and hidden resources that must have been required for Soviet payloads to be placed in orbit. Sheldon conjectures that their approximate 1.4 lead in weight sent into orbit implies that the total Soviet space program must have been at least as large as that of the United States in the mid-1960's. Sheldon contends that because the Soviet GNP is slightly less than one-half of the United States GNP, the space program must have comprised about 2 percent of the Soviet GNP in the mid-1960's, i.e., approximately double the relative share of the United States space program.[44]

SUMMARY

The public interest served by NASA's goals is an integral part of many traditional fields of American public policy: economic welfare, defense, international relations, and social welfare. The rate and direction of these public policies evolve through the complex democratic process, which, in turn, evolves from the public and private interactions in the market place, the political arena, and the legal system. Although the space program is a national goal, its relative importance vis-á-vis other national goals (urban development, equality, education, transportation, environmental quality, etc.) cannot be put into a clearly defined national goals structure. To date, no means has been developed that measures accurately the "value consensus" of the American people. Unlike many other public policies, such as those relating to labor or to agriculture, the space program has no organized interest group or lobby to strengthen its cause before the American public and before the Congress.

Although NASA was created under the Eisenhower Administration, that administration gave little sympathy to the space program. In 1957, when the first Russian Sputnik was launched, spending by the United States for space objectives was well below the 1954 level. President Eisenhower believed the manned lunar landing objective was "nuts." The attitudes of Presidents Kennedy and Johnson toward the space program clashed with the philosophy of Eisenhower. Both Kennedy and Johnson believed that sizable federal expenditures for space, and for research and development in general, would yield high returns to the American people.

The economic environment of the late 1950's and early 1960's was conducive to expanded federal spending on research and development. The period of the late 1950's and early 1960's was that of dampened economic growth. It was not until late 1965 that the gap between what the nation was actually producing and what the nation could produce was actually closed. Between 1956 and 1965 unrealized output (i.e., from idle resources) amounted to an estimated $250 billion. This unrealized output was equivalent to all federal spending for defense between 1963 and

1967; to about 10 space programs between 1957 and 1967; to about 15 federal natural resource programs between 1957 and 1967; and to about 20 federal education programs during the decade.

With the defense buildup for Vietnam, fiscal management became more difficult and the federal budget less flexible. Increases in defense spending and also "uncontrollable" civilian spending, such as social security, veterans' benefits, Medicare, and interest on the national debt, reduced the relative size of discretionary spending in the federal budget.

Coupled with an economic environment favorable to increased federal spending in the early 1960's was the fact that the Soviet Union was accomplishing many spectacular events in space. By the end of 1964, the USSR had logged a total of 407 hours of manned spacecraft flying hours, compared with 54 hours for the United States. In addition, the Soviet Union was the first to have a live animal in space, the first to have a man (and woman) in space, the first to have a multi-manned crew, and the first to have extra-vehicular activities. By the end of 1965, the Soviet program was becoming less spectacular. With the exception of the flight in 1967 when cosmonaut Komarov lost his life, the USSR had no manned space flights between March 1965 and October 1968.

There can be no doubt that between 1957 and late 1965 domestic and international factors exerted a favorable public reaction to, and Congressional appropriations for, a sizable space program. The reverse was true after late 1965.

NOTES

1. An analysis of the relation between individual and group values and an estimate of the direction of change for an inventory of fifteen American values since 1900 was made by Robin M. Williams, a sociologist. The values in the inventory include: (1) activity and work; (2) achievement and success; (3) moral orientation; (4) humanitarianism; (5) efficiency and practicality; (6) science and secular rationality; (7) material comfort; (8) progress; (9) equality; (10) freedom; (11) democracy; (12) external conformity; (13) nationalism and patriotism; (14) individual personality; and (15) racism and related group superiority. See Robin M. Williams, "Individual and Group Values," *The Annals of the American Academy of Political and Social Science*, Vol. I, *Social Goals and Indicators for American Society* (May, 1967), pp. 20–37. Also see Eltin E. Morisson, *The American Style* (New York: Harper, 1957), and Andrew Shonfield, *Modern Capitalism* (New York and London: Oxford University Press, 1965).

2. U.S. President, Presidential press conference, April 3, 1963.

3. Jerome B. Wiesner, *Where Science and Politics Meet* (New York: McGraw-Hill, 1965), especially Chap. 2, "Science and Presidential Leadership," pp. 13–18.

4. Edwin Diamond, *The Rise and Fall of the Space Age* (New York: Doubleday, 1964), pp. 30–32.

5. U.S. Congress, Senate, Committee on Aeronautical and Space Sciences, Staff Report, *Soviet Space Programs, 1962–1965; Goals and Purposes, Achievements, Plans, and International Implications* (Washington, D.C.: U.S. Government Printing Office, 1966), p. 3.

6. Amitai Etzioni, *The Moon-Doggle, Domestic and International Implications of the*

Space Race, especially Part II, "Slowing the Economy" (New York: Doubleday, 1964), pp. 72–112.

7. Gilbert Burck, "Hitching the Economy to the Infinite," Chapter 6 in *The Space Industry: America's Newest Giant* (Englewood Cliffs, N.J.: Prentice-Hall, 1962), pp. 84–85.

8. Leonard Silk, "The Impact on the American Economy," in *Outer Space: Prospects for Man and Society*, ed. by Lincoln P. Bloomfield (Englewood Cliffs, N.J.: Prentice-Hall, 1962), pp. 76–78.

9. Even Drew Pearson gave NASA a kind word for the first manned lunar landing shot. See Drew Pearson, "Apollo: Government-Industry Miracle," *Washington Post*, July 15, 1969, p. B-11.

10. Hillier Krieghbaum, *Science and the Mass Media* (New York: New York University Press, 1967), pp. 72–73.

11. Eva Mueller discusses the usefulness of personal interview surveys as a tool of public finance in "Public Attitudes Toward Fiscal Programs," *Quarterly Journal of Economics*, 77, 2 (May, 1963), 210–35.

12. In addition to the Harris and Gallup polls that frequently appear in many newspapers, other surveys about the space program include: U.S. Congress, House, Committee on Science and Astronautics, *A Survey of Views of Leading Industrial Executives on the National Space Program*, 90th Cong., 2d Sess., 1968; and Edward F. Furash, "Business Review of the Space Effort," *Harvard Business Review*, 41, 5 (September/October, 1963), 14–52.

13. Trendex, Inc., "A Public Opinion Survey: Attitudes Toward the Government Space and Moon Program" (New York: Trendex, Inc., October, 1968). The question asked by Trendex probably was biased in favor of a "yes" reply about the manned lunar landing. The question was worded to include the phrase, "a very competitive race to be first in space exploration."

14. The debate over the manned lunar landing program and the priority that should be given to that program was an equally hot topic in Washington after President Kennedy made his request that a manned lunar landing be accomplished before 1970. "Myopia in Space," editorial, *Industrial Research* (June, 1963).

15. Some of the statements about national goals are the result of research efforts undertaken by Presidential commissions. Other statements come from private organizations and research institutes. See, for example, The President's Materials Policy Commission, William S. Paley, Chairman, *Resources for Freedom*, 5 Vols. (Washington, D.C.: U.S. Government Printing Office, 1952); J. Frederick Dewhurst and Associates, *America's Needs and Resources* (New York: Twentieth Century Fund, 1955); Committee for Economic Development, *Budgeting for National Objectives; Executive and Congressional Roles in Program Planning and Performance* (New York: Committee for Economic Development, 1966); "National Purpose," 5 parts, *Life*, May 23, May 30, June 6, June 13, June 20, 1960.

16. The President's Commission on National Goals, *Report, Goals for Americans: Program of Action in the Sixties* (Englewood Cliffs, N.J.: Prentice-Hall, 1960).

17. Leonard A. Lecht, *The Dollar Cost of Our National Goals* (Washington, D.C.: National Planning Association, 1965), p. 8.

18. Albert D. Biderman, "Social Indicators and Goals," in *Social Indicators*, ed. by Raymond A. Bauer (Cambridge, Mass.: M.I.T. Press, 1966), pp. 79–86.

19. National Planning Association, *Program Planning for National Goals*, A joint statement by members of the NPA Board of Trustees, Standing Committees, and the National Committee on America's Goals and Resources, Washington, D.C., July, 1968.

20. Arthur Smithies, *The Budgetary Process in the United States* (New York: McGraw-Hill, 1955), pp. 229–32.

21. Assuming that resources could have been transferred from space to other civilian programs and that the resources also would have been employed, B. W. Augenstein estimates that the complete termination of the space program and release of space funds for the entire period between 1960 and 1969 would have financed only a single year's growth of one of three kinds of civilian programs. The three nonspace civilian goals include: (1) agriculture and natural resources, commerce and transportation, housing and community development, and education and manpower; (2) health and welfare; and (3) federal aid directly to the poor. B. W. Augenstein, "Spend for Space, but . . . ," *Christian Science Monitor*, August 9, 1969, pp. 14–17.

22. Section 3(a) of the Employment Act of 1946 requires the President to report annually about economic developments occurring each year and to recommend policy actions that promote balanced growth of the economy. Section 4(d) of the Act requires the Council of Economic Advisers to make an annual report about economic conditions to the President.

23. Herbert Stein, "The Drive for the Big Surplus," in *The Fiscal Revolution in America* (Chicago: University of Chicago Press, 1969), pp. 346–71.

24. *Economic Report of the President*, January, 1961, pp. 55–60.

25. *Economic Report of the President*, January, 1962, p. 83.

26. For a discussion of the several ways of measuring the difference between the actual and potential gross national product, see U.S. Department of Commerce, *Long-Term Economic Growth, 1860–1965* (Washington, D.C.: U.S. Government Printing Office, 1966); *Economic Report of the President*, together with the *Annual Report of the Council of Economic Advisers 1962, 1966, and 1968;* James W. Knowles, *The Potential Economic Growth of the United States*, Study Paper No. 20, Study of Employment, Growth, and Price Levels, Joint Economic Committee (Washington, D.C.: U.S. Government Printing Office, 1960); and Arthur Okun, "Potential GNP: Its Measurement and Significance," *The 1962 Proceedings of the Business and Economics Section, American Statistical Association*, Minneapolis, September 7–10, 1962.

27. Although potentially productive manpower and equipment were idle because of the insufficient spending, there is no precise way to measure how much total spending would have had to be to eliminate unemployment and idle plant and equipment. The relationship between higher output and reduced unemployment varies from firm to firm, from industry to industry, and from region to region. In addition, the skills of available workers are not perfectly transferable among alternative jobs, and, of course, some of the unemployed have no skills. However, on the basis of changes in the gross national product and changes in the rate of unemployment during the postwar period, the Council of Economic Advisers estimates that a 3 percent increase in real GNP is usually accompanied by a reduction in the global rate of unemployment of 1 percentage point. Changes in the real GNP and in the rate of unemployment between 1958 and 1959, 1962 and 1963, and 1964 and 1965 roughly approximate this estimate. For a discussion of the relationship between changes in the real GNP and changes in rates of unemployment, see U.S. Department of Labor, "Economic Growth and Employment Opportunities," *Manpower Report of the President*, 1966 (Washington, D.C.: U.S. Government Printing Office, 1966), pp. 28–31.

28. *Economic Report of the President*, 1962, pp. xxv–xxvi.

29. *Economic Report of the President*, January, 1969, p. 299.

30. For a discussion of the reliability of stories that many Russians have died in orbit, see Charles S. Sheldon, *Review of the Soviet Space Program* (Washington, D.C.: U.S. Government Printing Office, 1967), pp. 52–53.

31. *Ibid.*, pp. 11–45.

32. Joseph G. Whelan, Leonard N. Beck, Leon M. Herman, Domas Krivickas, and Armins Rusis, *Soviet Space Programs, 1962–1965: Goals and Purposes, Achievements, Plans, and International Implications*, Staff Report prepared for the Senate Committee on Aeronautical and Space Sciences, December 30, 1966, pp. 4–11.

33. For example, see Alexander G. Korol, *Soviet Research and Development: Its Organization, Personnel, and Funds* (Cambridge, Mass.: M.I.T. Press, 1965), pp. 211–14.

34. Leon M. Herman, "Soviet Economic Capabilities for Scientific Research," *Soviet Space Programs, 1962–1965: Goals and Purposes, Achievements, Plans, and International Implications,* Staff Report prepared for the Senate Committee on Aeronautical and Space Sciences, December 30, 1966, p. 418.

35. *Ibid.*

36. Organization for Economic Co-operation and Development, *Reviews of National Science Policy: United States* (Paris: OECD, 1968), p. 32.

37. For a more thorough discussion of the dollar-ruble rate of exchange, as well as other difficulties involved with making comparison between the U.S. and Soviet economies, see Alexander G. Korol, Chapter 7, "Soviet R & D Expenditures," in *Soviet Research and Development,* pp. 200–210; U.S. Congress, Joint Economic Committee, *Comparisons of the United States and Soviet Economies* (Washington, D.C.: U.S. Government Printing Office, 1959); and Abraham S. Becher, "Comparisons of the United States and USSR National Output: Some Rules of the Game," *World Politics,* 13, 1 (October, 1960), pp. 99–111.

38. Data collected by the National Science Foundation show that the number of engineers employed in the United States totaled 900,000 in January, 1962. See U.S. Congress, Senate, Statement by Jacob Perlman, *The Role and Effect of Technology in the Nation's Economy, Hearings,* before the Senate Select Committee on Small Business, 88th Cong., 1st Sess., June, 1963, pp. 168–70. The NSF data, however, are not tied to a series for the Soviet Union.

39. Organization for Economic Co-operation and Development, *The Research and Development Effort in Western Europe, North America and the Soviet Union* (Paris: OECD, 1965), p. 140.

40. Korol, *Soviet Research and Development,* p. 55.

41. Herman, "Soviet Economic Capabilities for Scientific Research," note 34 *supra,* p. 411, and pp. 419–20.

42. Korol, *Soviet Research and Development,* pp. 224–27.

43. In part, the lack of comparability between Soviet and Western national income statistics results from the fact that the Soviet definition excludes services, whereas the Western definition includes services. Soviet and Western statistics also differ because Soviet consumer goods are overpriced in relation to Soviet capital goods. The overpricing, as compared with Western countries, results from the fact that Soviet consumer goods bear a high turnover tax. Organization for Economic Co-operation and Development, *The Research and Development Effort in Western Europe, North America and the Soviet Union,* pp. 123–25.

44. Sheldon, *Review of the Soviet Space Program,* pp. 83–84.

2 Space Goals and the Evolution of a Federal Agency

In this country, the process of developing economic policy frequently is tied to activities that have other than economic objectives. This is not to suggest that there are no independent economic policies in the United States. Since the end of the nineteenth century, a national economic policy has been the maintenance of competition, implemented by the antitrust laws. The country has economic policies directed toward agriculture, labor, public utilities, and transportation. The Employment Act of 1946 is an economic policy designed to maintain economic stability and growth.

But defense and space policies do have significant economic effects. Throughout the history of this country, the federal government has been making economic policy when trying to achieve other objectives. When the public official formulates policy for NASA, he knows that his decisions will primarily affect space objectives. Secondarily, the public official knows that he is also playing a role in the formulation of economic policy. Subsequent chapters analyze the economic effects of space policies on employment, economic stability, the concentration of economic power, national economic growth, and the economic change in NASA-oriented communities. This chapter looks at the decision-making process that ultimately leads to these economic consequences. The major decisions made by NASA concern the selection of programs, the choice of programs for NASA installations, and the selection of business firms to work under contract for the space agency.

THE SELECTION OF NASA PROGRAMS

Perusal of government documents and Congressional testimony reveals little evidence of cohesive program planning on the part of NASA in its early years. NASA's first five Semi-Annual Reports to Congress (through the end of fiscal 1960) provide descriptions of the development and accomplishments of about a half-dozen broadly based space programs: manned space flight, space probes, satellites, data tracking systems, launch vehicles, engine developments, and nuclear applications of space technology. The sixth Semi-Annual Report was the first report that explicitly mentioned program objectives—a recapitulation of those set forth by President Kennedy in his famous May 25, 1961, special message to Congress requesting that a manned lunar landing become a national goal.

The lack of explicit program plans no doubt resulted from the reactive (i.e., to Soviet accomplishments) nature of the U.S. space program and from the fact that NASA inherited from other government agencies the most important space programs that had evolved in the postwar period.

Immediate Postwar Period

With a few exceptions, the National Advisory Committee for Aeronautics (NACA) and the several military departments focused most of their advanced research on the design and development of jet-propelled aircraft immediately after World War II. It appears to have been the consensus among scientists working for NACA and for the Department of Defense that full-scale missile development held a low-order priority among military and scientific objectives. From Wallops Island, Virginia, NACA was launching a few self-propelled guided missiles. NACA's Ames Research Center in California developed the "blunt nose cone principle," thereby solving the problem of re-entry for a ballistic missile warhead. Later the same kind of nose cone was used on ICBM's and on manned space capsules (Mercury).

In the late 1940's, of course, the Department of the Army's team, led by Wernher von Braun, was working on the Hermes project. This work was the antecedent to the Army's space booster programs for the development of the Redstone, the Jupiter, and the Juno. The Navy had limited funding for the development of sounding rockets designed to gather information about the regions in which rockets and rocket-payloads would operate. Space programs per se (mainly satellites) got virtually no official approval during the immediate postwar period. Satellite projects were not deemed worth funding because their military value was not readily apparent.[1]

The International Geophysical Year and Project Development

Space projects inherited by NASA were principally those to be developed for the International Geophysical Year—July 1957 through December 1958. Although the IGY was international in scope, there was little coordination among participating nations. Under the auspices of the National Academy of Sciences, the United States IGY Committee recommended on March 14, 1955 a plan to develop a satellite. The satellite was to be operational by the end of 1958.

Although the three military services were directed by top Department of Defense (DOD) leadership to coordinate efforts in the development of the United States satellite for the International Geophysical Year, three different satellite proposals materialized. The Army Ordnance Corps proposed Project Orbiter, a four-stage rocket based on the Redstone missile, capable of placing a 5-pound payload in orbit. As a result of its sounding rocket program, the Navy had its Viking rocket. For its satellite project, the Navy proposed use of the Viking rocket for the first of a three-stage launch system. The Air Force wanted to use the Atlas missile as the first of a two-stage system.

In August 1955, the Research and Development Policy Council of DOD assigned project responsibility to the Army. The Council delegated the development of the satellite to the Navy. In September 1955, the decision was reversed; the Navy proposal was adopted and named Project Vanguard.[2]

As indicated in Chapter 1, the Soviet Union celebrated the International Geophysical Year by launching Sputnik I on October 4, 1957. In reaction, the Army Ordnance Corps was given permission to join the satellite program, and it launched the first United States satellite on January 31, 1958. After several failures, the Navy successfully launched Vanguard on March 17, 1958. That launch was actually nine months ahead of the schedule set in 1955.

Most of NASA's early programs owe their parentage to components of DOD and NACA projects. From the Navy, NASA got the Vanguard project and the IGY tracking network. NASA's Delta booster, used to launch unmanned payloads, employed improved Vanguard second and third stages. From the Army, NASA inherited the Jet Propulsion Laboratory. The Air Force transferred to NASA its funds and program for the development of a 1½ million-pound thrust engine. During the months when the Army and Navy were attempting to deploy the first United States satellite, NACA research engineers at Langley and Wallops Island were expanding studies for a manned satellite program. NASA inherited from NACA's special Space Task Group at Langley the beginning of Project Mercury, as well as other NACA projects for propulsion systems, spacecraft configuration, guidance and control systems, and other space systems.[3]

Fiscal year 1959 was three months old when NASA was established. Over 75 percent of its $339 million funding level consisted of projects and monies from NACA (30 percent) and the Department of Defense (46 percent). In two supplemental bills, Congress appropriated about $83 million directly to NASA.[4] In addition to projects and funding, NASA inherited from NACA an in-house laboratory test research capability. From DOD, NASA inherited a systems-oriented contractor network with a strong research and production capability.

NASA's First Program Plans

In its first year, NASA drastically revised its original spending and program plans. About half of the funding the agency inherited for fiscal year 1959 was earmarked for space probes, rockets, and for scientific satellites. The same allocation held for NASA's proposed R & D budget for fiscal 1960, planned during late 1958. These early programs contained little or no provision for the development of large launch vehicles. In October 1958, the Department of Defense agreed to transfer the Huntsville ABM team that was working on large boosters to NASA. At that time NASA inherited the Saturn Program, an outgrowth of the Army's Jupiter project. NASA also reduced its allocation of funds for scientific investigations of space and applied the money to vehicle development (the Scout, Delta, Vega, and Centaur).

After revision, NASA's $521 million budget plan for fiscal year 1959 and 1960 (combined) was allocated as follows: 21 percent of funds for manned space flights; 28 percent for scientific satellites; 27 percent for vehicle development; 14 percent for engine development; and 10 percent for other space programs, including the maintenance of the Jet Propulsion Laboratory and tracking and data acquisition systems.[5]

NASA's Ten-Year Plan

Congressional rumblings began by the time NASA was two years old. Many Congressional observers believed that the space program was confused and disjointed because there seemed to be no definite space goals. Much of the Congressional concern arose from disagreements over what should be the respective military and civilian roles in space.[6]

In response, NASA prepared a long-range plan. One version of the final product was classified "secret" and another version "confidential." The highlights of the Ten-Year Plan, however, were presented to Congress during the 1960 budget hearings.[7] The plan and its estimated cost for a ten-year period were modest. Funding for NASA projects was estimated to average about $1.5 billion per annum. Most conspicuously absent from the plan was a manned lunar landing between 1960 and 1970. Instead, such a landing was proposed for the decade of the 1970's. Much of the plan was directed toward developing NASA's operational capability for large launch vehicles. In July 1960, the House Committee on Science and Astronautics issued a report requesting NASA to make the manned lunar landing a goal for the decade of the 1960's. The committee also asked that NASA establish a priority for the development of the large rocket engine that would be required for the lunar landing.[8]

The National Aeronautics and Space Council

Title II of the Space Act provides for a Space Council to advise the President in his planning, developing, and coordination of a comprehensive national space program. Membership on the Council was originally composed of the President, the Secretary of Defense, the Secretary of State, the NASA Administrator, the AEC Chairman, one other public representative, and up to three persons from private organizations. At NASA's inception, one of the principal responsibilities of the Space Council was to insure coordination between NASA and the Department of Defense. During the first years of NASA, the Space Council participated in the discussions and the formulations of plans for the transfer of DOD programs and facilities to NASA.

The members of the House Committee on Government Operations apparently believed that the Space Council, during its first two years of operation, did little or nothing to coordinate DOD and NASA programs. In its thirteenth report, the House Committee pointed out that during the first two years of its existence, the Space Council had neither appointed an Executive Secretary and a supporting staff, nor had it formulated a comprehensive and integrated space program. Each had been provided for in the Space Act. The Government Operations Committee reported that the Space Council had failed so dismally in its efforts to coordinate civilian and military space functions and plans that President Eisenhower asked Congress to abolish the Council.[9] Congress did not respond to that request.

President Kennedy took a different view of the Space Council. As President-elect he appointed an ad hoc committee on space, chaired by Jerome B. Wiesner. Although that committee made no specific recommendations for space programs, it advised that the Space Council be revitalized. Accordingly, President Kennedy sought and obtained amendments to the Space Act to change the structure of the Space Council. The Council's chairmanship was transferred from the President to the Vice President, and Edward C. Welsh was appointed Executive Secretary of the Council.

Almost immediately after it was revamped, the Space Council, under the enthusiastic chairmanship of Lyndon Johnson, had to devise a plan for a United States space program in reaction to the first Russian manned space flight in April 1961. To repeat, NASA's Ten-Year Plan called for a lunar program in the post-1970 period.

In addition to the national goal of a lunar landing within the decade of the 1960's, President Kennedy's May 25, 1961 special message to Congress requested other specific space objectives—faster development of the Rover nuclear rocket, accelerated work on communications and weather satellite systems, and development of a new military solid-propellant booster.[10]

With the manned lunar landing decision, NASA for the first time had objectives that captured the imagination and sympathy of most Americans, and certainly of Congress. Projects Gemini and Apollo became synonymous with the space agency.

Role of the National Academy of Sciences

The National Academy of Sciences, established by Abraham Lincoln more than a century ago, serves the federal government in an advisory capacity. The Space Science Board of NAS was established in June 1958 to assist in the development of space research programs after the end of the International Geophysical Year. One function of the Space Science Board has been to sponsor space science summer study groups. Participants at these conferences have included scientists in a wide range of academic disciplines, as well as persons representing government agencies engaged in the formulation of space policy (mainly the Department of Defense, the Atomic Energy Commission, the National Science Foundation, and the National Aeronautics and Space Administration).

The first summer study was conducted at the State University of Iowa in 1962. The work of that group, published in 1963, resulted in a 16-chapter tome entitled *A Review of Space Research*. A second summer study took place at Woods Hole, Massachusetts in 1965. The efforts of the 1965 summer group were directed toward program planning in the post-Apollo period.[11] The research of a third summer studies program, also at Woods Hole during 1967 and 1968, was on benefits derived from alternative space programs. The results and recommendations of the 1967–1968 summer group will be discussed in Chapter 3.

On the whole, NASA agreed that the research outlined in the reports of the Space Science Board should be undertaken. At the same time, however, NASA officials believed that only a fraction of the space projects and programs outlined could be implemented.[12] For example, the Iowa summer study group estimated that the United States would spend about \$10.6 billion for space research in fiscal 1967 and about \$12.5 billion by fiscal 1969,[13] whereas spending for space actually peaked at

$5.9 billion in fiscal 1966. By fiscal 1969, space spending amounted to $4.2 billion, about one-third of that estimated by the Iowa group in 1962.

The 1962 space science study group assigned as the top priority goal for the United States space program the investigation of extra-terrestrial life. The group also called for the manned exploration of Mars and for the construction and operation of new satellite programs.[14] A relatively low-cost recommendation made by the group was that NASA employ scientist astronauts on its manned space flight programs. Accordingly, NASA appointed five scientist astronauts in June 1965. In all, the 1962 group made about 130 different proposals for space projects and programs.

In contrast with its predecessor study group, the 1965 summer studies group in its report shifted space program priority emphasis away from manned experiments to unmanned scientific experiments. The group of scientists gathered at Woods Hole recommended that NASA put primary emphasis on the unmanned exploration of Mars and secondary emphasis on Venus and the other major planets. That group believed that unmanned experiments would result in the most significant advances in the exploration of space during the 1965–1985 period.[15] The 1965 summer studies group did not publish projections of federal spending for the space objectives that it outlined.

The Apollo Period

The manned lunar landing decision brought solidarity to the space program. For example, in fiscal 1959, NASA was spending about $11 million, i.e., 6 percent of its total R & D funds, on the development of large thrust engines. At that time, NASA had no definite missions for the giant engines. NASA's first Semi-Annual Report stated that its engine program would provide a booster large enough to launch experiments and payloads weighing several tons. Table 2–1 displays data that show the overwhelming role played by the Manned Space Flight Program. Research and development funding for the manned moon project rose from 25 percent of total spending in 1959 to 71 percent in fiscal 1967, amounting to about two-thirds of NASA's R & D budget during the decade. Although not shown separately, the greatest proportion (over half) of R & D funding for manned space flight was for the development of the Apollo spacecraft and for the development of the Saturn V launch vehicle.

For the most part, NASA's space science and applications programs are the different satellite projects that became household words to millions of Americans because of television. NASA's unmanned lunar and planet space probe projects are also administered by the Office of Space Science and Applications. The Explorer and Pioneer series are physics and astronomy projects.

The first United States space probe, which discovered the Van Allen

TABLE 2–1: SPACE PROGRAM DEVELOPMENT: FUNDS FOR RESEARCH AND DEVELOPMENT, FISCAL 1959–1968 (*MILLIONS OF DOLLARS*)

PROGRAM DEVELOPMENT	1959	1960	1961	1962	1963	1964	1965	1966	1967	1968	TOTAL	PERCENT
Manned space flight	50	119	296	538	1,502	2,718	2,961	3,200	3,024	2,789	17,197	66.2
Mercury	39	83	105	15	[a]	[a]	[a]	[a]	[a]	[a]	242	.9
Gemini	[a]	[a]	[a]	55	289	419	308	197	15	[a]	1,283	4.9
Apollo	11	36	191	452	1,185	2,278	2,629	2,941	2,923	2,538	15,184	58.5
Apollo applications	[a]	[a]	[a]	[a]	[a]	[a]	[a]	51	80	251	382	1.5
Advanced manned missions	[a]	[a]	[a]	[a]	11	21	24	10	6	[a]	72	.3
Space science and applications	129	148	260	438	620	708	688	716	576	549	4,832	18.6
Physics and astronomy	6	21	35	79	113	160	149	143	130	145	981	3.8
Lunar and planetary exploration	12	20	61	112	170	191	188	221	184	141	1,300	5.0
Launch vehicles[b]	28	60	115	182	216	242	251	236	149	124	1,603	6.2
Meteorological satellites[c]	1	8	14	27	42	69	31	32	42	[a]	266	1.0
Space applications	4	4	18	23	35	89	64	78	71	99	485	1.9
Advanced research and technology	14	22	32	103	264	317	332	289	269	313	1,955	7.5
Nuclear	3	4	7	27	69	79	57	58	53	54	411	1.6
Chemical propulsion	7	3	1	23	16	22	35	42	36	66	228	.9
Tracking and data acquisition	4	16	30	66	122	191	253	231	271	275	1,459	5.6
Total[d]	197	348	672	1,301	2,459	3,926	4,363	4,531	4,245	3,925	25,967	100.0

[a] No funding during the fiscal year.

[b] Excludes the Saturn series.

[c] Includes the Echo, Telstar, Relay, Syncom, Tiros, and Nimbus series.

[d] Includes other programs, principally Office of Administration, Technology Utilization, and University Affairs.

Source: National Aeronautics and Space Administration.

Radiation Belt in February 1958, was part of the Explorer series. Under its physics and astronomy programs, NASA has put three kinds of observatories into orbit: the Orbiting Solar Observatory (OSO); the Orbiting Geophysical Observatory (OGO); and the Orbiting Astronomical Observatory (OAO). NASA's physics and astronomy projects accounted for about 4 percent of the agency's total R & D spending during the last decade; the Centaur, Scout, and Delta have been the principal launch vehicles for this series.

Despite urging by the scientific community, NASA's lunar and planetary exploration series have not changed (i.e., increased) markedly since mid-1960. Those projects accounted for about 9 percent of NASA's R & D budget in 1961, compared with 4 percent in 1968. As a matter of fact, Congress took an independent view; in 1966, funds for the unmanned exploration of Mars (Voyager) were re-allocated so as to delay the anticipated launch from 1971 to 1973.[16] To be sure, Congress did this because of pressing Vietnam commitments.

The Surveyor soft lunar landing spacecraft was also part of the NASA lunar and inter-planetary program series. Because of the concentration on the manned lunar landing objective, NASA's Space Science and Applications program dropped from about 65 percent of NASA's R & D funding in fiscal 1959 to 14 percent in 1968; in dollars, the program authority quadrupled during the decade.

NASA's advanced aeronautics and space research projects have been fairly consistently funded at between 7 and 9 percent of NASA's R & D budget. Projects in this program office have ranged in diversity from NASA's participation in the national supersonic and hypersonic transport projects to the short-take-off and landing (STOL) and vertical-take-off and landing (VTOL) aircraft; and from studies on the sonic boom to a flight-weight nuclear engine (NERVA).

Pre-Apollo II Planning for Post-Apollo Programs

Programs promulgated by NASA in the mid- to late 1960's, like those in the early years of its existence, lacked Congressional and public appeal.[17] With the completion of NASA's Mercury shots in 1963, Congressional and public enthusiasm for the space program began to diminish from the heights reached during late 1961 and early 1962. For all practical purposes, Project Mercury was completed in May 1963 with Gordon Cooper's 22-orbit flight. Because of long-lead preparation, NASA's next manned flight did not take place until March 1965. Between fiscal 1961 and 1963, NASA's R & D budget had risen from about $700 million to $2.5 billion, and NASA's entire budget had risen from about $900 million to $3.5 billion during the same period. Despite reduced enthusiasm for the space program, the NASA budget continued to grow by about $2 billion during the following two years.

As early as the spring of 1964, President Johnson made inquiries on criteria and priorities for space missions that might follow those approved for the decade of the 1960's.[18] Emilio Daddario, member of the House Committee on Science and Astronautics, believed that Congress had become increasingly concerned with NASA's inability to look to the future after the manned lunar landing.[19]

It is difficult to know the reasons for the lack of long-range planning on the part of NASA during the mid- to late 1960's. Vietnam, no doubt, holds part of the explanation, as does the tragic Apollo accident. But as early as August 1965, James E. Webb, then Administrator of NASA, told Congress that:

> . . . what is necessary now is not a firm decision to undertake specific large new projects, but continued support and funding of the already ongoing scientific and technical efforts, and a rapid assimilation into the stream of our national policy making of opportunities and benefits that can be derived from the Gemini, Apollo, orbiting observatories, Nimbus, and other programs.[20]

The statement does not imply that NASA scientific advisory groups or industry did not examine alternative space goals for the post-Apollo period. On the contrary, the possible projects discussed were many. It does mean, however, that at that time there was no agreement among policy-making parties on the goals that should be pursued in the decade of the 1970's and beyond. It also means that NASA did not release schedules of space milestones to be accomplished, together with rough estimates of relative costs.[21]

Considerations Given to Future Space Objectives. In response to the President's 1964 request to evaluate space priorities, NASA established a Future Programs Task Group. The consensus of that group was that because of the heavy budget demands of established programs, such as Gemini, Apollo, Lunar Orbiter, and Surveyor, priority should be given to short-run planning unless the NASA budget was increased. The Task Group reported that unless national needs changed, large new mission commitments should be deferred.[22]

During the first half of 1966, the House Subcommittee on NASA Oversight studied the matter of future space objectives, seeking a delineation of national space goals in the 1970's. The subcommittee made inquiries to NASA management at headquarters, NASA center directors, aerospace industrial management, and the Department of Defense. The results of the study (a 450-page public document) revealed a wide spectrum of views. The only general agreement pertained to the urgent need to make positive and bold decisions on future national space goals at the earliest possible time.[23] The subcommittee urged that criteria be set forth for near-term (1967–1976) and long-term (1976–1985) program planning. For near-term planning, the subcommittee recommended the application of existing technology for such programs as a manned earth-orbital system, unmanned flybys of the planets, and unmanned scientific investigations of the solar system. The subcommittee called upon NASA to step up the evolution of its advanced programming.

Adjusting to Budget Cuts. At times, immediate program planning becomes more imperative than short-term and long-term planning. NASA, along with other government agencies, was required to make program adjustments in response to the Revenue and Expenditure Control Act of 1968. NASA requested $4.4 billion for fiscal 1969 and Congress authorized $4.0 billion for that year. NASA then imposed a $3.8 billion ceiling on itself to accommodate its expected share of total federal expenditure reductions.

The decrease in NASA's funding from the President's budget request amounted to about $500 million for fiscal 1969. About two-thirds of that reduction was absorbed by the Apollo Applications Program. Under NASA's revised budget plan, unmanned lunar and planetary exploration projects and the nuclear rocket program were also drastically reduced. The manned lunar landing program was not affected much by the revamping of NASA's budget.

THE SELECTION OF NASA INSTALLATIONS

With the exception of the research centers transferred to NASA from other government agencies in the late 1950's, NASA's decisions on the selection of field installations are an example of de facto cost-effectiveness analysis. As Chapter 3 brings out, there is nothing new about rational decision-making in the federal government. What is new is the somewhat more formal and uniform framework that planning, programming, and budgeting (PPB) provides for the government decision-making process. To be sure, NASA did not use a formal PPB system when it selected the locations of installations. However, NASA site selection teams established priorities and criteria, and they evaluated the relative merits of hundreds of possible locations for the field centers that eventually were established. The ultimate decision for the location of two NASA research centers (Houston, Texas, and Cambridge, Massachusetts) exemplifies the influence of political maneuvering that frequently takes place in Congress. This is true especially for the Electronics Research Center in Cambridge, which was established only after prolonged and heated Congressional debate.

Facilities Transferred to NASA

The beginning of NASA's acquisition of facilities and properties was contained in the Space Act. Title III, Sections 301 and 302, of the Act provided for the termination of all functions performed by the National

Advisory Committee for Aeronautics (NACA). Title III also specified that the duties and properties of that agency be transferred to the new National Aeronautics and Space Administration. Upon inception, NASA inherited the five principal research installations of NACA, the oldest of which had been established during World War I.

The facilities transferred from NACA included Langley Research Center, Ames Research Center, Lewis Research Center, Flight Research Center, and Wallops Station. With the exception of Wallops Station, each of those field installations is (1969) under the jurisdiction of NASA's Office of Advanced Research and Technology. The Office of Space Science and Applications administers the rocket launch and tracking facilities at Wallops Station, Virginia.

The Space Act also authorized NASA, in broad and general terms, to acquire, construct, improve, repair, operate, and maintain research laboratories, testing sites, and related facilities necessary for the conduct of its mission.[24]

In April 1958, several months before NASA was actually established, President Eisenhower directed NACA and the Department of Defense to arrange for the transfer of necessary DOD facilities to the new agency and to cooperate in the joint use of facilities.[25] This executive directive ultimately led to the transfer to NASA of the government-owned, contractor-operated Jet Propulsion Laboratory at the California Institute of Technology in December 1958, and to the transfer of the Army's ballistic missile facilities in July 1960. Responsibility for the Jet Propulsion Laboratory rests with NASA's Office of Space Science and Applications and responsibility for the Marshall Space Flight Center is with the Office of Manned Space Flight.

For perspective, Table 2-2 shows the organizational structure of NASA's fourteen research centers. The information, applicable in 1969, gives the location, the principal research function, the date of original establishment, and the date each center was acquired by NASA.

Goddard Space Flight Center

Just one month after President Eisenhower signed the Space Act and before the actual establishment of the National Aeronautics and Space Administration, Maryland newspapers announced on August 1, 1958, that the first "new" NASA field center would be located in Beltsville, Maryland. About 550 acres of the Beltsville Agricultural Test Center of the United States Department of Agriculture were transferred to NASA. The center, named after a pioneer in rocket technology, Robert H. Goddard, is located 15 miles northeast of Washington, D.C. After the initial transfer, NASA acquired an additional 640 acres in Greenbelt, Maryland.

An important reason for establishing this new field installation to supplement those centers transferred from NACA was to expedite and to accommodate the absorption of the Navy's Washington, D.C.-centered Vanguard project. Although the Vanguard scientific satellite project was transferred to NASA by an Executive Order on October 1, 1958, the move into a permanent building at the Goddard Space Flight Center did not take place until late 1960. During the interim period, civilian personnel working on the Vanguard project used the facilities of the Naval Research Laboratory in Washington, D.C.

As Table 2–2 shows, the Goddard Space Flight Center is responsible for the research necessary for unmanned spacecraft and sounding rocket experiments. Some of the meteorological and communication satellites developed at Goddard include: Tiros, Nimbus, Relay, and Syncom. Two world-wide satellite tracking networks are housed at the Goddard Space Flight Center. These global communication network facilities link NASA's tracking stations in the United States to the rest of the world.

The Space Nuclear Propulsion Office

Inter-agency cooperation on the space program extended beyond the transfer of programs and facilities from various military and civilian departments to NASA. In 1960, NASA and the Atomic Energy Commission entered into an agreement to establish the joint NASA-AEC Space Nuclear Propulsion Office. The headquarters office of SNPO is located at the Atomic Energy Commission facility in Germantown, Maryland. The Space Nuclear Propulsion Office has an Albuquerque field extension office that maintains liaison with AEC's Los Alamos Scientific Laboratory. Another field extension office in Cleveland maintains liaison with NASA's Lewis Research Center. In addition, a Nuclear Rocket Development Station was established at AEC's Nevada Test Station at Jackass Flats.

The Space Nuclear Propulsion Office was initially established to develop compact energy sources for space experiments and explorations beyond those required for the manned lunar landing mission. Currently, SNPO is engaged in activities for the development of nuclear propulsion systems, as well as other systems that generate large amounts of electrical power using nuclear reactors.[26]

The Manned Spacecraft Center

Within five months after President Kennedy's request in May 1961 that the United States commit itself to a successful manned lunar landing, NASA established four new field locations to assist in meeting that objective. In August 1961, it was announced that Cape Canaveral (now Cape Kennedy) had been selected as the launch site for the Manned Space

TABLE 2–2: THE LOCATION AND MISSION OF NASA'S RESEARCH CENTERS

RESEARCH CENTER	NASA OFFICE AND MISSION	COMMENT
Langley Research Center Hampton, Va.	Office of Advanced Research and Technology: Space and aeronautic research on vehicle configuration, materials, and structures.	Established in 1917 by NACA. Transferred to NASA at inception. Initial support for Project Mercury.
Ames Research Center Moffett Field, Calif.	Office of Advanced Research and Technology: Research in such physical sciences as environmental physics; guidance and control systems; and aeronautics magnetic fields.	Established in 1940 by NACA. Transferred to NASA at inception. Initial support for Project Mercury.
Lewis Research Center Cleveland, Ohio	Office of Advanced Research and Technology: Research on advanced propulsion and power generation systems.	Established in 1941 by NACA. Transferred to NASA at inception. Responsible for the operation of the Nuclear Propulsion Test Station of Plum Brook in Sandusky, Ohio.
Flight Research Center Edwards, Calif. (Mojave Desert)	Office of Advanced Research and Technology: Manned flight research within and outside the atmosphere.	Established in 1947 by NACA. Transferred to NASA at inception.
Electronics Research Center Cambridge, Mass.	Office of Advanced Research and Technology: Research on space and aeronautical electronic components and systems.	Established, after heated Congressional debate, by NASA in 1964. NASA announced the closing of ERC in December 1969.
Space Nuclear Propulsion Office Germantown, Md.	Office of Advanced Research and Technology: Development and test work on components and systems used in nuclear rockets.	Established by NASA in 1962. Manages the joint NASA-AEC nuclear rocket program. The principal test area is located in Jackass Flats, Nev.
Wallops Station, an island in Accomack County, Va.	Office of Space Science and Applications: Rocket-launch facilities and tracking station.	Established by NASA in 1945. Transferred to NASA at inception and expanded in 1959 to include the Chincoteague Naval Air Station.
Goddard Space Flight Center Greenbelt, Md.	Office of Space Science and Applications: Dual mission of development of unmanned satellite programs and maintenance of two world-wide satellite tracking networks.	NASA's first "new" space flight center, established in 1958.

Jet Propulsion Laboratory California Institute of Technology, Pasadena, Calif.	Office of Space Science and Applications: Multifarious R & D activities for lunar and deep space probes.	Established by the War Department in 1936. Transferred from the Army to NASA in 1958. JPL is a government-owned, contractor-operated installation.
Marshall Space Flight Center Huntsville, Ala.	Office of Manned Space Flight: Develop heavy launch vehicles and space transportation systems for the Manned Space Flight Program.	Initial facilities transferred from the Army Ballistic Missile Agency in July 1960. The Center is responsible for operations at the Michoud Assembly Facility, the Mississippi Test Facility, and the Slidell Computer Facility.
Kennedy Space Flight Center Brevard County, Fla.	Office of Manned Space Flight: Checkout and launch of manned and unmanned space vehicles.	Established by NASA in August 1961. The Air Force's Eastern Test Range, established in 1950, is also located at Cape Kennedy (formerly Cape Canaveral).
Manned Spacecraft Center Houston, Tex.	Office of Manned Space Flight: Design and development of manned spacecraft; astronaut crew training; and space flight mission control.	Established by NASA in September 1961. The center is responsible for operations at the White Sands Test Facility.
Michoud Assembly Facility New Orleans, La.	Office of Manned Space Flight: Design, fabrication, and assembly of launch vehicles.	Built in 1942 as a shipyard, taken over by the Army in 1951 for the manufacture of tank engines, and transferred to NASA in September 1961. Michoud is a government-owned, contractor-operated facility.
Mississippi Test Facility Hancock County, Miss.	Office of Manned Space Flight: Acceptance testing of rocket engines and launch vehicles.	Established by NASA in October 1961. The Mississippi and Tennessee Rivers connect MTF with the Michoud facility and with the Marshall Space Flight Center. MTF is government-owned and contractor-operated.

Flight Program; during September 1961, NASA began preparations for the establishment of the Manned Spacecraft Center in Houston, and for the acquisition and conversion of the unused government-owned Michoud manufacturing plant in New Orleans; and, one month later, NASA announced that it would establish a vehicle test facility in southwestern Mississippi (Hancock County).

As indicated, NASA inherited from NACA the Manned Space Flight Program, popularly known as Project Mercury. As the pace of the space program intensified, NASA became concerned over possible inadequacies of its training facilities located at Langley. On July 7, 1961, Robert C. Seamans, then Associate Administrator, sent a memorandum to the Director of Manned Space Flight Programs requesting that surveys be made for the selection of a site for the Manned Space Flight Laboratory. In response to the July memorandum, Abe Silverstein, then Director of Manned Space Flight Programs, outlined site selection criteria. The first item set forth by Silverstein was a "general geographic guideline": "In view of the present concentration of NASA installations on the East and West Coasts and in order to extend to a new section of the country an opportunity to participate and contribute to the space effort, preference will be given to a mid-continent location."[27]

According to the record of a meeting of the site survey team held in Washington in August 1961, the survey team agreed not to question the general geographic guideline. The record of that meeting does not make it clear, however, whether the geographic criterion would be included as a factor in the site selection decision-making process. It is impossible to know the relative importance of at least three factors in the final decision on the establishment of the Manned Spacecraft Center in Houston: (1) a desire on the part of NASA officials to broaden the geographic base of the nation's research and development effort; (2) the actual advantages that Houston itself offered; and (3) possible political pressures.

Although the political aspect in the decision-making process is shrouded in secrecy, considerable speculation exists. The speculation arises from the fact that several key political positions were held by Texans at that time. Also, Lyndon B. Johnson was then Vice President of the United States and Houston Congressman Albert Thomas was then Chairman of the House Independent Offices Subcommittee. In 1961, that subcommittee was responsible for handling NASA appropriations.[28] Not unexpectedly, NASA officials emphatically denied that improper political pressures had been brought to bear on the site selection decision.

In addition to the geographic criterion, the Silverstein memorandum outlined factors that should be considered in selecting the site of the proposed Manned Spacecraft Center. The criteria were listed in three orders of priority: "essential," "important," and "desirable." The site selection committee subsequently incorporated most of the "important" factors into the "essential" group. The following factors were considered by the site selection committee in its evaluation.

Essential Criteria

Transportation. The location must provide the capability to transport, by barge, large and cumbersome space vehicles. The location should also have both an all-weather commercial jet service airport and a Department of Defense air base installation capable of handling high performance military aircraft for use by astronauts in maintaining flight capability.

Climate. To speed construction and to reduce facility costs, the area should have a mild climate to permit out-of-door work throughout the year.

Local Industry Support. The area should be a well-established industrial complex to support research and development activities and to fabricate prototypes of spacecraft.

Labor Supply. To permit rapid construction of facilities without incurring premium labor costs, the area should have a well-established supply of competitive construction contractors and a good supply of workmen in the building trades.

Community Facilities. To enhance the recruitment and retention of scientific personnel, the site should be located close to a culturally attractive community, with facilities for higher education.

Area. In addition to a minimum of 1,000 acres, the site should have semi-isolated adjacent acreage for further development.

Other Essential Factors. The location should have public utility systems capable of providing minimum specified amounts of electrical power, water, and telephone service to the center.

Desirable Factors

Impact on the Area. The proposed facility should be compatible with any existing regional development plans and the community should possess sufficient flexibility to absorb increased population and expansion of industrial and transportation facilities.

Site Development and Operating Costs. In site selection, consideration should be given to the relative costs that will be incurred for site development, and ultimately costs of operation.

Interim Facilities. The site should have reasonably adequate facilities to house about 1,500 people on a temporary basis.[29]

The committee reviewed and evaluated eighteen possible site locations. Table 2–3 presents information about the site selection committee's evaluation of each of the eighteen locations.[30] As a result of the committee's review, the number of locations warranting further investigations

TABLE 2–3: EVALUATION OF POSSIBLE SITES FOR THE MANNED SPACECRAFT CENTER

POSSIBLE LOCATIONS	CRITERIA						SCORE		
	TRANSPORTATION								
	JET AIRPORT	DOD AIR BASE	MILD CLIMATE	COMMUNITY CULTURAL AND EDUCATIONAL FACILITIES	ADEQUATE COMMUNICATIONS SYSTEM	ADEQUATE WATER SUPPLY	NUMBER "YES"	NUMBER "UNCERTAIN"	NUMBER "NO"
Baton Rouge, La.	No	Yes	Yes	Yes	Yes	Yes	5	0	1
Charleston, S.C.	No	Yes	Yes	Yes	Uncertain	Yes	4	1	1
Corpus Christi, Tex.	Uncertain	Yes	Yes	No	Yes	Yes	4	1	1
Hampton, Va.	Uncertain	Yes	Yes	No	Uncertain	Yes	3	2	1
Houston, Tex.	Yes	Yes	Yes	Yes	Yes	Yes	6	0	0
Jacksonville, Fla.	Yes	Yes	Yes	No	Yes	Yes	5	0	1
Los Angeles, Calif.	Yes	Yes	Yes	Yes	Yes	Yes	6	0	0
Memphis, Tenn.	No	Yes	Yes	No	Yes	Yes	4	0	2
Miami, Fla.	Yes	Yes	Yes	No	Yes	Yes	5	0	1
Mobile, Ala.	No	Yes	Yes	No	Yes	Yes	4	0	2
New Orleans, La.	Yes	Yes	Yes	Yes	Yes	Yes	6	0	0
Portland, Ore.	Yes	No	Yes	No	Yes	Yes	4	0	2
Santa Barbara, Calif.	No	Yes	Yes	Yes	Yes	Yes	5	0	1
San Diego, Calif.	Yes	Yes	Yes	Yes	Yes	Yes	6	0	0
San Francisco, Calif.	Yes	Yes	Yes	Yes	Yes	Yes	6	0	0
Savannah, Ga.	No	Yes	Yes	No	Yes	Yes	4	0	2
Seattle, Wash.	Yes	Yes	Uncertain	Yes	Yes	Yes	5	1	0
Tampa, Fla.	Yes	Yes	Yes	No	Yes	Yes	5	0	1

Source: Adapted from the minutes of the meeting of the site selection committee, Washington, D.C., August 11, 1961.

was reduced to ten: Baton Rouge, Houston, Jacksonville, Los Angeles, Miami, New Orleans, Santa Barbara, San Diego, San Francisco, and Tampa. If the geographic criterion had been applied, only Baton Rouge, Houston, and New Orleans would have qualified as having a mid-continent location. Houston and New Orleans were both given the maximum number of evaluation points.

NASA Administrator James E. Webb announced on September 19, 1961, that the new Manned Spacecraft Center would be located in Houston. Site selection, Mr. Webb maintained, was influenced only by NASA's decisions on the locations of other facilities required for the accelerated Manned Space Flight Program.[31]

Michoud Assembly Facility

Because of the widely diverse requirements, site selection for the plant that was to manufacture NASA's launch vehicles was quite different from that of the Manned Spacecraft Center in Houston. The vehicle manufacturing plant selection committee began the evaluation task on August 4, 1961. The assignment of the committee, given by the Associate Administrator, Abe Silverstein, was to recommend selection of a suitable and existing manufacturing plant for NASA's launch vehicle program.[32]

The committee's selection of the Michoud Assembly Facility was based on requirement analyses of possible government-owned or contractor-owned plants. Before establishing the criteria for selection, the committee set forth six basic assumptions. They were: (1) program schedules would not permit construction of a new facility; (2) the fabrication and assembly of present and future NASA vehicles would require a substantial manufacturing area; (3) minimum production capacity would be based on the immediate requirements of the present program, but plant capacity should exceed immediate requirements to provide for future expansion; (4) work at the plant would be of a development nature in the initial phase, in contrast with straight assembly line production; (5) all launches would be from the Atlantic Missile Range; and (6) all transportation of completed units would be by water.

Proceeding from these assumptions, the committee outlined the criteria for plant selection. The committee also distinguished between those that were "essential" and those that were "desirable."

Essential Factors

Accessibility. The plant must be on year-round navigable water, with a minimum width of 100 feet and a minimum depth of 10 feet; and must also have access to rail transportation.

Availability. The plant must be available to NASA by September 30, 1961.

Size. The plant must have a minimum of 675,000 square feet, with a high bay area of 350,000 feet (overhead clearance of at least 40 feet and width of 90 feet); 200,000 square feet of engineering and administrative space must also be available.

Utilities. To attain adequate cleanliness, the plant must be air-conditioned. Public utilities had to be able to supply the minimum specified amounts of electric power and water.

Community Support. An urban area capable of absorbing 3,000 employees and their families must be near the plant.

Foundation. The plant must have a sub-surface foundation capable of supporting unusually heavy machinery.[33]

Desirable Criteria

Desirable features included: proximity to an urban area with available housing, shopping, educational, religious, recreational, cultural, and commercial airport facilities. The committee deemed it desirable that the plant be so located relative to the Marshall Space Flight Center and to the Atlantic Missile Range that travel would require less than two weeks by barge. Location in an area with a diversified manufacturing base and an available supplementary labor force was also considered advantageous by the committee.

56 Possible Locations

The U.S. Army Corps of Engineers, along with various government agencies, provided the plant selection committee with data on possible manufacturing facilities. The committee evaluated 56 possible plants in 23 different states. For most plants, insufficient size (including a high bay area) was the main disqualifying factor. Almost three-fourths of the evaluated plants could not provide either adequate floor space or overhead clearance for the production of NASA's gigantic launch vehicles. Proximity to test and launch areas precluded selection of plants located on the Pacific Coast and also those in the central states that were not located on navigable rivers. About 25 percent of the plants simply were not available for occupancy by the September 30, 1961 deadline established by NASA. Lack of capability for future expansion disqualified about 25 percent of the possible plants.

The Michoud plant, located about 15 miles east of downtown New Orleans, has a history of reconversion and deactivation dating from the early years of World War II. The plant was initially designed for the manufacture of ocean-going ships. Auxiliary facilities, such as canals and pilings for shipways, were constructed before the plant itself. The government changed its plans in the early 1940's, and the facility was completed as a plant for the manufacture of plywood airplanes. After the production and test of two prototypes, the plywood aircraft was deemed unsatisfactory. Shortly after that, the government closed the Michoud plant, and it remained deactivated until the beginning of the Korean War. At that time, the Chrysler Corporation leased the facility for the production of engines for Sherman and Patton tanks. Chrysler modified the existing production facilities and also installed an air-conditioning system. Although the plant was deactivated between 1954 and 1961, the government had maintained it for standby use.

NASA Launch Vehicle Test Site

The ad hoc committee for the selection of NASA's launch vehicle test site issued its report only a few days after the vehicle manufacturing plant site selection committee made its report. The test site criteria were formulated in conjunction with those established for the vehicle manufacturing site selection procedure.

For example, the committee members recognized that it would be ideal to locate the manufacturing, test, and launch facilities at one master location. At that time, the principal NASA launch area was located at the Atlantic Missile Range in Brevard County, Florida, but lack of sufficient land precluded the establishment of a three-function master site at that location. The committee believed, however, that time lags could be reduced by locating the manufacturing and test sites in the same general area.[34] The requirement for continuous liaison among technical personnel during the manufacture and test stages of the program also made proximity of the manufacturing and test sites desirable. Blast and sound level hazards resulting from testing, however, added the additional requirement that the site be located in a remote area. The committee believed that wide separation of the facilities probably would have an adverse effect on costs, delivery schedules, and the reliability of space vehicles. Therefore, a "critical distance"[35] factor between the manufacture and test sites became an important criterion in the site selection process.

In addition to the objective of proximity to the manufacturing site, the committee set forth other criteria, distinguishing those that were essential from those that were desirable.

Some Essential Factors

Isolation. Public safety required that the test facility be located in an isolated area. Based on calculations, an eight to ten mile minimum distance to densely populated areas was established. Another essential was that all lands within a 125 decibel accoustical boundary be acquired and that human habitation be prohibited within a 110 decibel boundary (i.e., acquisition of 8,100 acres and inhabitation restrictions for 151,000 acres).

Accessibility. The site had to be accessible to the ocean by a navigable channel having a minimum depth of 10 feet and width of 100 feet. Also, the site had to be located within 15 miles of a public highway.

Climate. To permit year-round operations, the climate had to be mild.

Desirable Criteria

Foremost of the second-order criteria was that of a maximum barge trip of two weeks from the test site to the Marshall Space Flight Center and to the Atlantic Missile Range. Other desirable criteria were: a maximum of 50 miles between the test site and the manufacturing facility; scheduled airline services for the surrounding communities; a diversified industrial base; an available labor force for facility support operations; and adequate shopping, educational, religious, and cultural facilities.

30 Possible Sites

Utility companies, the Army Corps of Engineers, the Air Force Civil Engineers, and Navy Public Works Offices provided NASA's site selection team with a list of 30 possible sites for the space vehicle test facility. Between August 9 and August 12 and on August 17, 1961, the committee made site review trips.[36] As a result of the findings the committee judged that six sites warranted further evaluation. The six sites were: Brownsville, Texas; Corpus Christi, Texas; Cumberland Island, Georgia; Elgin Air Force Base, Florida; New Orleans, Louisiana; and the Pearl River Basin, Mississippi.

The committee made a comparative evaluation of the six possible sites. The evaluators used a point-rating system, 300 points being allocated to construction cost criteria and 300 points being allocated to operation costs. For facilities construction, only those costs that would vary significantly among the locations were considered. Costs that would be incurred for such facilities as test tower structures and buildings were not included. Although such costs would also vary between the possible sites, their exclusion probably did not distort the analysis because each of the possible locations is in an isolated area in the southern part of the United States.

Among those costs considered, land acquisition, foundation support, and access construction costs (i.e., canals and bridges) were the largest components of the total. Estimated costs and the number of points assigned were inversely related.[37] The high cost, low-point site was located in the vicinity of New Orleans (40 miles southeast). The location was a delta swamp, only two feet above sea level. Construction costs for foundations and for road access were markedly higher than similar costs at other possible locations.

The weighting system reflected the committee's belief that there should be proximity between the manufacturing and test sites. Proximity to the Michoud Assembly Facility accounted for more than half of the total number of points assigned to all operations. Lack of proximity to the manufacturing facility and to the Marshall Space Flight Center eliminated the Texas locations and Cumberland Island. The report of the committee acknowledged the fact that any weighting system that combines intangibles is arbitrary. The report leaves little doubt that selection of the Pearl River area was planned with the selection of the Michoud Assembly Facility. The committee believed that it was imperative to select a site in the vicinity of New Orleans to insure personnel liaison during production and test operations. The New Orleans delta site, another possibility, was less satisfactory because of much poorer construction and working conditions.

The Mississippi Test Facility, as it was later named, is located 38 air miles from New Orleans. The facility is located in Hancock County, with much of the acoustic buffer zone in Pearl River County.

Cape Canaveral Launch Site (Cape Kennedy)

As indicated earlier, most of the Army Ballistic Missile Agency's launch facilities located at Cape Canaveral were transferred to NASA under provisions of the Eisenhower directive of April 1958. With the decision to pursue an intensive manned space flight program, however, NASA had to build facilities to launch large vehicles. One possible site, of course, was Cape Canaveral. In late June 1961, a launch site survey committee was established to evaluate potential sites. The site survey team, a joint NASA-DOD undertaking, was headed by Kurt H. Debus, then Director, NASA Launch Operations Directorate, and Major General L. I. Davis, USAF, then Commander, Air Force Missile Test Center.

This launch site survey committee evaluated launch requirements for vehicles with sufficient payload capability to accomplish a manned lunar landing and return by direct flight. In the foreword of its report, the committee stated that the study results were preliminary and subject to change as the Manned Space Flight Program requirements became firmly defined. The committee analyzed launch requirements for an all-liquid NOVA vehicle and for a solid-propellant first-stage NOVA vehicle. The committee released the results of its study on July 31, 1961. Almost a year later, on July 11, 1962, NASA announced its decision to use the lunar orbit rendezvous technique for the manned lunar landing. The decision for the lunar orbit rendezvous mode, in preference to the direct flight or the earth orbit rendezvous mode, is discussed in Chapter 3.

Ground Rules Set for the Committee

A June 1961 memorandum sent by the Associate Administrator, Robert C. Seamans, to General Davis and to Dr. Debus established the ground rules for the committee evaluation of possible launch sites. The committee was directed to concern itself solely with launch operations. The committee was to assess and evaluate possible launch site locations, land acquisition requirements, spacecraft and launch vehicle preparation facilities, vehicle launch facilities, and other requirements for a large vehicle launch site. Given the ground rules, the committee made its site appraisal on the basis of seven principal criteria: costs, accessibility, schedule, location, launch vehicle impact hazard, overhead flight hazard, and use of existing facilities. For each criterion, evaluations were made for both the all-liquid NOVA vehicle and for the solid-propellant first-stage NOVA vehicle.

With a few exceptions, the basic launch complex for both kinds of vehicles was essentially the same, consisting of vertical assembly buildings, pad and sub-pad facilities, operations support buildings, and facilities for propellants. Hence, there was little difference in the estimated costs resulting from changes in vehicle configuration at any given launch site. A given cost estimate, however, pertained to either the solid or the liquid vehicle rather than for a launch complex capable of launching both solid- and liquid-propellant vehicles. The committee stated that the costs of a dual capability launch complex would be considerably greater than any estimate given.[38]

The Alternative Costs of Possible Sites

Table 2–4 displays data on the committee's estimates of capital and operating costs for the launch complex required for a solid-propellant first-stage NOVA vehicle at eight possible locations.

Construction cost estimates for the Christmas, Hawaii, and Mayaguana Island locations were markedly higher than those for other sites. Christmas Island is located about 1,150 miles south of the Hawaiian Islands. At the time, the Island was used by the British Atomic Weapons Establishment. Construction of a launch complex at Christmas Island would have required the importation of materials and of labor because of an insufficient indigenous population. Mayaguana Island, a British possession located about 550 miles south of Cape Canaveral, also lacked a skilled as well as unskilled available labor supply. Neither Christmas nor Mayaguana Island had adequate housing and community support facilities. The committee reported that the Island of Hawaii could provide both labor personnel and community facilities. Because of material imports, construction costs were estimated to be about 1.4 times higher in Hawaii than on the West Coast of the United States. Mayaguana Island is in the hurricane belt, and the Island of Hawaii, of course, has volcanic hazards.

The lower land acquisition costs for an off-shore Cape Canaveral location, compared with those for an on-shore location, were estimated to be more than offset by higher construction costs. The higher construction

TABLE 2–4: COST COMPARISON FOR POSSIBLE LAUNCH SITES TO SUPPORT THE MANNED LUNAR LANDING PROGRAM

POSSIBLE LAUNCH SITES	ESTIMATED CAPITAL COSTS[a] (MILLIONS OF DOLLARS)				ESTIMATED OPERATING COSTS
	LAUNCH FACILITIES[b]	RANGE SUPPORT FACILITIES[c]	REAL ESTATE[d]	TOTAL	BASE = 100[e]
Brownsville, Tex.	$ 699	$202	$21	$ 922	128.1
Cape Canaveral, Fla. (on-shore)	694	123	60	877	110.7
Cape Canaveral, Fla. (off-shore)	844	126	22	992	122.9
Christmas Island, Pacific Ocean	1,607	852	[f]	2,459	473.7
Cumberland Island, Ga.	699	194	15	908	126.3
Mayaguana Island, Atlantic Ocean	1,121	905	[g]	2,026	223.7
South Point (Ka Lae), Haw.	1,170	410	9	1,589	275.2
White Sands Missile Range, N.M.	699	147	[h]	846	100.0

[a] Cost estimates are for solid-propellant first-stage vehicles.

[b] Estimated cost of facilities and buildings for launch and missile support.

[c] Estimated cost of facilities and buildings for range support.

[d] Cost figures are for the minimum amount of land required.

[e] The estimates are projected annual mission and range operating costs for fiscal 1966. The low-cost area (White Sands) is set to be the base; other areas are shown in relation to the low-cost area.

[f] Insufficient time for evaluation and agreement by committee.

[g] Subject to negotiation because Mayaguana Island is a British possession.

[h] Estimated to be $200,000.

Source: Adapted from Department of Defense and National Aeronautics and Space Administration: *Phase I: Joint Report on Facilities and Resources Required at Launch Site to Support the NASA Manned Lunar Program* (Debus-Davis Report) July 31, 1961.

costs for the possible off-shore site resulted from the need to construct additional structural support for the vertical assembly and launch building, higher utility costs, and additional costs for the construction of a breakwater and a causeway. Because water and weather conditions have a greater adverse effect off-shore than on land, the operating costs and maintenance costs of an off-shore location also would be higher than those incurred at an on-shore site.

In terms of money outlay only for the construction and operation of a launch complex, the White Sands Missile Range in New Mexico clearly would have been the least-cost site. The committee estimated that land acquisition cost would have amounted to about $200,000, compared with $15 million for a Cumberland Island site and $60 million for the on-shore Cape Canaveral site. Table 2–5 provides qualitative information on the committee's appraisal of White Sands, compared with other possible sites,

TABLE 2-5: COMPARATIVE ANALYSIS OF POSSIBLE LOCATIONS FOR LAUNCH SITE TO SUPPORT THE MANNED LUNAR LANDING PROGRAM

EVALUATION CRITERIA	COMMITTEE COMMENT
Cost	On basis of costs alone, advisable to eliminate Christmas Island, Hawaii, and Mayaguana.
Accessibility	All sites except White Sands Missile Range have water accessibility and permit barge transport from assembly and test points to launch area.
Schedule	Mission and range support facilities can be constructed in time to support expected flight schedule at each proposed site.
Location	Cumberland Island and Brownsville would require intermittent closing of intracoastal waterway. Christmas Island, Hawaii, and Mayaguana would create logistic problems; proximity of Christmas Island to equator would provide some technical advantages.
Launch vehicle impact hazard	For Brownsville and White Sands Missile Range the launch azimuth would have to be limited because several large cities[a] would be located within a larger fan area. No serious land impact hazard exists for the other sites.
Overhead flight hazard	For Brownsville and White Sands Missile Range, large portions of the United States would be flown over through the second burnout stage. Some sizable population centers[b] might be endangered by possible abort during first- or second-stage burning. No serious overflight hazard exists for the other sites.
Use of existing facilities	Economies that might be derived from use of existing equipment, facilities, and technical experience would favor White Sands Missile Range or Cape Canaveral, Fla., as the possible launch vehicle site.

[a] Dallas, Austin, Galveston, and Houston would be subject to booster impact hazard from a White Sands launch with a wide azimuth.

[b] For example, Tampa, Palm Beach, Miami, New Orleans, and Dallas.

Source: Adapted from Department of Defense and National Aeronautics and Space Administration: *Phase I: Joint Report on Facilities Required at Launch Site to Support the NASA Manned Lunar Landing Program* (Debus-Davis Report), July 31, 1961.

for non-cost criteria. The disadvantages of White Sands included: the requirement that vehicle manufacturing facilities be located near the Missile Range because of the lack of water accessibility; the launch azimuth[39] would have to be limited to avoid possible launch impact hazard; and some sizable population centers (Atlanta, Birmingham, New Orleans, Dallas) might be endangered by overflight hazards. The committee believed that an East Coast or a Pacific island site would reduce hazards to the continental United States from possible abort during the first or second stage burning of the launch.[40]

The Debus-Davis Report was submitted to the Associate Administrator on July 31, 1961. Unlike the reports of the other site evaluation committees, the report did not make an explicit recommendation that

Cape Canaveral (on-shore site) be selected as the location from which to launch America's manned space vehicles. Information in the committee's report indicated that the site had cost advantages over all sites except White Sands. In addition, it ranked high on other evaluation criteria. NASA's announcement of the selection of Cape Canaveral came on August 24, 1961. In September 1961, NASA asked Congress to authorize the purchase of 80,000 acres of land on Merritt Island, Florida to expand the Cape Canaveral launch facilities. Although Congress adjourned before the requested authorization was enacted, NASA, using its reprogramming authority, transferred funds from its research and development account to its construction of facilities account and began to acquire the land before the end of that calendar year.[41] Two years later, in December 1963, Cape Canaveral was renamed the John F. Kennedy Space Center.

Electronics Research Center

Selection of Boston for the site of the Electronics Research Center caused more political speculation and furor than did selection of any of the other new research centers established by NASA. Congressional hearings indicate that Albert J. Kelley, then Director of NASA's Electronics and Control Group, recommended to James E. Webb that such a center be established early in the fall of 1962. Kelley had been directed to devise a plan to meet NASA's growing requirements in electronics research. According to Kelley's Congressional testimony, NASA examined three possible ways to strengthen the agency's electronics capability: (1) the establishment of several diffused electronics research groups at existing centers; (2) establishment of an electronics research subcenter at the Langley Research Center; and (3) establishment of a new center.[42]

According to testimony, NASA initially discussed the proposed Electronics Research Center, at the Boston location, with the Bureau of the Budget in October 1962. Discussions about the proposed center were kept within the Executive Department until the proposal was incorporated into the 1964 Presidential Budget Request. In Congressional correspondence, James E. Webb stated that the (then) proposed electronics research center was held so closely that he (Webb) doubted whether any member of Congress knew about the matter until it was discussed at press briefings the day before the President sent his budget to Congress.[43]

The controversy over possible political favor from NASA must have arisen from the fact that a number of delegates from Massachusetts held key Congressional positions at that time. During the second session of the 87th Congress, John W. McCormack was Speaker of the House. Joseph W. Martin, former minority leader, as well as former Speaker of the House, was a member of the House Committee on Science and Astronautics. Both Edward P. Boland and Silvio O. Conte were members of the House Appropriations Committee. President Kennedy, of course, was in the White House.[44]

The fiscal 1964 appropriation authorized funds for the establishment of an Electronics Research Center by NASA. Congress, however, made the authorization contingent upon the preparation, by NASA, of a detailed study on the need for, the nature of, and the geographic location of the proposed center.[45] NASA's report was submitted to Speaker of the House John McCormack, on January 31, 1964.

Essential Criteria

The location criteria set forth by the Electronics Research Center site selection committee differed sharply from those outlined by other site selection committees. The primary mission of the Electronics Research Center was to be the management of a joint NASA-university-industry program for advanced research in the development of electronic components and subsystems, rather than the manufacture, test, and launch of gigantic space vehicles.

The four criteria set forth by the Electronics Research site selection committee were: (1) proximity to educational institutions that have proven high-level research capabilities in science and engineering; (2) proximity to an area with industrial research and development capabilities in electronics; (3) proximity to an established, research-oriented, scientific and engineering community; and (4) availability of rapid and convenient air transportation facilities with connections to the principal space flight centers as well as to other parts of the country.

After preliminary investigations, the committee dropped a fifth criterion, proximity to adequate housing, cultural, religious, and recreational facilities. The committee believed that any area that could provide research-oriented graduate educational facilities also would have such community support facilities.

Table 2–6 provides information about the Electronics Research Center site selection committee's evaluation of nine different locations. On the grounds of an apparent imbalance in the geographic distribution of federal funds for research and development and an avowed attempt on the part of NASA to build a broadly based national research and development network, James E. Webb directed the committee to exclude from its investigations any areas in or near established NASA centers.[46] By this directive, possible locations such as Baltimore, Los Angeles, and San Francisco were excluded from consideration.

The committee devised an academic rating system for its first and prin-

TABLE 2–6: EVALUATION OF POSSIBLE SITES FOR A NASA ELECTRONICS RESEARCH CENTER

EVALUATION CRITERIA	POSSIBLE LOCATIONS									
	ANN ARBOR-DETROIT	BOSTON	CHICAGO	MINNE-APOLIS	NEW YORK	PHILADEL-PHIA	PITTS-BURGH	PRINCE-TON	ST. LOUIS	RANKING LOCATION
Academic strength in electrical engineering and physics										
Master's degrees conferred (weighted average)[a]	147	551	123	60	780	338	128	72	40	New York
Ph.D. degrees conferred (weighted average)[a]	28	103	37	13	55	25	43	28	16	Boston
Electronics R & D environment										
Small businesses (number)	23	175	113	60	242	54	12	37	7	New York
Research laboratories (number)	32	120	89	19	282	46	45	34	8	New York
R & D employees (number)	7,100	14,700	5,600	3,900	44,100	13,300	4,100	11,000	4,500	New York
General R & D environment										
Professionals in R & D laboratories (number)	1,700	5,000	3,000	2,100	22,700	6,200	2,100	5,000	1,900	New York
Institute of Radio Engineers membership (number)	700	4,200	1,900	600	10,100	3,600	400	2,100	400	New York
Electronic engineers, U.S. 1960 Census (number)	3,600	6,000	8,100	2,300	21,300	7,600	2,600	N.A.	2,300	New York
National Science Foundation list of scientists (number)	1,800	6,000	6,300	2,400	13,500	4,200	2,400	N.A.	1,700	New York
Technical meetings held annually (number)	27	99	81	20	149	97	26	11	12	New York
Availability of air transportation	good	good	good	good	good	good	good	good	good	all satisfactory
Housing, cultural, and recreational facilities	good	good	good	good	good	good	good	good	good	all satisfactory
Number of times ranking location	zero	one	zero	zero	nine	zero	zero	zero	zero	

[a] The average number of degrees conferred in electrical engineering and physics between 1960 and 1962, weighted on a 2 to 1 basis in favor of electrical engineering.

Source: Adapted from National Aeronautics and Space Administration, *Electronics Research Center,* Report of the U.S. House Committee on Science and Astronautics (Washington, D.C.: U.S. Government Printing Office, 1964), Table p. 2, no page number given.

cipal criterion. The educational environment of possible sites was to be judged by averaging the numbers of graduate degrees awarded by accredited academic institutions, within a 30-mile radius of proposed sites, in electrical engineering and in physics between 1960 and 1962.[47] According to the evaluation system, the Massachusetts Institute of Technology ranked second to Stanford University in doctoral degrees awarded in electrical engineering, and second to the University of California (Berkeley) in the number of degrees awarded in physics. The decision to exclude those areas already having NASA centers eliminated the California institutions and placed Boston first. The fact that the New York area ranked ahead of Boston for the educational criterion at the master's degree level was disregarded because the facilities were scattered over too wide an area. The New York area was defined to include: New York City, Jersey City, New Rochelle, the New Jersey Oranges, Staten Island, and Long Island.

To evaluate the strength of the possible sites in research and development in electronics, the committee used information provided in the eleventh edition of the joint National Academy of Sciences and National Research Council's publication, *Industrial Research Laboratories in the United States*. The committee also used the *List of Small Business Concerns Interested in Performing Research and Development* (4th edition) prepared by the Small Business Administration. As Table 2–6 shows, Boston ranked second to the New York City area on the basis of numbers of small businesses, research laboratories, and research and development personnel, but well ahead of the other seven locations listed. The general research and development environment of possible sites was judged by the number of different kinds of technical personnel working in each area, and by the annual number of professional and technical meetings held there.[48] As measured by this criterion, Boston again ranked behind the New York metropolitan area. On some professional membership lists, Chicago and Philadelphia also ranked ahead of Boston.

In 1964, after submitting its site evaluation report to Congress, NASA began to acquire land and construct its Electronics Research Center. The President's Budget requested $5 million for fiscal 1964. After extensive hearings, Congress authorized $3.9 million for the initial costs of the facility. The Electronics Research Center, located about one and a half miles west of the center of Boston, is immediately north of the Massachusetts Institute of Technology and one and a half miles from Harvard University. By the end of 1967, four buildings had been constructed at the Electronics Research Center, with a capital investment amounting to $11.5 million.[49] In December 1969, NASA announced its plan to close the Electronics Research Center as an economy measure.

CONTRACTOR SELECTION AND SPACE SYSTEMS PROCUREMENT

The question of the most efficient organizational arrangement between the government, industry, and the nonprofit world for the development of large missile systems has been a controversial public issue since early 1945. At that time, the Joint Chiefs of Staff established a Special Committee on Guided Missiles to examine and to monitor the nation's guided missile program. In deciding on its institutional arrangements, NASA could, and did, draw on the experience of six research-oriented predecessor agencies: the Atomic Energy Commission, the International Geophysical Year Program of the National Science Foundation, the National Advisory Committee on Aeronautics, and the three military agencies in the Department of Defense.

NASA'S PREDECESSOR AGENCIES

Table 2–7 presents data that show the varying degrees of contractor participation in the research and development programs of six government agencies in 1957. The information is also given for 1966. At one end of the spectrum was the Atomic Energy Commission, with about 2 percent of its R & D funds allocated to an in-house effort in 1957. Historically, the Atomic Energy Commission has relied on universities and industry to manage its facilities and to do its research work. At the other end of the spectrum was the National Advisory Committee on Aeronautics, with only 2 percent of its research performed under contract in 1957. Although a few universities and nonprofit research institutions participated in the special National Science Foundation Program for the International Geophysical Year, most of the program was carried out by public employees in government-owned facilities.

The Arsenal System of Army Ordnance

Despite the fact that the Department of the Army has traditionally owned and operated its munitions factories, industry received about 90 percent of the Army's funds for missile procurement in the late 1950's. The percentage was somewhat lower for other R & D programs, with industry performing about 55 percent of all R & D for the Army. Testifying before Congress, Major General John B. Medaris stated that the Army arsenals were, in fact, research and engineering centers, concerned primarily with the design, development, and test of prototypes. Once accepted, new weapons systems were turned over to industry for production in quantity. In the late 1950's, the Army did not seek much help from industry for technical direction or for systems engineering. Rather, the Army Ballistic Missile Agency assumed the responsibility of "weapons systems manager." The Army, however, did have a long-term con-

TABLE 2–7: FEDERAL OBLIGATIONS FOR RESEARCH AND DEVELOPMENT, FISCAL 1957 AND 1966

AGENCY	TOTAL (MILLIONS OF DOLLARS)	PERCENT OF TOTAL			
		INTRAMURAL	EXTRAMURAL		
			PROFITS	UNIVERSITIES	NONPROFITS
		Fiscal 1957			
Department of Defense	$1,884	54	37	7	2
Army	468	55	32	10	2
Navy	579	50	42	8	1
Air Force	800	54	38	6	2
National Advisory Council on Aeronautics	55	98	[a]	2	[a]
International Geophysical Year[b]	13	69	[a]	23	8
Atomic Energy Commission	420	2	64	29	5
		Fiscal 1966			
Department of Defense	$7,024	27	63	6	4
Army	1,585	36	59	3	1
Navy	1,602	46	45	7	2
Air Force	3,342	14	76	4	5
National Aeronautics and Space Administration	5,050	17	78	4	1
Atomic Energy Commission	1,212	2	55	39	4

[a] Less than 1 percent.

[b] Administered as a special program of the National Science Foundation.

Sources: National Science Foundation, *Federal Funds for Science, VII*, pp. 68–71, and *Federal Funds for Research, Development, and Other Scientific Activities, XVI*, pp. 112–13.

tract with the Jet Propulsion Laboratory of the California Institute of Technology. The Jet Propulsion Laboratory played an important role in the development of the Army's missile capability. For example, the Jet Propulsion Laboratory developed the first surface-to-surface ballistic missile. As indicated in the last section, the Jet Propulsion Laboratory, under contract, was transferred to NASA in late 1958.[50]

Navy Procurement Methods

Like the Army, the Navy has an in-house technical capability for the design and development of systems based on experience gained in its shipyards. During the late 1950's, about 50 percent of all Navy research was performed in government-owned facilities. Also akin to the Army Ballistic Missile Agency in Huntsville is the Naval Ordnance Test Station at China Lake, California. Another similarity in the institutional arrangements of the two agencies is the long-term contract that the Navy has had with the Applied Physics Laboratory of the Johns Hopkins University. The Applied Physics Laboratory performed the initial research and development work for a number of naval weapons and missile systems and passed the results along to industrial manufacturers. Although the Navy Special Projects Office performed the systems management function for a number of projects in the 1950's, including the Polaris, the Navy did let a number of systems management contracts with business firms during the late 1950's.[51]

Contractor Support and the Air Force

In the development of its missile program through the late 1950's, the Air Force relied on industry to develop and to produce its missile systems.[52] At that time, the technical management resources of the Air Force were provided by one company, Space Technology Laboratories, Inc., formerly a division of the Ramo-Wooldridge Corporation. Space Technology Laboratories was responsible for the technical development of the Atlas, the Thor, the Titan, and the Minuteman missiles. Each of the missile systems was under prime contract with a different contractor (Convair, Martin, Douglas, and Boeing, respectively). The role of the four primes was to assemble and test the missile systems, with assistance from Space Technology Laboratories.[53]

NASA's Procurement Policy

The National Aeronautics and Space Act of 1958 granted to NASA broad authority and flexibility in procurement actions. Section 205 (b) (5) gave NASA's Administrator the authority "to enter into . . . and perform such contracts . . . or other transactions as may be necessary to the conduct of its work and on such terms as it may deem appropriate."

The pattern of contractor participation that evolved during the last decade has some characteristics of the systems procurement policy of the Army, the Navy, and the Air Force. NASA performs the technical management of its programs, including the complete conduct of flight tests, at its research centers. This basic organization roughly resembles that of the Army Ordnance arsenal system. Yet NASA relies more heavily on industry for the design and manufacture of its systems than does the Army or the Navy. As Table 2–7 shows, NASA's R & D contracts with business firms were about the same percentage as those of the Air Force in fiscal 1966. The urgency of developing a United States manned space flight capability and the rapid increase in spending for space after mid-1961 was reflected in a large decrease in the relative size of intramural R & D spending. In fiscal 1961, NASA spent about 25 percent of its R & D funds internally, compared with 15 percent in 1962 and 10 percent in 1963.

Selection of NASA Contractors

Shortly after its inception, NASA procured the services of McKinsey and Company, management consultants. That concern investigated NASA's contracting policies and industrial relationships. In October 1960, McKinsey and Company presented its report to T. Keith Glennan, then Administrator of NASA. In two sentences, McKinsey and Company described the totality of NASA's contractor selection process. To be sure, the same description also applies to any other agency of the federal government. "No aspect of NASA's job is more politically sensitive than its contracting process. In substantial part, this political sensitivity arises out of the large volume of contracts being let and their significance to individual contractors and to the communities in which their plants are located."[54]

The substance of the sentences, of course, is that it is really not possible to find out how NASA actually selected its half-dozen or so principal contractors. The information can only be gleaned from Congressional testimony, NASA correspondence in defense of its actions, and from research studies made by academic groups. This does not mean that the NASA machinery for contractor selection cannot be fully described; it can, in fact, be described in great detail. What it does mean is that the corporate information accompanying contract bids is held as privileged. This, in turn, means that the criteria used for contractor selection are not in the public domain.

For example, in 1959, NASA awarded a $102 million contract to North American Aviation for the development of a 1.5 million-pound thrust, single-chamber rocket engine. Congress wanted to know how and why North American Aviation won the contract. Representative Overton

Brooks requested Glennan to provide the House Committee on Science and Astronautics with NASA documents that supported the selection of North American Aviation. Although Glennan was willing to release some documents to Congress, he was not willing to supply the one key document—the report of the Source Evaluation Board. Representative Brooks sought the aid of the General Accounting Office to obtain the report of NASA's Source Evaluation Board, to no avail. In defending that report as privileged information, Glennan obtained the support of President Eisenhower. The President, as a matter of executive privilege, did not release the requested documents to either the GAO or to the House Committee. Virtually the same controversy flared again in 1960 over the selection of the McDonnell Aircraft Corporation to build the Mercury spacecraft. The same resolution was also made.[55]

NASA's Source Evaluation Board

For large contracts[56] NASA establishes a Source Evaluation Board to assist in the selection of the contractor. NASA procurements, of course, differ from many other government procurements for several reasons: (1) NASA's equipment must be virtually failure-proof for man-rating (i.e., space vehicles safe for man); (2) NASA's procurement is for low-volume, high-cost hardware; and (3) NASA's requirements encompass all the risks and uncertainties associated with the acquisition of sophisticated R & D systems. This means that cost is only one, and not necessarily the most important, criterion evaluated by the Source Evaluation Board.

Each board, consisting of about eight technically cognizant members, first develops, approves, and releases a Request for Proposal (RFP) and a Source Solicitation List. After establishing its set of weighted criteria, the board evaluates and ranks all acceptable proposals. The formal evaluations made by the board contain much information and consist of many pounds of paper. The findings of the Source Evaluation Board, however, are only facts and are used only as guides in the selection process. It is the Source Selection Official that makes the final decision.

For expected procurement awards of $5,000,000 or more, the Administrator of NASA and other high NASA officials (usually the Deputy Administrator and one or two Associate Administrators) are the Source Selection Officials. For field procurements expected to be in excess of $1,000,000 but less than $5,000,000, the appropriate field installation director assumes responsibility as the Source Selection Official. When a procurement action of similar size originates from headquarters, the cognizant headquarters program director becomes the Source Selection Official.[57]

Major steps in the procedure of the Source Evaluation Board (SEB) can be summarized as follows: (1) SEB review of the statement of work and SEB development of qualification and evaluation criteria; (2) SEB establishment of the preliminary source list; (3) SEB review of the request for proposals; (4) issuance of the request for proposals; (5) holding proposals conference; (6) SEB establishment of the scoring or rating system to be used; (7) technical reviews of proposals to SEB; (8) SEB review and evaluation of technical reports and first ranking of proposals; (9) SEB review of contractor experience gained from available government data, from visits to plants, from interviews with company officials, and from correspondence with companies in the competition; (10) SEB final ranking of proposals; (11) SEB written report on its findings and oral presentation to the Source Selection Official; and (12) contract selection statement signed by the Source Selection Official.

The Economic Roles of 93 Business Firms Working on the Space Program

Since its inception, NASA has had prime or subcontractual arrangements with at least 20,000 different organizations.[58] To understand the roles played by these contractors, it is useful to look at the economic functions they perform for NASA. Peck and Scherer outlined a convenient classification for analyzing the economic roles of defense and space contractors.[59] They suggest the following quintet: systems firms, subsystems firms, overflow procedures, components and parts producers, and materials makers. For NASA, it is necessary to add another economic function: that of the service support firms.

Because it is impossible to classify all 20,000 of NASA's contractors, the grouping that follows is limited to 93 business firms that provided employment data to the joint NASA/DOD Economic Information System in June 1966.[60] Each one of the business firms had at least 10 employees working on space projects in mid-1966.

A given distinction between a subsystem and a major subsystem, or for that matter, between a minor subsystem and a major component, need not be universally accepted. Furthermore, the activities of a given firm can overlap between two or more economic functions, or even the entire spectrum. Because the manned lunar landing project is NASA's largest program, the classification is based on the primary manned space flight activity of the company during fiscal 1966. The space program has been shrinking and also changing since 1966. Therefore, the classification is not necessarily applicable after 1966. For those companies doing little or no work for the Manned Space Flight Program, the major NASA activity determines the classification.

The Systems Firms

The space systems firms contract to deliver complete space systems, such as a launch vehicle or a guidance system. Typically, these firms engage in research and development work as well as in production and assembling activities, and they are responsible for the delivery of the complete space system. Although they are prime contractors for particular systems, these firms also do subcontract work for other systems firms or even subsystems firms. For example, North American Aviation (now North American Rockwell) was NASA's largest subcontractor as well as its largest prime contractor during fiscal 1966.

TABLE 2–8: NASA'S SPACE SYSTEMS FIRMS

NAME OF COMPANY	ECONOMIC FUNCTION
Ball Brothers Research	Orbiting Solar Observatory (non-MSF)
Boeing Company	Launch vehicle – S-IC stage of Saturn V; lunar orbiter (non-MSF); technical support; program integration
Chrysler Corporation	Launch vehicle – First stage Saturn IB
Douglas Aircraft[a]	Launch vehicles – Saturn S-IVB stage; Delta (non-MSF)
Fairchild Hiller	Pegasus (non-MSF); engineering services
General Dynamics	Launch vehicles – Atlas; Little Joe II; Centaur (non-MSF)
General Motors	Guidance and navigation system; Apollo spacecraft
Grumman Aircraft Engineering	Lunar module; Orbiting Astronomical Observatories (non-MSF)
Hughes Aircraft	Surveyor; Syncom, applications technology satellites (non-MSF)
International Business Machines	Saturn instrument unit; technical services
Martin Marietta	Gemini launch vehicles – Titan II; payload integration for Apollo applications
McDonnell Aircraft[a]	Mercury, Gemini spacecraft; airlock
North American Aviation	Command and service module; launch vehicle – S-II stage of Saturn V; Saturn launch vehicle engines

[a] Before merger of McDonnell Aircraft and Douglas Aircraft.

Source: Compiled from data in the Joint NASA/DOD Economic Information System and from NASA's Fiscal 1966 Procurement Report.

Table 2–8 shows NASA's principal systems contractors. Although much of the work undertaken by Hughes Aircraft Company is under subcontract from the Jet Propulsion Laboratory, its responsibility for the Surveyor (the soft unmanned lunar landing module) warrants placing the company among the space systems firms. The Boeing Company and International Business Machines are also included among the systems firms. For these firms the overlap among the economic functions of system producer and support services is clear. The Boeing Company is engaged in program integration for the manned lunar landing project and International Business Machines operates the real-time computer at the Manned Spacecraft Center in Houston, Texas.

The Subsystems Firms

The role of these firms is to provide major subsystems, such as engines, earth landing systems, and stabilization systems. Although most subsystems firms are subcontractors, some are also prime contractors. Those firms under prime contracts provide such subsystems as mission simulators or operations display devices directly to NASA. Table 2–9 lists NASA's principal subsystems contractors and the subsystems that the companies have provided to NASA contractors and also to NASA.

In this grouping TRW, Inc., is classified as a subsystem firm because of its work for the Grumman Aircraft Engineering Company on the descent engine for the lunar module. The company also provides such technical services as trajectory and systems analyses for the Manned Space Flight Program at the Manned Spacecraft Center in Houston. In view of its NASA-wide mission, TRW could have been grouped among the systems firms. That company has been responsible for several unmanned satellite projects, including Mariner, Pioneer, and the Orbiting Geophysical Observatory (OGO).

The Overflow Producers

Using the description of Peck and Scherer, overflow producers are almost exclusively subcontractors. The task of the overflow producers differs from that of the subsystems firms because the technical R & D competence of the overflow producer is the same as that of the contracting firm. The overflow producer gets NASA business because of insufficient capacity in the plants and laboratories of the contracting firm. Firms that let contracts to the overflow producers are the major systems firms and the principal subsystems firms. Slightly more than 10 percent of NASA's contractors supplying data to the Economic Information System in June 1966 were overflow producers. Table 2–10 lists the names of NASA's overflow producers and also gives the names of the firms letting contracts to these overflow producers.

The Components and Parts Producers

These firms provide such components as valves, pressure tanks, actuators, and gauges directly to NASA under prime contract, or indirectly through NASA's contracts. Although these companies provide parts for various space systems, they are not necessarily small business firms. A glance at Table 2–11 shows that some of the giants in American industry are included among the suppliers of components and parts. Borg-Warner, Cutler-Hammer, Fairchild Camera and Instrument, Hercules, International Telephone and Telegraph, Lear Siegler, Litton Industries, Texas Instruments, Varian Associates, and Western Electric were all among the 500 largest industrial corporations in 1966.

TABLE 2-9: NASA'S SUBSYSTEMS CONTRACTORS

NAME OF COMPANY	SUBSYSTEM PROVIDED BY COMPANY
American Machine & Foundry	Vehicle simulator; Apollo docking test device
AVCO Corporation	Heatshield for Apollo spacecraft
Beech Aircraft	Cryogenic storage tanks for Apollo spacecraft and lunar module
Collins Radio	Communications and data systems for Apollo spacecraft
Conrac (Gianinni Controls)	Apollo propellant quantity gauging systems
Curtiss-Wright	Turbojet combusters
Dynamics Corp. of America	S-Band antenna for Apollo tracking ships; lighting systems for lunar module
General Precision	Apollo mission simulator
General Signal	C-Band transponders for Apollo spacecraft
General Telephone & Electronics (Sylvania)	Transportable ground stations
General Tire and Rubber (Aerojet General)	Propulsion systems
Goodyear Tire & Rubber (Goodyear Aerospace)	Propellant tanks; temperature controls
Honeywell, Inc.	Stabilization and control systems; inertial reference guidance systems
Marquardt Corp.	Reaction control system for Apollo spacecraft
Motorola, Inc.	Various electronic subsystems for major space systems firms
Northrop Corp.	Earth landing systems
Radio Corp. of America	Communications system for lunar module; various radar subsystems
Raytheon Company	Computers for guidance and navigation systems
Ryan Aeronautical	Radar altimeter and velocity sensor for Surveyor
Sanders Associates	Operational display system for Marshall Space Flight Center
Schlumberger, Ltd. (Electro-Mechanical Research)	Telemetry pulse code modulating ground system
Signal Oil & Gas (Garrett Corp.)	Environmental control systems
Standard Kollsman Instrument	Optical subsystems for guidance and navigation system
TRW, Inc.	Lunar module descent engine
Textron, Inc. (Bell Aerospace Corp.)	Lunar module ascent engine
Thiokol Chemical	Solid-propellant engines; retro engine for Surveyor
United Aircraft	Fuel cells; liquid propulsion engines; LM environmental control system; space suits
Westinghouse Electric	Television camera for lunar module

Sources: Compiled from data in the Joint NASA/DOD Economic Information System and from NASA's Fiscal 1966 Procurement Report.

TABLE 2-10: FIRMS PERFORMING THE OVERFLOW FUNCTION FOR NASA CONTRACTORS

NAME OF COMPANY	OVERFLOW FUNCTION
Aeronca, Inc.	Honeycomb panels for Apollo spacecraft (North American Aviation)
American Bosch Arma	Control assemblies for Apollo spacecraft (Grumman)
Bogue Electric	Motor generators for Saturn V (General Electric)
Cook Electric	Recoverable camera capsule (North American Aviation); experiments (University of Wisconsin)
Electronic Communications	Specification changes for Saturn V (Douglas)
Emerson Electric	Receiving stations; Electrical equipment (General Electric and Chrysler)
Federal Mogul Bower	Bellows for launch vehicles; ducting (Boeing, Chrysler, and North American Aviation)
International Harvester (Solar Division)	Heat exchangers; ducting, fuel lines (Boeing and North American Aviation)
Murdock Machine and Engineering	Honeycomb panel for lunar orbiter (Boeing)
Rohr Corp.	Heat shield panels for Saturn V (Boeing)
Sun Shipbuilding and Drydock	Test chambers (Northrop)

Sources: Compiled from data in the Joint NASA/DOD Economic Information System and from NASA's Fiscal 1966 Procurement Report.

The Materials Makers

These companies provide basic materials, such as fiberglass, high-quality steel, aluminum, and titanium, for the nation's space program. Because the plants of these firms are geared toward civilian production, the Joint NASA/DOD Economic Information System does not have information about many materials makers. None of the companies used in this sample, i.e., those with at least 10 employees working on space projects, are materials makers. Procurement awards lists, however, show that prime and subcontract awards are let to a number of materials producers. For example, DuPont de Nemours supplies fibrous potassium to the Boeing Company; Union Carbide supplies liquid helium to the Grumman Aircraft Engineering Company; American Cryogenics supplies liquid oxygen to the Marshall Space Flight Center in Huntsville; and the Borden Company supplies separators for silver oxide to the Goddard Space Flight Center in Beltsville.

The Service Firms

As indicated, the firms in this group provide engineering and other scientific services to NASA and to NASA contractors. Another kind of firm is that which provides maintenance and operation services. For the most part, these firms work under prime contract. In this grouping, Lock-

TABLE 2-11: FIRMS PROVIDING COMPONENTS AND PARTS FOR THE NATION'S SPACE PROGRAM

NAME OF COMPANY	COMPONENTS PRODUCED
Babcock Electronics	Relays
Borg-Warner	Pumps
Cutler-Hammer	Frames; antennas; amplifiers
EDO Corp.	Sensor probes
Excello Corp.	Gimbal actuators
Fairchild Camera & Instrument	Optical and photographic parts
General Instrument	Test equipment; simulators
Hercules, Inc.	Solid-propellant motors
International Telephone and Telegraph	Connectors; wire; receivers
Lear Siegler	Couplings; gyroscopes; cameras; filters; transformers
Litton Industries	Resolvers; antenna equipment
Menasco Mfg.	Valves; tanks; checkout systems
Moog, Inc.	Actuators
Packard-Bell Electronics	Computer parts; checkout systems
Perkin-Elmer	Optical and electronic sensors
Pneumodynamics	Hatch actuators; shock actuators; value actuators
Raymond Engineering Laboratories	Various electronic parts
Sangamo Electric	Tape recorders and parts
Texas Instruments	Various electronic parts
United Industrial	Pyrotechnics
Varian Associates	Various electronic parts
Western Electric	Communications equipment

Sources: Compiled from data in the Joint NASA/DOD Economic Information System and from NASA's Fiscal 1966 Procurement Report.

heed Aircraft and LTV, Inc., are included among the firms that provide services. Again because of overlap in economic function, they could be grouped among the systems firms: Lockheed for the Agena and LTV for the Scout launch vehicle. Table 2–12 shows 19 other firms that have contributed to the space program by providing services to NASA and to NASA contractors.

NASA has come under attack from Congress, the Bureau of the Budget, the Civil Service Commission, the General Accounting Office, and from the National Federation of Federal Employees for its use of service support contractors. During the rapid build-up of the space program, the use of service support contractors seems to have been acceptable to each of the organizations; not so after Congress tightened the purse strings on NASA in the mid-1960's. The use of service support contractors became especially controversial when James E. Webb announced in the spring of 1968 that 800 civil servants would be separated from the Marshall Space Flight Center in Huntsville because of insuf-

TABLE 2-12: FIRMS PROVIDING SERVICES TO THE SPACE PROGRAM

NAME OF COMPANY	SERVICE
AT & T (Bellcomm)	Systems engineering
Avondale Shipyard	Barge repairs
Atlantic Research	Various engineering and scientific studies for major space systems firms
Bechtel Corp.	Technical support for Kennedy Space Center
Bendix Corp.	Technical support for Kennedy Space Center launch operations
Brown Engineering	Technical support for Kennedy Space Center and Marshall Space Flight Center
Control Data	Computer services for centers and for major space systems contractors
Ford Motor Co. (Philco-Ford)	Maintenance and operation for Manned Spacecraft Center Mission Control Center
General Electric	Apollo integration; checkout and reliability; engineering services; computers
Hayes International	Technical support for Kennedy Space Center and for Marshall Space Flight Center
Itek Corp.	Technical support for Goddard Space Flight Center
Mason Rust	Maintenance and operation for Michoud Facility
LTV, Inc.	Computer software for Kennedy Space Center and Slidell Computer Facility
Lockheed Aircraft	Computer software for Manned Spacecraft Center
Scientific Data Systems	Computer services
Spaco	Engineering services
Sperry Rand	Computer services for Jet Propulsion Laboratory; support for Astrionics Laboratory, Marshall Space Flight Center
Vitro Corp.	Engineering and test support for Marshall Space Flight Center
Westinghouse Air Brake Corp. (Melpar)	Various engineering studies for Goddard Space Flight Center, Marquardt, General Electric, and General Motors

Sources: Compiled from data in the Joint NASA/DOD Economic Information System and from NASA's Fiscal 1966 Procurement Report.

ficient funds. At the same time, NASA retained the support of a number of its service contractors located in Huntsville.

According to the Comptroller General of the United States, the interest of the General Accounting Office in NASA's use of service support contractors is twofold: (1) does NASA properly implement applicable laws and regulations (principally procurement and civil service)? and (2) is there an effective means to decide, on the basis of cost analysis, whether NASA should contract for services or perform the services in-house?[61] In civil service jargon, support service is an operation subordinate to the function of an agency which, even though performed under contract, could be performed within an agency using its civil servants. As indicated, NASA's support service contracts cover a broad range of activities—from simple housekeeping chores, such as security guards and mailroom activities, to sophisticated engineering and computer activities.

Putting it simply, the position of NASA is that during the rapid initial build-up of the space program, support service contracts were necessary to implement work planned by government personnel. During this phase of the program, according to NASA, there was an insufficient supply of available in-house personnel. As the program matured, support service contracts continued to provide NASA with flexibility. The product mix of required skills changed continually as the program went from research to development, and from production to test and launch. In addition, essential personnel were required in different geographic locations as the program passed into a new phase—from Huntsville and New Orleans to Cape Kennedy.[62] NASA cost analyses also show that the government and contractors incur essentially the same costs. The most important direct costs for contractors and for the government are wages, employee benefits, materials, and other supplies.[63]

Those opposing NASA's point of view, of course, contend that service support contracts are more costly, and that contractor employees displace civil servants. In terms of cost analysis, the problem is one of computing indirect costs and opportunity costs.

As in other controversies of this sort, the disputants usually have data and information that support their respective points of view. In August 1968, the House Committee on Government Operations recommended that the Bureau of the Budget issue specific criteria for making cost comparisons between functions performed by civil service personnel and those performed by service support contractors.[64] Because the matter is of relatively recent origin and because the data are not conclusive, the issue of NASA's support services contracts has not been resolved (as of 1969). Moreover, the issue probably will not be resolved until both NASA's funding position and the future direction of the space program are determined by Congress.

Success at Winning R & D Contracts

Edward B. Roberts, of the Massachusetts Institute of Technology, made a study of the profile of R & D contract-winning companies and of the principal criteria on which contracts are awarded. Although the results of Roberts' research are not universally applicable to all government contracting, his findings provide insight into that complex process.[65]

Most of the data used by Roberts are from the files of two Department of Defense installations. Initially NASA was included in his sample, but at the time (beginning in 1962), information from NASA's contractors was not sufficiently complete to warrant tabulation. The data used included 100 procurement competitions of DOD, ranging in size from \$100,000 to \$2,000,000. The contracts were let between January 1960 and June 1964.

One of Roberts' findings was that two-thirds of the winning companies had performed technical work for the initiator of the contract. In contrast, only one-third of the losers had prior contractual experience. According to two-thirds of the winners, they specifically designed their contract proposals to satisfy what they believed to be the technical preferences of the client. The winners also stated that they frequently addressed the contents of proposals to a particular person or to a given group. One-third of the losers pursued the same kind of behavior. Roberts' data show that companies winning R & D contracts have greater prior contract experience and better technical understanding than do companies losing procurement awards.

Another set of findings related to the determinants of contractor selection. Roberts found that about 75 percent of the procurement awards went to the company ranked by the government to be most competent technically. Roberts' "technical rank data" support the widely held impression that superior technical competence is a key factor in the award of R & D contracts. Cost is another determinant in contractor selection. Frequently, government procurement officers first make a determination of technical competence and they examine only the cost aspects of proposals of the technically acceptable bidders. If only one company is deemed technically qualified, the selection is predetermined unless the government cancels the procurement request. Despite the importance of technical competence, Roberts' data also show that the "low" bidder usually received more contract awards than any other bidder.

In another study, Roberts examined the role of follow-on contracts in government-sponsored research and development.[66] The data came from follow-on questionnaires to the companies that provided the information discussed above. The purpose of the second study was to correlate the contractor's view of his schedule, cost, or technical problems with the receipt of follow-on production contracts. Roberts found that the managers of companies with the greatest anticipations for large follow-on contracts encountered the most severe performance problems. He attributes this to the possibility that the initial R & D contracts must have been written with unrealistic time schedules, cost estimates, and technical performance goals. Another finding was that only one of the 18 new contracts in his survey was a production contract. To Roberts, this suggests that either (1) few production opportunities arise from R & D projects; or (2) that his two-year lag was insufficient to show the application of R & D to production.

Other studies indicate that the latter explanation is the correct one.[67] Edward Greenberg, using data from the Department of Defense list of 100 military Prime Contract Awards (1959–1960) and the DOD list of 500 military Prime Contracts for Experimental, Development, Test, and Research Work (1960–1964), finds a strong lock-in effect. With a two- to five-year time lag, Greenberg's data show that the R & D contractors also tend to be the companies receiving the production contracts.

Burdens in the Procurement Process

Contractor selection by NASA and also by DOD must have been made more difficult by social and economic objectives incorporated in various national laws. Some of the burdens on contractual arrangements include: minimum wages, maximum hours, nondiscrimination in employment, excess profits, "Buy American," use of small business, and use of firms in labor surplus areas. In addition to these, of course, there are strong political pressures for states to have a "just" share of procurement contracts. Despite laws and regulations specifying that contracts placed with small businesses or in depressed areas shall not result in higher costs, it is difficult to imagine that these programs do not increase procurement costs. If nothing more, additional administrative costs are incurred.[68]

NASA keeps close records on its procurement awards to small business. Although its definition varies slightly by industry, the Small Business Administration generally considers a manufacturing firm as a small business if it has fewer than 500 employees. NASA also insures that its largest contractors know the share of subcontracts going to small business. Of course, if a small business gets one or two generous procurement awards, it is frequently lifted out of the small business class. Classic examples of this kind of corporate growth include Ramo-Wooldridge, Litton Industries, and Raytheon. A list of companies with rapid corporate growth of more recent origin would include Scientific Data Systems, Varian Associates, Edgerton, Germeshausen, and Grier, and Sanders Associates.

NASA procurement reports show that between 1963 and 1967 about 7 percent of the dollar value of procurement awards went to small business. In terms of the number of procurement actions, however, NASA awarded almost two-thirds to small businesses during the period. NASA's largest business contractors report that between 1961 and 1967 about 38 percent of the value of their subcontracts were awarded to small businesses.

The federal government has a policy of encouraging the award of contracts in areas with persistent unemployment. The United States Department of Labor defines substantial unemployment as a rate of unemployment in excess of 6 percent of the labor force. With a high rate of national unemployment, it was not difficult for NASA to succeed in placing a large number of contracts in labor surplus areas during its early years. For example, NASA's fiscal 1962 procurement report shows that about 40 percent of the dollar value of its procurement awards went to labor surplus areas, whereas in fiscal 1967, with the economy operating close to full-employment, only 2 percent of the value of NASA's awards were placed in labor surplus areas.

Kinds of Contracts

With some variation, NASA negotiates three kinds of contracts: the fixed-cost, the cost-plus, and the incentive contract. Typically the fixed-cost contract results from advertised competitive bids (sealed bids), with NASA usually selecting the low-bid contractor. Most of NASA's fixed-cost contracts have been for the construction of facilities. Government-research—oriented agencies contend that the advertised competitive bid does not lend itself to research and development procurement. Because of this, fixed-cost contracts have comprised a relatively small proportion (about 10 percent) of the total value of NASA's procurement awards during the last decade.

NASA used cost-plus contracting as its principal procurement tool until late 1963. The main advantage of the cost-plus contract is that it permits maximum freedom to experiment with advanced technologies. This is true because the government commits itself to reimburse the contractor for all allowable costs specified in the contract.[69] On the disadvantage side, the cost-plus contract is purported to encourage the understatement of target cost. The R & D contract has been characterized by uncertainty and by large overruns. NASA officials are proud of the agency's shift away from the cost-plus contract to the incentive contract. In fiscal 1962, cost-plus contracts accounted for about 84 percent of the total value of NASA's procurement awards. By fiscal 1967, cost-plus contracts had dropped to 20 percent of procurement awards. The major shift took place between fiscal 1965 and 1966 when cost-plus contracting dropped from about 70 to 40 percent of awards.

The de-emphasis on cost-plus contracting resulted in a large increase in the relative importance of incentive contracts—from less than 2 percent of total awards in fiscal 1962, to 15 percent in fiscal 1965, to 68 percent in fiscal 1967.[70] With the incentive contract, fees might be below or above target amounts. The fee depends upon performance under a weighted three-way trade-off between cost, performance, and schedule. For the major contractors on the Gemini Program, the incentive parameters were weighted about 44 percent for cost, 43 percent for performance, and 13 percent for schedule.[71] With the incentive contract, the contractor presumably participates in program cost underruns and cost overruns.[72] Officials in NASA believe that the incentive contract also serves as a management tool because it forces corporate attention to the trade-off between the incentive parameters when work statements require change.[73]

SUMMARY

The advent of any new major government agency produces many changes that permeate the economic and social structure of the nation. The rate and direction of the socio-economic effects result from policy decisions that determine the kind and size of programs, the locations of facilities, and the organizational structure under which mission objectives are accomplished.

NASA's early projects came from the National Advisory Committee on Aeronautics (NACA), the three military departments, and the International Geophysical Year Program. NASA's inheritance, however, was relatively small because of the lack of much military and scientific interest in space projects during the early post-World War II period. It is clear from NASA's first Semi-Annual Reports to Congress that the Space Agency did not formulate definite program goals in its early years. In 1960, NASA estimated that its program costs would amount to $1.5 billion per annum during the decade of the 1960's. Although NASA's funding plans were relatively small, the scientific community recommended that NASA undertake dozens of different space programs during the decade of the 1960's. One group of scientists, meeting under the auspices of the National Academy of Sciences, estimated that federal spending for space research would amount to more than $10 billion per annum by the end of the sixties. Yet NASA's early budgets did not include sufficient funding for a manned lunar landing during the decade. At its peak, spending by NASA reached only $5.9 billion during fiscal 1966 and was less than half the amount projected by some in the scientific community.

President Kennedy's request that NASA accomplish a manned lunar landing by 1970 gave NASA at least three things: (1) a cohesive goal-oriented plan; (2) a definite mission for large-thrust engines that the agency had been developing; and (3) a program that captured the interest of the public and the Congress. Between 1959 and 1967, spending for the Manned Space Flight Program rose from 25 percent to 75 percent of NASA's R & D budget.

Because of the long-lead time for most hardware used for space programs, Congress, in the mid-1960's, began to ask NASA to formulate space priorities for the 1970's and beyond. NASA's response to requests for program plans was to insist that it was not timely to undertake large new projects because of the sizable budget demands of established programs. NASA's reluctance to provide Congress with long-range space goals in the mid to late 1960's resulted from the agency's preoccupation with program adjustments for existing projects caused by (1) the tragic Apollo accident in January 1967; (2) the increase in federal spending for Vietnam; and (3) the Revenue and Expenditure Control Act of 1968.

NASA's largest field facilities were established as a direct consequence of President Kennedy's decision that the nation commit itself to the goal of a manned lunar landing. The decision about the locations for NASA's manned space flight field installations reflects a combination of three factors: (1) weighing the relative costs and advantages of dozens of possible sites; (2) an acknowledged attempt by NASA to widen the national R & D base; and (3) a logistics requirement for navigable waterways between sites to permit barge transportation for gigantic equipment. The Gulf Coast states became the principal beneficiaries of NASA's final decision for the locations of its facilities. After evaluating 56 possible plants, NASA chose the Michoud Facility in New Orleans as its vehicle assembly facility. The Pearl River Basin in southwest Mississippi was chosen over 29 other locations for NASA's vehicle test facility. Cape Canaveral (now Cape Kennedy) was deemed superior to 7 other possible sites. Amid speculation about political favor from NASA, the agency selected Houston for the home of its Manned Spacecraft Center in preference to 18 other cities.

The third important factor in the evolution of NASA as a federal agency was the development of a government-industry-university team for the conduct of work. The R & D management experience of NASA's predecessor agencies markedly influenced the structure of the space agency's procurement system. Like NACA, the Navy, and the Army Ordnance arsenal system that existed in the late 1950's, NASA performs the technical management function for most of its programs at its research centers; and following the pattern of the Air Force and also of the Atomic Energy Commission, NASA relies more heavily on industry for the design and manufacture of its space systems than the Army, Navy, or NACA. As a result, about 90 percent of those working on space projects have been employed by contractors and about 10 percent by the federal government. In industry, the economic functions of thousands of different contractors have been to provide NASA with systems, subsystems, components, parts, materials, as well as a line of support services ranging from simple housekeeping chores to sophisticated engineering and computer activities. Because corporate information accompanying contract bids is privileged information, it is difficult to know whether NASA actually selected its leading contractors on the basis of least-cost and greatest technical competence. Research studies, based on questionnaires sent to contract-winning companies and contract-losing companies, show that the government usually selects the technically superior as well as the low bidding firm. In view of this fact, the anecdote that swept Washington immediately after the first manned lunar landing may be judged fictitious rather than factual. According to the story, mission commander Armstrong, right after take-off, advised his two fellow astronauts, Aldrin and Collins, to be on their toes at all times by stating: "Remember guys, this space vehicle was built by the lowest bidding contractors."

In the belief that incentive contracts improve contractor performance in a three-way trade-off between cost, performance, and schedule, especially when program requirements and work statements change, NASA shifted from the use of cost-plus contracts to the use of incentive contracts in the mid-1960's. By the late 1960's, about 70 percent of total contract awards let by NASA were under incentive contracts. The effects of NASA's contractor selection process on the geographic concentration of R & D facilities and on the competitive environment will be discussed in Chapter 8: Performance—Concentration, Monopoly, and Economic Control.

NOTES

1. U.S. Congress, House, Committee on Government Operations, *Report on Government Operations in Space: An Analysis of Civilian-Military Roles and Relationships,* 89th Cong., 1st Sess., June 4, 1965, pp. 22–27.

2. *Ibid.,* pp. 27–29.

3. Loyd S. Swenson, Jr., James M. Grimwood, and Charles C. Alexander, *This New Ocean: A History of Project Mercury,* NASA Historical Series, SP-4201 (Washington, D.C.: U.S. Government Printing Office, 1966), pp. 93–94.

4. Robert L. Rosholt, *An Administrative History of NASA, 1958–1963* (Washington, D.C.: U.S. Government Printing Office, 1966), pp. 58–59.

5. *Ibid.,* pp. 72–76.

6. U.S. Congress, House, Committee on Government Operations, *Report on Government Operations in Space,* p. 56.

7. U.S. Congress, Senate, *NASA Authorization, Fiscal 1961, Hearings,* before the Senate Committee on Aeronautical and Space Sciences, 86th Cong., 2d Sess., 1960, pp. 19–22.

8. U.S. Congress, House, *Report: Space, Missiles, and the Nation,* 86th Cong., 2d Sess., 1960, p. 55.

9. U.S. Congress, House, Committee on Government Operations, *Government Operations in Space,* pp. 56–57.

10. *Ibid.,* pp. 60–61.

11. National Academy of Sciences-National Research Council, Space Sciences Board, *Space Research: Directions for the Future* (Washington, D.C.: National Academy of Sciences, 1965).

12. Charles M. Atkins, *NASA and the Space Science Board of the National Academy of Sciences,* NASA Historical Series, HHN-62 (Washington, D.C.: National Aeronautics and Space Administration, September 1966), p. 57.

13. National Academy of Sciences-National Research Council, *A Review of Space*

Research, Publication No. 1079 (Washington, D.C.: National Academy of Sciences, 1962), Vol. 16, pp. 27–28.

14. Heather David, "Three-Pronged Role Urged for NASA," *Missiles and Rockets,* January 14, 1963, p. 19.

15. National Academy of Sciences, *Space Research: Directions for the Future,* pp. 5–12.

16. Joseph E. Karth, "The Congress, Public Policy, and Science," *TRW Space Log* (Spring, 1966), p. 57.

17. The time frame covered in this section does not include the programs outlined by President Nixon's Space Task Group. The report of that group, given to the President in September 1969, will be discussed in Chapter 10 on the future profile of the space program.

18. U.S. Congress, House, Committee on Science and Astronautics, *Summary Report: Future Programs Task Group, A Report by the National Aeronautics and Space Administration to the President,* with a Foreword by James E. Webb, 89th Cong., 1st Sess., April, 1965, p. v.

19. Emilio Q. Daddario, "Congress Faces Space Policies," *Bulletin of the Atomic Scientists* (May, 1967), p. 14.

20. U.S. Congress, Senate, *National Space Goals for the Post-Apollo Period, Hearings,* before the Committee on Aeronautical and Space Sciences, 89th Cong., 1st Sess., 1965, p. 5.

21. It is easy to understand why officials in government agencies are more than a little reluctant to present public estimates of costs. Once Congress has been given a cost estimate, be it ever so rough, Congress wants agency officials to stay within that cost estimate. Costing research and development projects is more uncertain than costing most other projects. NASA has been forced to justify to Congress the magic $20 billion figure for the cost of the Manned Space Flight Program for years.

22. Note 18, *supra.* Committee on Science and Astronautics, *Summary Report: Future Programs Task Group,* pp. 3–6.

23. U.S. Congress, House, Subcommittee on NASA Oversight of the Committee on Science and Astronautics, *Staff Study: Future National Space Objectives,* 89th Cong., 2d Sess., 1966, pp. III–XVI.

24. Section 203 (b) (3).

25. Rosholt, *An Administrative History of NASA, 1958–1963,* p. 38.

26. Harold B. Finger, "Nuclear Energy in the Space Age," *The United States in Space* (Washington, D.C.: NASA, Office of Scientific and Technical Information, 1963), pp. 71–72. Also, National Aeronautics and Space Administration, Facilities Management Office, *Facilities Data* (Washington, D.C.: National Aeronautics and Space Administration, January 1968), p. I–20.

27. An undated memorandum from the Director of Manned Space Flight Programs to the Associate Administrator.

28. See Rosholt, *An Administrative History of NASA,* pp. 213–14; Stephen B. Oates, "NASA's Manned Spacecraft Center at Houston, Texas," *Southwestern Historical Quarterly* (January 1964), pp. 350–75; and Swenson, Grimwood, and Alexander, *This New Ocean,* pp. 390–91.

29. Adapted from text of the minutes of the meeting for selection of the manned space flight laboratory, August 11, 1961, Washington, D.C.

30. During a meeting of the committee, the Assistant Director for Congressional Relations requested the site selection committee to evaluate several additional locations as the possible site of the proposed manned spacecraft center. They included: Shreveport, Louisiana; Bogalusa, Louisiana; Houma, Louisiana; Fort Polk, Louisiana; Victoria, Texas; and Bradenton, Florida.

31. Swenson, Grimwood, and Alexander, *This New Ocean,* p. 391.

32. Files, National Aeronautics and Space Administration, Washington, D.C.

33. *Ibid.*

34. *Ibid.*

35. The committee deemed that distances greater than that which an automobile can travel in about one hour would have progressively larger impacts on costs, schedules, and reliability.

36. Files, National Aeronautics and Space Administration, Washington, D.C.

37. The point rating formula was: obtain the estimated costs at each location, obtain the low-cost and the high-cost area, and then linearly normalize the points assigned to each location. The magnitude of the cost difference between lowest and highest amounted to a factor of slightly more than 3.5.

38. Department of Defense and National Aeronautics and Space Administration: *Phase I Report: Joint Report on Facilities Required at Launch Site to Support the NASA Manned Lunar Landing Program,* July 31, 1961, pp. 10–11. Hereafter cited as Debus-Davis Report.

39. The horizontal direction or bearing.

40. Debus-Davis Report.

41. Rosholt, *An Administrative History of NASA,* p. 215.

42. U.S. Congress, House, Committee on Science and Astronautics, *1964 NASA Authorization, Hearings,* before the Subcommittee on Space Sciences and Advanced Research and Technology of the House Committee on Science and Astronautics, 88th Cong., 1st Sess., March–May 1963, pp. 2206–7.

43. Letter, James E. Webb to John J. Williams (Delaware), March 31, 1963, *1964 NASA Authorization Hearings,* pp. 2222–23.

44. Joint Committee on Printing, *Congressional Directory, 87th Cong., 2d Sess.* (Washington, D.C.: U.S. Government Printing Office, 1962).

45. U.S. Congress, House, *National Aeronautics and Space Administration Authorization Act of 1964,* Sec. (d) (2), 88th Cong., 1st Sess., 1963, H.R. 7500.

46. National Aeronautics and Space Administration, *Electronics Research Center, Report to the U.S. House Committee on Science and Astronautics* (Washington, D.C.: U.S. Government Printing Office, 1964), pp. X–XI.

47. The average number of degrees conferred in electrical engineering was multiplied by two and added to those conferred in physics. The committee rejected, as unfeasible, the possibility of ranking universities by the numbers of published research papers or of Nobel Prizes credited to faculty members. *Ibid.,* Appendix D, no pagination.

48. Data used by the committee came from the National Academy of Sciences—National Research Council; the Institute of Radio Engineers; the U.S. Bureau of the Census; and the National Science Foundation.

49. National Aeronautics and Space Administration, Facilities Management Office, *Facilities Data,* January 1968.

50. U.S. Congress, House, Committee on Government Operations, *Organization and Management of Missile Programs,* 86th Cong., 1st Sess., September 2, 1959, H.R. 1121, pp. 50–51.

51. *Ibid.,* pp. 51–52.

52. U.S. Congress, Senate, Testimony of General Bernard A. Schriever, *Investigation of Governmental Organization for Space Activities, Hearings,* before the Senate Committee on Aeronautical and Space Sciences, 86th Cong., 1st Sess., 1959, p. 425.

53. Not unexpectedly, the Air Force received much criticism for the part it let Space Technology Laboratories play in the development of its missile systems capability. For a succinct description of the controversy, see U.S. Congress, House, Committee on Government Operations, *Eleventh Report: Organization and Management of Missile Programs,* 86th Cong., 1st Sess., September 1959, pp. 81–100. For a less concise (803 pages) but a more vivid presentation, see U.S. Congress, House, Committee on Government Opera-

tions, *Organization and Management of Missile Programs,* 86th Cong., 1st Sess., February and March 1959. Also, U.S. Congress, House, Committee on Government Operations, *Report; Air Force Ballistic Missile Management—Formation of the Aerospace Corporation,* 87th Cong., 1st Sess., May 1961.

54. McKinsey and Company, Inc., Washington, New York, Chicago, London, San Francisco, and Los Angeles, Management Consultants, *An Evaluation of NASA's Contracting Policies, Organization, Performance* (Report to Dr. T. Keith Glennan, October 28, 1960), pp. 1–4.

55. Rosholt, *An Administrative History of NASA*, pp. 99–102.

56. The minimum figure changes from time to time, but usually ranges between $1,000,-000 and $5,000,000.

57. Willard I. Zangwell, *Top Management and Selection of Major Contractors at NASA* (internal NASA discussion paper, August 1968). Also National Aeronautics and Space Administration, "A Reply to Senator Gaylord Nelson," (document in the office of NASA's Historian, 1965). (mimeographed)

58. The number of different contractors has been estimated to be as high as 80,000.

59. Merton J. Peck and Frederic M. Scherer, *The Weapons Acquisition Process: An Economic Analysis* (Cambridge, Mass.: Graduate School of Business Administration, Harvard University, 1962), pp. 114–16.

60. The purpose and usefulness of the Joint NASA/DOD Economic Information System will be discussed in greater depth in Chapter 5, which deals with measures of impact.

61. U.S. Congress, House Testimony of Elmer B. Staats, *Support Service Contracts, Hearings,* before the House Committee on Government Operations, 90th Cong., 1st Sess., June 21, 1967, pp. 2–3.

62. U.S. Congress, House, Testimony of Harold B. Finger, *Support Service Contracts, Hearings,* before the House Committee on Government Operations, 90th Cong., 1st Sess., June 21, 1967, pp. 61–93.

63. U.S. Congress, House, Committee on Government Operations, Testimony of Harold B. Finger, *A Cost Profile for Support Services, Hearings,* before the House Committee on Government Operations, 90th Cong., 2d Sess., April 1968, pp. 3–32.

64. U.S. Congress, House, Committee on Government Operations, *Report: Criteria for Support Service Contracts,* 90th Cong., 2d Sess., August 5, 1968, pp. 4–5.

65. See Edward B. Roberts, *Questioning the Cost Effectiveness of the R & D Procurement Process* (Research Program on the Organization and Management of Research and Development, Sloan School of Management, September 1965, Paper No. 144–65). Also, Edward B. Roberts, *Marketing and Engineering Strategies for Winning R & D Contracts* (Sloan School of Management, July 1965, Paper No. 128–65).

66. Edward B. Roberts and William H. Dyer, *The Role of Follow-on Contracts in Government-Sponsored Research and Development* (Alfred P. Sloan School of Management, Paper No. 269–67, July 1967).

67. Edward Greenberg, "Some Aspects of the State Distribution of Military Prime Contract Awards," *Review of Economics and Statistics,* XLVIII, 2 (May 1966), 205–10; and Edward Greenberg, *Relationships between R & D Contracts and Production Contracts,* (NASA Economic Research Program, Washington University, St. Louis, Missouri, June 1967).

68. Stanford Research Institute presents a fairly comprehensive analysis of burdens in the procurement process. See George T. Hayes, Project Manager, *The Industry-Government Aerospace Relationship* (Menlo Park, Calif.: Stanford Research Institute, 1963), especially Appendix D, Vol. II, pp. 39–55.

69. Clarence H. Danhof, *Government Contracting and Technological Change* (Washington, D.C.: Brookings Institution, 1968), pp. 252–53.

70. Procurement award data came from National Aeronautics and Space Administra-

tion, *Annual Procurement Report, Fiscal 1967* (Washington, D.C.: National Aeronautics and Space Administration, 1967), pp. 12–17 and pp. 66–67.

71. U.S. Congress, Senate, Testimony of George E. Mueller, *NASA Authorization for Fiscal Year 1967, Hearings,* before the Senate Committee on Aeronautical and Space Sciences, 89th Cong., 2d Sess., February and March 1966.

72. On the rather heroic assumption that contractors act to maximize short-run profits, it has been argued that contractors can derive higher fees from incentive contracts than from cost-plus contracts because they can manipulate multiple contractual arrangements. It is contended that contractors tend to prefer incentive contracts to cost-plus contracts and that incentive contracts usually result in cost over-runs rather than under-runs. See Charles E. Bradley and Clayton C. McCuistion, *Contract Decision-Making and Incentive Fee Contracts* (unpublished research report to NASA, December 1965).

73. Mueller, *NASA Authorization for Fiscal 1967,* p. 172.

3 Economic Efficiency and Planning, Programming, and Budgeting in NASA

In many governmental groups in Washington, it is fashionable to be conversant with cost-effectiveness analysis, variously known as planning-programming-budgeting, cost-benefit analysis, program budgeting, and cost-utility analysis.[1] Broadly conceived, cost-effectiveness analysis is no more than the attempt to assure efficiency in the conduct of business by government, and in this context, by the federal government. Cost-effectiveness analysis is a process of identifying and selecting among alternative goals and objectives, and then quantifying and comparing the costs and the benefits derived from each of the alternatives. Cost-effectiveness analysis amounts to an attempt to emulate the resource allocation procedure of the private economy in the public sector of the economic system. Quantitative estimates of alternative costs and benefits are used as substitutes for the resource allocation mechanism of the price system.

Because the sections that immediately follow on the historical origins of planning-programming-budgeting, levels of application, and methodological difficulties are a digression from the main stream of thought, it is suggested that those with a knowledge of the principles of economics and of the use of cost-effectiveness analysis skip to the sections on case studies of the application of cost-effectiveness analysis in NASA, which begins with a discussion of NASA's lunar landing orbital rendezvous decision.

THE CURRENT INTEREST IN PPBS

The desire to apply cost-effectiveness analysis to the nonmilitary government departments and agencies stems from the widespread use of the method in the Department of Defense. The major experimental use of planning-programming-budgeting in the Department of Defense dates from the appointment of Robert S. McNamara as Secretary of Defense in 1961. At that time, a group of economists who had specialized research capabilities in the field were appointed to senior management positions in the Department of Defense. These were the so-called "whiz kids" who worked for the Department of Defense. Many of those accepting government service had acquired their technical capabilities while employed at the nonprofit RAND Corporation in Santa Monica, California. Charles J. Hitch was appointed Assistant Secretary of Defense (Comptroller) and Alain C. Enthoven was appointed as Assistant Secretary of Defense (Systems Analysis). Much to the displeasure of some senior military officers, these appointments gave experts in the field the opportunity to apply the results of their economic research to the financial and management problems of the Department of Defense. Throughout their stay at the Department of Defense, Hitch and Enthoven always emphasized the fact that their advisory functions related to matters of economy and efficiency in defense planning; the functions of the Joint Chiefs of Staff as principal military advisers to the Secretary of Defense and to the President were not affected by their appointments.[2]

On August 25, 1965, President Johnson issued a directive to cabinet members and to agency heads establishing a new government-wide planning and budgeting system. The objective of the directive was simple; but meeting the requirements of that objective has been, and will continue to be, far from simple. According to President Johnson, the then newly established program was nothing more than the application of the most modern management tools in government to insure that a finer life can be brought to every American at the least possible cost. The new program was designed, "to insure sounder judgement through more accurate information; to pinpoint those things we ought to do more, and to spotlight those things we ought to do less; to make our decision-making process as up-to-date as our space-exploring equipment."[3]

President Johnson and the Bureau of the Budget hoped that successful implementation and operation of a government-wide planning-programming-budgeting system would result in five major benefits to the country by enabling the government and the nation to:

1. identify our national goals with precision and on a continuing basis;
2. choose among those goals the ones that are most urgent;
3. search for alternative means of reaching those goals most effectively at least cost;
4. inform ourselves not merely on next year's costs, but on the second, and third, and subsequent year's costs of our programs; and
5. measure the performance of our programs to insure a dollar's worth of service for each dollar spent.[4]

In a message to the Congress on March 17, 1967 concerning the quality of American government, President Johnson stated that the new system would not result in decisions. Rather, the system would provide information for the governmental decision-making process. For the system to be successful in identifying alternatives, each federal department was required to:

1. develop its objectives and goals, precisely and carefully;
2. evaluate each of its programs to meet these objectives, weighing the benefits against the costs;
3. examine, in every case, alternative means of achieving these objectives; and
4. shape its budget request on the basis of this analysis, and justify that request in the context of a long-range program and financial plan.[5]

THE HISTORICAL ORIGINS OF PPBS

Although application of a planning-programming-budgeting system in the government might conceivably be revolutionary, the idea certainly is not new. As a field in theoretical economics, cost-effectiveness analysis owes its origin to the theory of public finance, which in turn, is part of that branch of economic analysis known as welfare theory. Welfare analysis also plays an important role in international economic theory. Economists have long sought to identify the conditions necessary to achieve the optimum level of economic welfare. In the jargon of welfare economics, the optimum level of welfare is achieved when the marginal rate of substitution for goods and services is equal to the marginal rate of technical transformation for all products. In less technical terms, this means that the optimum welfare level is achieved when it is impossible to make someone better off without making someone else worse off in both exchange and production.[6]

Outside of the field of pure economic theory, the use of cost-effectiveness analysis in the decision-making process has been dated from the Garden of Eden (Genesis 2 and 3).[7] It is scarcely debatable that the decisions made by Adam and Eve in the Garden of Eden were based on a gross under-estimation of costs and a gross over-estimation of benefits. Along more practical lines, in the Rivers and Harbors Act of 1902, Congress explicitly required that cost-benefit studies be made to assist the decision-making process. In the private sector, Donaldson Brown, former Chairman of the Finance Committee of General Motors, incorporated the PPBS method in the 1924 corporate budget and financial plan of General Motors.[8]

For water resource projects, the United States Army Corps of Engineers began to make formal economic evaluations of projects on the basis of costs and benefits in the late 1920's. By the late 1930's the use of this method for project justification became a widespread practice in the Department of Agriculture, the Bureau of Reclamation of the Department of the Interior, as well as in the Army Corps of Engineers.[9]

In the early fall of 1941, shortly before Pearl Harbor, David Novick developed a Production Requirements Plan that attempted to solve wartime priority and allocation problems. By late 1942, the War Production Board, by means of the Controlled Materials Plan, employed an unsophisticated version of cost-effectiveness analysis.[10] Other examples of the use of cost-effectiveness analysis during World War II include: assessment of the optimum number of ships put in a convoy; the spacing between ships in convoys; the ratio of escort vessels to cargo carriers; and the optimal aiming of aircraft guns, bombs, depth charges, and torpedoes.[11]

LEVELS OF APPLICATION

Conceptually, cost-effectiveness models can be constructed for at least four levels in the governmental decision-making process: (1) the allocation of resources between the public and the private sector; (2) the allocation of resources within the public sector; (3) the allocation of resources between programs within a given goal or federal agency; and (4) the least-cost means of achieving a given program objective.

Allocation of Resources Between the Private and the Public Sector and Within the Public Sector

The highest tier of theoretical pertinence of cost-benefit analysis relates to the achievement of optimum efficiency in the allocation of resources between the public and the private sector. This problem is one of the main concerns of economic welfare theory and the theory of public finance. To solve this problem, the analyst, or the economist, would have to find a way to measure all social wants—private and public. In the language of economists, what must be derived is a "social welfare function." This lofty, ivory-tower concept would measure the true, and agreed-upon, economic, social, political, and ethical value judgments and preferences of all people in our society. The social welfare function may be useful in pure economic theory and logic, but even those using the concept believe that current social research methods are not sophisticated enough to give empirical content to a social welfare function, making it a useful tool for policy formulation. Some economists, and more recently, some political scientists, believe that if individuals can evaluate their social wants and if a social welfare function can be derived, that function probably will take the form of some kind of voting process that approximates true public preferences.[12]

The optimum allocation of resources between programs and federal agencies within the public sector presents as complex a question as the allocation of resources between the public and the private sector. The optimum allocation of resources between as disparate fields of public endeavor as space, defense, health, education, transportation, and public housing also requires the application of a "social welfare function." As indicated in Chapter 1, in this country the establishment of national goals and their priority structure is a function of the market place, the legal system, and the political process. Presidential and Congressional action, or inaction, ultimately determines the choice between public programs. The Congressional decision, of course, is based on a voting system. Little reflection is required, however, to know that our present political process does not represent an agreed-upon economic, social, political, and ethical value structure (i.e., the true social welfare function). The difficulty is that of measuring and comparing the range of social costs and benefits of public activities that have greatly different objectives. There is no way to know which is more valuable: going to the moon; youth rehabilitation through the Job Corps program; the reduction of air-pollution; or the elimination of traffic congestion. This comparison does not, of course, even address the question of whether such problems are solvable by the application of the same resources or the terms on which any possible transformation could be made.

Even if a trade-off in terms of social worth could be determined, other kinds of trade-off problems exist because resources used in the production of one kind of public program are not, in fact, perfectly transferable to other kinds of public programs. Because resources are not perfect substitutes, transfer among programs usually takes place at increasing cost and decreasing benefit. It is naive to believe that Wernher von Braun's space team would be as efficient in solving problems arising in urban ghettos or in Appalachia as it has been in solving problems associated with the exploration of space. For optimum efficiency within the public sector, it is necessary to measure the rate at which costs and benefits change when resources are re-allocated.

Resource Allocation for a National Objective and the Least-Cost Means of Achieving that Objective

Cost-effectiveness analysis makes its greatest contribution within a well-defined and quite narrow field. Although they shade into one another, these narrower fields constitute the third and fourth levels in the governmental decision-making process. Given the goal and the amount of military preparedness required, optimum resource allocation would require the application of an operational "military worth function" to choose among competing programs. For the space program it would be determination of a "space worth function"; for health programs it would be a "health worth function." But even within more narrowly defined fields of public endeavor, it is frequently, if not usually, political criteria rather than economic criteria that determine the allocation of resources to a specified national goal. It was Congress, at the request of the President, that decided that this country's principal space program would be the manned lunar landing project. For about half of NASA's life that objective absorbed between 60 and 70 percent of the space budget. Similarly, Congress decided that part of the defense budget would be allocated to an anti-ballistic missile shield for the nation.[13] It might be argued that each of these objectives is a higher-order political goal (i.e., in the second tier of the governmental decision-making process) and that cost-effectiveness analysis is not conceptually applicable because the Manned Space Flight Program and the anti-ballistic missile system accomplish such different public objectives.

Even the severest critics of the effectiveness of cost-effectiveness analysis as a tool for decision-making in the federal government should admit, however, that after program objectives have been defined, they should be achieved at the lowest possible cost. This is the most limited concept of cost-effectiveness analysis; yet it still assumes that those working in the federal government can make correct trade-off analysis between cost, quality, and time. Although not in the formal framework of cost-effectiveness analysis, the next chapter discusses the ways in which one program office in NASA, the Office of Manned Space Flight, makes trade-off decisions between cost, quality, and schedule.

DIFFICULTIES OF METHOD

In addition to conceptual difficulties that preclude the use of cost-effectiveness analysis for all decisions on resource allocation, there are a number of problems in method. The most important difficulties encountered by analysts applying the cost-effectiveness technique include: the selection of appropriate discount rates and amortization periods; changes in the quality of public goods and services; the unpredictable nature of the results of research and development; the continually changing nature of international events; and the allocation of joint costs and benefits in multi-purpose projects.

Selection of Discount Rates and Amortization Periods

Because most costs and benefits resulting from public investments occur through time, it is necessary to select appropriate discount rates and amortization periods. The costs and benefits of projects requiring appreciable time, whether public or private, must be "discounted." As used here, discounting has nothing to do with the commercial notion of discounting, i.e., the reduction of the commercial price of a commodity.

Discounting to an economist means the equalization of a future to a present value. Human nature dictates that the enjoyment of benefits in the present is preferable to the enjoyment of the same benefits in the future. For example, a young couple is willing to incur a debt burden well in excess of the purchase price of a house to enjoy home ownership right after marriage rather than 15 years later. The purchase price of the house less the annual stream of interest payments on the mortgage over the amortization period is the "discounted" value of the house. Discounting of public undertakings is no different.

When comparing government projects, an incorrect choice of discount rates and amortization periods would distort the allocation of resources and would reduce economic welfare. For example, the present value of $1,000 50 years in the future and discounted at 6 percent is $54; discounted at 3 percent the present value is $228. The present value of $1,000 ten years in the future and discounted at 6 percent is $558; discounted at 3 percent the present value is $774.[14] Use of an excessively low discount rate, coupled with an excessively short time frame, could result in the acceptance of uneconomic projects. For example, if the economy were operating at full employment, resources would be drawn from more highly valued projects to less desirable ones.

There is agreement among economists that the proper discount rate for cost-benefit studies should reflect the opportunity cost of the use of resources on the proposed project. Opportunity cost, as used here, is the cost of not being able to put the resources to work on alternative projects. Assuming full employment, this means that if resources are used to build highways, the same resources are not also available to build airports. In this case, the opportunity cost is the loss of benefits that would have resulted from the construction of airports. During World War II, when a mission commander selected one target to be destroyed, the opportunity cost of that selection was the stream of dangers still existing from those targets not destroyed. If resources are transferred from the production of goods and services by business into the production of a public highway system, the opportunity cost of that transfer is the foregone income that would have accrued to the business sector. The lowest possible return to the business sector, and also the smallest measure of the opportunity cost, would be the prevailing bond rate of interest. To be sure, the appropriate discount rate for cost-benefit studies varies by time and by place. During times of widespread unemployment, the opportunity cost of using resources in the public sector can be zero. This is true because if the resources remain idle, they would not yield an alternative return.[15]

In 1967, the Comptroller General of the United States conducted a survey of discounting techniques used by federal agencies in evaluating future programs. Of the twenty-three agencies surveyed, ten departments stated that they actually used the discount technique in evaluating programs for fiscal year 1969. The discount rates as well as the rationale for their selection varied widely. The rates ranged from a low of 3.0 percent for Job Corps programs in the Office of Economic Opportunity to a high of 12.0 percent for some projects sponsored by the United States Department of the Interior. The rationale for the rate used by the Office of Economic Opportunity was that it reflected the "rate of return on a safe investment." The rationale used by Interior was that its rate was "representative of average capital return in the private sector."[16] In the spring of 1968, the Bureau of the Budget established the uniform discount rate of 10 percent for all federal departments. The Bureau also stipulated that cost-benefit analyses be tested using the alternative discount rates of 7.5 percent and 12.5 percent in cases when decisions would be changed by small changes in the discount rate. The Bureau of the Budget directive provides uniformity, but it does not necessarily reflect the true opportunity cost of various federal programs. This is true particularly for those projects that employ resources that would remain idle unless used by the government.

Changes in the Quality of Goods

The objective of all productive activity in the government sector is to satisfy public or social wants. Hence, changes in the quality of public goods and services affect the stream of benefits that accrue from public spending. The installation of improved highway surfaces, at no increase in real cost, for example, results in "more" highway and higher welfare for the users. Similarly, a change in the aesthetic appearance of a public road yields psychic benefits, or losses if the change is negative, to those using the road. Because the notion of quality is one of tastes and subjective value judgments, economic analysis has not been successful in quantifying changes in quality. Unless two similar commodities have quality attributes such as size or weight, there is no objective way to incorporate differences in the quality of goods and services into cost-benefit studies.[17]

Research and Development

Another problem area in cost-effectiveness analysis is evaluation of research and development efforts, especially basic research. In the absence of federal support for research and development, the private sector probably would provide less than the "socially optimum" amount. By definition, the results of basic research are both unknowable and unpredictable. Obviously, cost-benefit criteria cannot be applied when benefits are unknowable and unpredictable. The Bureau of the Budget recognizes this fact. Funds for basic research usually are budgeted by level-of-effort-methods, which is largely a process of defining levels of effort in major scientific fields and then making value judgments about priorities.[18] Despite this fact, the National Science Foundation has not been exempted from the government-wide requirement that agencies make budget requests in a cost-effectiveness framework. Although not basic research, all NASA effort is classified by the Bureau of the Budget as research and development.

National Security Programs in Foreign Affairs

Certain government agencies, such as the Department of State, the United States Information Agency, the Agency for International Development, and the Central Intelligence Agency, are struggling to learn whether, and if so, how, the planning-programming-budgeting technique affects them. Cost-effectiveness analysis probably is of little use as an internal planning tool to these agencies. Most program objectives of the foreign affairs and security agencies are reactive policies, frequently resulting from day-to-day changes in the actions of foreign governments. In addition to continual change, many factors that affect foreign affairs and national security simply are not subject to realistic quantification.[19] Delineation and measurement of the costs and benefits of policy alternatives in diplomacy, if ever attainable, are far removed from the present planning-programming-budgeting process.

Joint Cost and Program Interdependence

The allocation of costs and benefits on multi-purpose projects presents a final conceptual difficulty. Trade-off analysis becomes opaque when programs are interdependent. The joint use of hardware and facilities for different guided missiles and space projects is a good example of program interdependence. Clearly, tracking networks and stations are an integral part of NASA's Manned Space Flight Program. But these facilities are also used for a half dozen or so different unmanned space objectives. The same is true of the launch vehicle complexes, and even the development of launch rockets. The problem of allocating joint costs arising from program interdependence is at least twofold. First, large nonrecurring costs for research, development, production, and construction must be allocated among different missions. As discussed in Chapter 2, the development of the large Saturn launch vehicle was started several years before the manned lunar landing became the principal space objective of NASA. The question that then arises is, "what proportion of those early R & D program costs incurred for large launch vehicles should be allocated to the Manned Space Flight Program in general and what proportion to any given mission?" In addition to the correct allocation of nonrecurring costs, recurring operational costs must be apportioned among the separate programs. Incorrect allocations of both kinds of costs, of course, distort cost-benefit ratios and can result in the general misallocation of resources.[20]

VIEWS ON COST-EFFECTIVENESS ANALYSIS

The spectrum of views on the usefulness of planning-programming-budgeting as a tool of public policy ranges from complete condemnation to unequivocal acceptance. Because most of those working in the field are cognizant of the conceptual and practical limitations of cost-effectiveness analysis, those most knowledgeable are cautiously optimistic about its future success.

In the military hierarchy no one has made a stronger public stand against the use of cost-effectiveness analysis for defense decision-making than Admiral Hyman G. Rickover. Testifying before Congress in the mid-1960's, Rickover stated:

> I think that cost-effectiveness studies are occasionally used to kill something—to kill it by studying it to death. The use of studies for this purpose is well-known in Washington. The question of nuclear propulsion for the carrier John F. Kennedy is a good example. . . . Making the John F. Kennedy nuclear powered would have added substantially to the capability of the fleet. However, the decision on this one ship was delayed over a year while the Navy attempted to respond to a request to undertake a comprehensive, quantitative study of whether the future Navy will, indeed, make full use of nuclear power. . . . A decision was finally made by the Department of Defense against putting nuclear propulsion in this carrier in order to avoid further delay in the construction of the ship. But is it really necessary to engage in cost-effectiveness studies on the whole future of the Navy before we can decide to put nuclear propulsion in a single ship?
>
> The problem of selecting weapons systems for development, engineering them to completion, introducing them into service, and training people to operate them requires extensive experience in the field under consideration. Without such experience it is easy to make the mistake of oversimplifying the factors that are taken into consideration and so to reach an incorrect decision.
>
> In the cost-effectiveness studies performed by the analysts, they compute numerical values for the effectiveness of nuclear power. However, before they make the calculation, they make certain simplifying assumptions in order to be able to do the arithmetic. These assumptions just happen to eliminate from consideration the principal military reason for wanting nuclear power in the first place.[21]

Alain C. Enthoven, who has been working in the field since the 1950's, first as a professional economist and then as Assistant Secretary of Defense, looks at cost-effectiveness analysis somewhat differently. Speaking about the technique, Enthoven states:

> . . . It is not physics, engineering, mathematics, economics, political science or military operations and yet it involves elements of all of the above disciplines.

But regardless of its make-up, the art of systems analysis—and it is an art—like the art of medicine, must be based on the scientific method, using the term in the broadest sense.

. . . Of course, neither science nor systems analysis is infallible. But infallibility is not being claimed; it would be worse than unscientific to do so. However, scientific method does have a self-correcting character that helps to guard science from persistence in error in the long run.[22]

Three other economists, David Novick, Bertram Gross, and Michael Spring,[23] each of whom has played an important role in implementing the use of cost-effectiveness analysis in the public sector, describe program budgeting as: "potentially the most significant management improvement in the history of American government."

One additional view about the analytical value of cost-effectiveness analysis warrants citation. It is the opinion of Henry S. Rowen, President of the RAND Corporation and former Assistant Director of the Bureau of the Budget. Discussing the minimum value of program budgeting, Rowen states:

If PPB leads to nothing more than each agency reexamining its objectives, then PPB would be a useful system. Objectives of government programs are often stated in such vague and nonoperational terms as to be little more than home-and-mother platitudes. At the other extreme, they are sometimes stated in such narrow and conventional terms as to rule out any consideration of relevant alternatives.[24]

RISKS OF USING COST-EFFECTIVENESS ANALYSIS

The risks that can be incurred by the use of cost-effectiveness analysis arise principally from abuse of the approach and from misunderstanding of what planning-programming-budgeting can and cannot accomplish.[25] Cost-effectiveness analysis is no more and no less than a management tool. Although it provides guidelines, it does not make decisions.

The greatest danger in using cost-effectiveness analysis is inherent in the approach. If the choice of the objective or the objectives to be maximized is wrong, the subsequent analysis must be misleading and useless as a guide for decision-making. This is true even if supporting analysis is flawless.

Ancillary to the wrong choice of our objective function is a risk that Senator Henry M. Jackson believes makes Congress skeptical about the usefulness of planning-programming-budgeting. That is the danger that some analysts, a year or two out of graduate school and without much wisdom and judgment, will have too much influence in making key decisions.[26] There is only one way to avoid this risk. The heads of government departments should not place those without sufficient background and training, whether they are PPB specialists or not, in key decision-making positions. Along less serious lines, it is possible that those federal employees (in the late 1960's more than one hundred annually) who train for PPBS work at universities will be frustrated when they return to their jobs. As economists working in the field know, it is hard to do good cost-effectiveness analysis. Although it is not difficult to acquire the set of techniques, it is difficult to gain sophisticated understanding of the relevant costs and benefits, and it is even more difficult to get the results of analyses implemented into policy decisions.

The third and last risk to be discussed is one that might result from the "oversell" of planning-programming-budgeting. Cost effectiveness analysis is not costless. To be sure, not all problems that lend themselves to cost-effectiveness analysis warrant such analysis. By mid-1969, the Bureau of the Budget had not established operational criteria on the cost-effectiveness of using the cost-effectiveness technique. The possibility of overuse of the tool can be gleaned from occasional proposals that Congress establish a sizable PPB group of its own, possibly within the Government Accounting Office, to check on the validity (by duplication) of studies made by Executive Departments.

THE USE OF COST-EFFECTIVENESS ANALYSIS IN NASA

The second part of this chapter presents five case studies on the use of cost-effectiveness analysis in NASA. The studies include: (1) NASA's lunar orbital rendezvous decision; (2) an investigation made by the Stanford Research Institute of the international costs and benefits of different management techniques for a manned orbital research laboratory; (3) an analysis of the economic benefits of satellite information systems made by the Planning Research Corporation; (4) an interesting attempt by the RAND Corporation to develop what might be called a "space worth function"; and (5) the results of a summer study effort sponsored by the National Academy of Sciences on the peaceful uses of satellite systems. The discussion in Chapter 2 of NASA's site and facility selection process could have been put into this section.

With the exception of one study, the choice of the technique to be used for the manned lunar landing, discussion of these studies does not imply either that NASA accepts the findings of the research groups or that NASA has been able to use the results of the studies in planning decisions. In addition, not everyone will agree that the case studies selected are good examples of the use of planning-programming-budgeting in the federal government. Some of the studies emphasize costs; others look mainly at benefits. None of the studies actually succeeds in developing an operational "space worth function"; nor do they put the role of the space program into a more general "social welfare function." All of the studies have one thing in common. They show that NASA has, in fact, looked at alternatives.

Of the use of cost-effectiveness analysis in the agency, NASA states:

> . . . Trade-off type studies relating to performance, costs, schedule and other pertinent factors are used in project planning and management, and in the planning and conduct of individual missions. We have not, however, applied specific discount rates. That is to say we have not attempted to express in terms of percentage discounts, the effect of future costs and benefits, although those economic factors are implicit in our consideration of alternative programs and assessment of priorities. Frankly, we find it extremely difficult to quantify such elusive economic considerations as they affect research and development efforts in the space environment.[27]

NASA'S LUNAR ORBITAL RENDEZVOUS DECISION

Slightly more than one year after President Kennedy urged that a manned lunar landing become a national goal, NASA decided how the objective should be accomplished. In the early 1960's, analysis of existing technology showed that three principal methods were possible: the direct ascent technique; the earth orbital rendezvous technique; and the lunar orbital rendezvous technique. As is well known now, NASA selected the third method. The selection of the lunar rendezvous technique is an excellent example of how a federal agency seeks to achieve economic efficiency (i.e., the fourth tier in the governmental decision-making process) in the accomplishment of a given objective.

An in-depth discussion of the selection process for the technique to be used for the manned lunar landing is limited by the fact that many of the documents used for the decision remain under national security restrictions. Although NASA has relatively few secret documents compared with the Department of Defense, it appears that once documents become classified, subsequent declassification is a lengthy process.

As in so many government decisions, selection of the lunar rendezvous technique was not without inter-organizational and inter-governmental disagreement. In May 1961, two study groups were organized to investigate matters relating to the manned lunar landing and to the profile of the nation's launch vehicle program. On June 16, 1961, the first task group, chaired by William Flemming, made its report. One conclusion of the Flemming Report was that a manned lunar landing would be possible within a 6½- to 8½-year period of time. That conclusion was based on an analysis of procurement lead times for hardware required for a direct ascent approach to the moon. The direct ascent technique called for the use of one large launch vehicle. The second task group made its report in December 1961. That study was a joint NASA and DOD effort, co-chaired by Nicholas Golovin and Laurence Kavanau. The Golovin Report suggested that the direct ascent approach be discarded because of technical schedule problems that would probably arise from the use of a single, gigantic launch vehicle. The Golovin group favored the earth orbital rendezvous technique.[28] Although there was no unanimity within NASA, the greatest censure over the final selection of the lunar rendezvous method came from Jerome Wiesner, then Scientific Adviser to the President. On technical grounds, Wiesner charged that the lunar orbital rendezvous decision was unsound. He suggested that either the direct ascent or the earth orbital rendezvous technique would be superior. On economic grounds, he argued that the lunar orbital rendezvous technique was a technological dead-end because of its limited value for other kinds of missions.[29]

Obviously, it can never be proved whether Wiesner and his staff of scientists were right. It is conceivable that the earth orbital technique would have been as successful as the lunar orbital rendezvous technique proved to be. NASA historians report that the lunar orbital rendezvous technique was selected after a million man-hours of engineering studies demonstrated that it would accomplish the lunar mission most cheaply. In NASA, Joseph F. Shea, Deputy Director of Systems, Office of Manned Space Flight, coordinated mode selection studies made by NASA field centers and by NASA contractors. NASA's news conference announcing the lunar orbital rendezvous decision was on July 11, 1962. Joseph Shea testified before the House Committee on Science and Astronautics the next day. Commenting on Shea's justification for the decision, Representative Ken Heckler stated:

> I think that this will go down as a classic in decision-making. History will tell us whether it is right. But I desire to compliment you, Dr. Shea, and all those concerned with the NASA decision, which is soundly based in the reasoning you have presented. The important fact is that you have reached a decision and have weighed the alternatives carefully.[30]

The Alternatives

The principal technical differences between the direct ascent, the earth orbital rendezvous, and the lunar orbital rendezvous techniques rested on three interrelated factors. The first factor was the launch vehicle requirement. The direct mode would have required the development of a vehicle significantly larger than the advanced Saturn. The advanced Saturn was to become known as the Saturn V. At that time, NASA had concepts for such a vehicle—the NOVA. Only the third stage of the NOVA was under contract in mid-1962. At that time, configurations for the NOVA and the advanced Saturn called for the same third stage. The NOVA would have been capable of putting almost twice as many pounds into earth orbit as the advanced Saturn.

The earth orbital rendezvous approach would have required the use of two Saturn vehicles, each about five-eighths as powerful as the NOVA. Two pieces of hardware, a spacecraft and a fuel tanker, would have been put in space separately because their combined weight would have exceeded the payload to orbit capability of the Saturn. After separate launch into earth orbit, the fuel tanker would rendezvous, dock, and transfer fuel to the spacecraft. Development of the fuel tanker would have been required. The lunar rendezvous technique required a single launch of the Saturn V from the pad at Cape Kennedy.

In mid-1962, NASA decided to postpone funding of the NOVA vehicle for at least a year or two and concentrate on development of more advanced stages of the Saturn. NASA officials hoped that concentrated development would permit deeper space penetrations, larger payloads in earth orbit, and larger payloads to the moon in the future, with the then existing basic vehicle configuration.

A second major consideration in NASA's technical trade-off analysis was the hardware to be used for the actual lunar landing. Both the direct ascent and the earth orbital rendezvous methods would have required the use of the command and service module (CSM) for the lunar landing; whereas the lunar orbital rendezvous technique called for the use of the lunar excursion module (LEM). Feasibility studies showed that the landing could be achieved with either module. The studies also showed that use of the command and service module probably would require wider use of television cameras to improve pilot visibility during the lunar landing. It was believed that use of the lunar excursion module would provide more glass area and better visibility because it would be designed specifically for operation in the vicinity of the moon and, therefore, would require no provision for re-entry into the earth's atmosphere. As is now well known, with the lunar rendezvous technique, the lunar excursion module separates from the command and service module for descent to the lunar surface.

The third principal technical trade-off was the rendezvous technique itself. The lunar orbital rendezvous technique required rendezvous and docking of the lunar excursion module with the command and service module during the return phase of the mission. The other two methods would not have required the rendezvous maneuver for return to earth. Based on technical studies, NASA officials believed that this maneuver would be less difficult than the actual lunar landing and probably no more difficult than rendezvous in earth orbit.

In terms of hardware, each of the techniques called for additional development. The direct ascent approach required the development of the NOVA, a vehicle with a payload-to-orbit capability substantially greater than the advanced Saturn. To launch the same payload, using two advanced Saturns for the earth rendezvous technique, would have required the development of a fuel tanker. The unique hardware for the lunar orbital rendezvous technique was the lunar excursion module, eventually nicknamed "the bug" or "the spider."

In terms of mission capability, the earth orbital rendezvous technique and rendezvous techniques in general were developed and practiced during the Gemini Program, as well as by the lunar excursion module and the command and service module during the Apollo 9 flight shortly before the actual manned lunar landing.

The Study Effort

The study groups assembled to help make the lunar mode decision were impressive. In addition to the coordinating and evaluating effort of the Systems Office of the NASA Headquarters Office of Manned Space Flight and the joint NASA/DOD Large Vehicle Planning group, study groups were formed at the Manned Spacecraft Center in Houston, the Marshall Space Flight Center in Huntsville, the Langley Research Center in Hampton, Virginia, and the Launch Operations Center at Cape Kennedy. To a lesser extent the Lewis Research Center in Cleveland, Ohio, and the Ames Research Center in Mountain View, California, also participated.

NASA let contracts for special studies to compare the advantages and disadvantages of the several lunar landing techniques. Some business firms with capabilities in space systems research provided NASA with unsolicited studies made at their own expense. Either paid or unsolicited contract studies were made by the Massachusetts Institute of Technology, the Boeing Company, Convair, Douglas Aircraft, Grumman Aircraft Engineering, Lockheed Aircraft, Ling-Temco-Vought, Martin, North American Aviation, Northrop, Republic Aviation, and by Space Technology Laboratories.[31]

Analysis of the Alternatives

From unclassified studies, it is known that the results of the various studies were used in a comprehensive probability analysis of the three techniques. Some of the variables put into the probability calculations included: overall mission success, overall mission safety, safety in the vicinity of the moon, reliability in the vicinity of the moon, guidance and navigation reliability in the vicinity of the moon, propulsion reliability in the vicinity of the moon, human factors reliability, life support equipment reliability, mechanical reliability, communications reliability, and crew and scientific equipment reliability in the vicinity of the moon.

The judgment of NASA was that the lunar rendezvous technique provided the highest probability of overall mission success, with essentially equal probability for mission safety, compared with the alternative modes. On the basis of schedule considerations, NASA officials believed that the lunar orbital rendezvous technique would accomplish the national goal of a manned lunar landing between a year to a year and a half earlier than the other techniques. Because the lunar orbital rendezvous mode required the least amount of technical development work beyond that existing in 1962, NASA officials estimated that the lunar-orbital rendezvous technique would cost between 10 percent and 20 percent less than the other modes. On November 7, 1962, NASA let the contract for the lunar excursion module to the Grumman Aircraft Engineering Company. With that contract, NASA's choice of the lunar orbital rendezvous technique for the manned lunar landing was confirmed.

SOCIAL BENEFITS FROM A MANNED ORBITAL RESEARCH LABORATORY

Although necessarily enshrouded with ethical value judgment, researchers do attempt to identify and even set priorities on social and political benefits and costs that arise from public investments. One such study was made by Stanford Research Institute in 1965.[32] The task of the research group was to compare the benefits and the costs of policy options in the management of a manned orbital research laboratory. The underlying assumption of the analysis was that the United States would, at some time in the future, develop and operate such a laboratory. This, in turn, assumed that a manned orbital research laboratory would become an integral part of the space program.

Possible policy options for the United States ranged from complete divestment of control and ownership of the laboratory to other governments, presumably through a United Nations organization, to permanent retention of control and ownership. The Stanford Research Institute group identified, but did not attempt to quantify, possible benefits and costs resulting from use of the laboratory under the various management schemes. As Table 3–1 shows, the study did not estimate military costs and benefits of alternative actions. Among the socio-political benefits listed in the table is prestige. It is almost unnecessary to state that there is no (good) measure of the increments and decrements in United States prestige and good will that would result from the selection of a given management option. Another benefit not susceptible to quantification that appears throughout the list is that arising from United States pre-eminence in science and technology.

This kind of socio-political benefit and cost analysis does in some way show that scientific and space projects have wide international ramifications. Except perhaps for the study format, there is nothing new about the SRI analysis. For years, diplomats have been charged with the responsibility of making value judgments about possible courses

TABLE 3–1: BALANCE OF SOCIO-POLITICAL BENEFITS AND COSTS UNDER DIFFERENT MANAGEMENT SCHEMES FOR A MANNED ORBITAL RESEARCH LABORATORY

OPTIONS IN MANAGEMENT AND USE	SOCIO-POLITICAL BENEFITS	SOCIO-POLITICAL COSTS
Dedication to mankind by deed to the United Nations	Prestige in world community; enhanced world cooperation in the regulation of peaceful uses of outer space; international good will.	Constraints on future actions and uses of the laboratory. Alienation of non-UN countries; alienation of some allies.
Dedication to a scientific group	Same as above; also prestige in the scientific community; reputation of United States as the patron of science.	Constraints on future actions and uses of the laboratory; alienation of technically backward countries; alienation of some allies.
Retention by the United States	Control; reservation of future options; support for the interests of allied nations.	Loss of propaganda advantage; alienation of neutrals.
Lease and loan of facilities	Lessee's good will; international good will for sharing; prestige in scientific community for sharing; adhesion of allies; initiation of projects not possible unilaterally.	Jurisdictional disputes; constraints on use.
Sharing of data	Same as above; another form of foreign aid; political leverage.	Alienation of nonparticipants.
Use limited to basic research missions	Advancement of learning; scientific prestige; unanticipated findings and applications.	Delay on alternative projects; foregone projects and other uses of scientific personnel.
Use limited to applied scientific missions	Improvements in: world food supply, transportation, health, safety, weather forecasting, and cartography; another form of foreign aid; prestige as benefactor of mankind.	Aggravation of some existing social problems; delay on alternative projects; foregone projects; other uses of scientific personnel.

Source: Adapted from Stanford Research Institute, *Priority Analysis of Manned Orbital Research Applications,* Vol. II (Menlo Park, Calif.: Stanford Research Institute, September, 1965), pp. 12–13.

of action to be pursued by the United States. Certainly this kind of socio-political analysis of benefits and costs underlies the United States ratification of the 1967 Treaty on Outer Space. With that treaty, the United States and about a dozen other countries established principles for the peaceful international exploration and use of outer space. The principal provisions contained in that treaty were suspension of any claims of national ownership, the protection of astronauts, and prohibition of placing nuclear or other weapons of mass destruction in outer space.

ECONOMIC BENEFITS OF SPACE STATION OPERATIONS

In April 1967, NASA's Advanced Manned Missions Program of the Office of Manned Space Flight let a contract to the Planning Research Corporation (PRC) to study the economic benefits of space station operations.[33] The main purposes of the study were (1) to develop a uniform method of cost-benefit analysis for use in evaluating earth-oriented operational space stations; and (2) to estimate the potential benefits to be derived from earth applications of satellite-assisted information systems during the period between 1970 and 1989. Essentially, NASA wanted information about the relatively near-term benefits that might accrue from the application of satellite-derived information to the solution of earth-related problems. This kind of information, provided for each major program, is essential if, as directed by the Bureau of the Budget, NASA is to use planning-programming-budgeting as a management tool.

NASA established a set of ground rules from which the research group was to work. The two most important assumptions set by NASA were technical and engineering constraints. First, the costs of launch vehicles, spacecraft, and mission support and control were not to be included on the cost side of the model. The rationale for this exclusion was that too much research time would have had to be used for developing an appropriate estimate of the interdependent use of hardware and the correct allocation of joint costs to earth-oriented satellite projects. Second, technical feasibility was to be limited to changes in sensor capabilities. The group was to assume that NASA launch vehicles, spacecraft, and mission support activities would be available from "off-the-shelf" use of existing hardware. The research group was also directed to concentrate on an in-depth study of a small number of possible applications of satellite-derived information rather than to develop a broad view of all possible socio-economic effects arising from the development and use of space platforms.

Because technical feasibility was limited to the development and use of appropriate remote sensors, the study group did not distinguish between any possible differences in costs or benefits that might result from the use of manned rather than unmanned earth-oriented space platforms.

Gross Estimate of Benefits from Satellite-Obtained Information

With the advice of several federal agencies concerned with the use of natural resources, such as the Departments of Agriculture and the Interior, and by scanning the technical and socio-economic literature, the PRC research group developed a list of 243 possible uses for satellite-obtained information. From that list, 21 areas were selected as possible candidates for case studies. All research and development uses of space sensor systems were excluded from selection because of the inherent

TABLE 3-2: ESTIMATE OF ANNUAL BENEFITS DERIVED FROM A SATELLITE-OBTAINED INFORMATION SYSTEM, 1970–1989: RANKING OF TWENTY-ONE SELECTED USES

FIELD OF POTENTIAL USE	ESTIMATE OF ANNUAL BENEFITS (MILLIONS OF DOLLARS)
Agriculture: Stress[a]	$27,000
Health: Diseases of man	16,300
Agriculture: Inventory and yield	11,300
Oceanography: Fishing	1,600
Health: Diseases of animals	1,400
Natural disasters	645
Hydrology	374
Health: Solid wastes	189
Resources management: Soil survey	115
Geography: Mapping	114
Government operations: Tax assessment	87
Search and rescue missions	57
Geophysics: Fuel and mineral detection	42
Health: Water pollution	19
Forestry	9
Oceanography: Nautical charting	9
Oceanography: Ship routing	7
Health: Air pollution	1
Oceanography: Monitoring of icebergs	[b]
Geophysics: Erosion and silting	[b]
Agriculture: Location of new land	[b]
Total	$59.3[c]

[a] Stress such as wheat rust.
[b] Less than one million dollars in annual benefits.
[c] Billions of dollars.

Source: Adapted from information provided in a report submitted by Planning Research Corporation to NASA in January, 1968, Vol. I, p. 16.

difficulties of forecasting and quantifying possible benefits from R & D. A rough estimate was made of the potential benefits that might be derived from satellite-obtained information for each of the 21 candidate study areas. The sensor requirement for each use was also identified. Table 3–2 presents the 21 candidates, along with an estimate of the gross value of benefits that might be obtained from the use of satellite information systems. The results of the PRC calculations suggest that the greatest benefits derived from the use of earth-oriented satellite systems probably will accrue to the agricultural sector of the world economy. Cost-benefit ratios were not computed for the 21 candidate study areas.

Selection of Five Case Studies

After review of the 21 possible areas for in-depth studies, the PRC research group selected five uses of satellite information systems that promised the realization of large benefits during the 1970's and the 1980's. Because of the economic importance of agriculture to Southeast Asia, and also to the United States, the researchers selected two agricultural cases: (1) rice inventory and yield, and (2) crop stress resulting from wheat rust. The group selected reduction in the incidence of malaria by low water control as an example of the application of satellite information systems to help solve health problems. A health problem was included because the PRC group wanted to apply their method to an estimate of nonmonetary benefits, such as prevention of illness and lives saved. For an application in the area of conservation, the group selected increased harvest of marine life (albacore tuna). Finally, the matter of snow pack and melted snow run-off was chosen as an application in the field of hydrology because of the growing importance of limited water resources and improved management of dams.

Table 3–3 displays the results of the study made by the Planning Research Corporation. Although not shown in the table, the total net benefits resulting from the five possible applications were estimated by the PRC group to amount to about $1 billion annually. The benefits generated from these applications were included in the estimate of benefits for the 21 candidate study areas.

The PRC research team believes that their findings support the economic feasibility of satellite-assisted information systems. This conclusion is based on the fact that the cost-benefit ratios displayed in the table are well above the traditional 1:3 cut-off point (defined by PRC). In addition to a high cost-benefit ratio, the PRC group believes that satellite-assisted information systems are more efficient (i.e., cheaper) than alternative systems.

TABLE 3-3: ESTIMATE OF COSTS AND BENEFITS RESULTING FROM SATELLITE-OBTAINED INFORMATION SYSTEMS: FIVE SELECTED EXAMPLES

FIELD OF POTENTIAL USE	TOTAL COST OF SATELLITE SYSTEM[a] (MILLIONS OF DOLLARS[c])	TOTAL NET BENEFITS[b] (MILLIONS OF DOLLARS[c])	TWENTY-YEAR (1970–1989) COST-BENEFIT RATIO[d]	TOTAL COST OF ALTERNATIVE SYSTEM (MILLIONS OF DOLLARS[c])
Agriculture: Stress from wheat rust	$5.9	$6,029.2	1:1018	$ 12.4
Agriculture: Increased production of rice from supply assessment	3.1	402.7	1:128	8.7
Conservation: Increased tuna fish catch	7.0	3,124.5	1:449	505.7
Health: Malaria control	5.4	1,603.1	1:296	[e]
Hydrology: Dam management	6.4	833.8	1:131	18.1

[a] Includes costs for space-borne sensors, data handling, and power and telemetry systems. Excludes costs of space hardware, launch, and mission support.

[b] Includes primary and secondary benefits less costs incurred by users and producers.

[c] Incremental dollars discounted at 6 percent for twenty years.

[d] Computed by PRC from unrounded data.

[e] Not reported by PRC.

Source: Adapted from information provided in a report submitted by Planning Research Corporation to NASA in January, 1968. Vol. I, p. 11.

Measurement of Benefits

For each of the case studies, the PRC group first identified the benefits that would probably result from more accurate and more timely predictions obtained from earth orbital satellite information systems. The group distinguished between primary and secondary benefits. Primary benefits were defined as (1) those resulting from increased efficiency in the use of resources; and (2) those benefits arising from the employment of previously unemployed resources. The dollar value of these benefits was computed in terms of an increase in the gross national products of the United States and countries in the rest of the world. The secondary benefits, as measured by the research group, were those arising from induced increases in income and employment via the multiplier effect.[34]

Measurement of Costs

In deriving cost data, the PRC research team estimated those costs of satellite-obtained information that were over and above those of existing information systems. Cost estimates were based on three components:

research and development cost, investment costs, and operation and maintenance costs. The PRC group contends that the fundamental criterion for the use of information obtained from satellite systems is the cost of the relevant alternative information system that provides equivalent benefits without the use of spacecraft. For many of the applications studied, the research group found that the relevant alternative system would be the installation of sensing devices in aircraft. For some of the 21 candidate study areas, the PRC team found no feasible alternative system to the space satellite system.

As a caveat, the PRC research group correctly advised NASA that the results of their research would change (i.e., the cost-benefit ratios would drop) if part of the cost of launch, spacecraft, and mission support operations were apportioned to satellite information systems.

STUDY PREPARED FOR NASA BY THE RAND CORPORATION

NASA let a contract to the RAND Corporation, Santa Monica, California, in 1967 to develop a method for analyzing the comparative costs and benefits of alternative space plans.[35] Although not specified in the contract, the purpose of the research was to develop a method that might yield a viable "space worth function." The study was limited to comparisons of various space plans. No attempt was made to solve the more difficult problem of comparing space programs with nonspace programs, i.e., to derive an operational "social welfare function." The research concentrated on an analysis of the trade-off between costs and benefits of two of the goals outlined in the Space Act: (1) space research to advance scientific knowledge; and (2) space research to enhance national security. To a much lesser extent the RAND group included in their study two other space objectives outlined in the Space Act: (3) obtaining economic benefits from space activities; and (4) space research to enhance national prestige.

Limitations of the Study

The RAND group clearly stated that their model was a first attempt to measure alternative United States space plans. Numerical results were presented for the sole purpose of demonstrating the use of the model; they were not an analysis of the relative merits of space programs considered by NASA. The space projects used in the model came from tentative plans outlined by a NASA headquarters steering group before the sharp NASA budget cut in the late 1960's. The RAND group examined the national security objectives of possible space programs in terms of cost savings to the Department of Defense rather than in terms of the military effectiveness of a given weapons system. No attempt was made to develop a "military-worth function" for possible future space programs. Finally, the RAND evaluation scheme did not include a forecast of space activities that might be taken by the Soviet Union.

Single-Goal versus Multi-Goal Planning

As a frame of reference, the RAND group outlined a simple trade-off scheme for single-goal planning. With single-goal planning, decision-makers would evaluate any given space project according to its separate contribution to each of the four space objectives. With single-goal planning, no attempt is made to analyze the joint costs and joint benefits that result from the interdependence of space objectives. The four space objectives for single-goal planning are displayed in Table 3–4. The table also shows the RAND group's evaluation of relevant programs, location in space of activities, technological requirements, relative costs of achieving each goal, and the amount of public interest that is likely to be attached to each goal.

TABLE 3–4: POSSIBLE COSTS AND BENEFITS RESULTING FROM SINGLE-GOAL PLANNING FOR SPACE PROJECTS

EVALUATION FACTOR	ADVANCEMENT OF SCIENCE	NATIONAL SECURITY	BENEFIT ECONOMY	NATIONAL PRESTIGE
Kinds of programs	Unmanned Automated	Manned and unmanned Automated	Unmanned Automated	Manned and unmanned Automated
Space location	Earth orbit, moon, and planets	Earth orbit	Earth orbit	Moon and planets
Technological requirement	Existing	Advanced	Existing	Advanced and existing
Relative costs	Low	Medium	Low	High
Public interest	Low	Low	Medium	High

Source: Adapted from S. H. Dole, H. G. Campbell, D. Dreyfuss, W. D. Gosch, E. D. Harris, D. E. Lewis, T. M. Parker, J. W. Ranftl, and J. String, "Methodologies for Analyzing the Comparative Effectiveness and Costs of Alternative Space Plans" (Santa Monica, Calif.: RAND Memorandum, RM 5656-NASA, August, 1968), Vol. 2, p. 11.

The RAND research team briefly evaluated the ramifications of each of the single-goal objectives in the following way: (1) A science-oriented program would contain little advanced technology and few, if any, manned space flights. The program probably would not have much public appeal. (2) A security-oriented space program would emphasize advanced technology as well as manned and automated spacecraft systems in earth orbit. Despite the fact that it would not have flights to the moon or the planets, it would probably be expensive. The security-oriented goal also is not likely to have much public appeal. (3) A space program oriented to economic benefits might reduce the role of NASA to that of underwriting and supervising high-risk programs of agencies and industries using space systems. Because of the high cost of manned space flight projects, economic benefit criteria probably would eliminate such flights from NASA's program plans. (4) A prestige-oriented space program would select goals designed to capture the interest of the world. Programs for prestige would emphasize advanced technology and also manned and unmanned automated flights to the moon and to the planets. Because the prestige goal program would meet virtually all space objectives, it probably would be costly.[36]

Although single-goal planning requires decision-makers to know the relative importance of each goal, the RAND group correctly believes that trade-off analyses are easier with separate measures of effectiveness. This contrasts with multi-goal planning. For multi-goal planning, measurement techniques must show the relative costs and benefits of space programs to more than a single goal. As an example, take the science objective as a base for multi-goal planning. Given a clearly defined scientific effectiveness per dollar outlay, the contribution of any space project to other goals could be measured by either a decrease in scientific effectiveness for equal cost, or by the same scientific effectiveness for added cost.

RAND's Method of Measuring Scientific Effectiveness

To measure scientific effectiveness, the RAND research group constructed what they call a Comprehensive Catalog of Scientific Objectives in Space (CCSOS). The catalog contains 1,033 space objectives in pure science, scientific research, exploration, and discovery that might be attainable within the next twenty years.

The 1,033 space objectives in turn provide information about five principal basic and applied scientific research objectives: (1) the earth and its environment; (2) extra-terrestrial life; (3) the solar system; (4) the universe; and (5) space as a laboratory. To give the catalog empirical content, the RAND team proposes to send out a series of questionnaires to groups of experts in various scientific fields.[37] The results of the questionnaires would yield a value apportionment factor for each cell in the model. The sum of the values assigned to each of the 1,033 scientific space objectives would equal 100 percent. To be sure, the value array would portray the combined opinions of those participating in the surveys at a point in time. The RAND research group proposes to change the value array as projects are completed and as others are added. The values assigned to the objectives (1,033 for this particular model) become the inputs for appraising scientific effectiveness. The measure of total effectiveness of a given space project is the extent, expressed as a fraction, to which that program provides scientific information to solve the various problems listed in the Comprehensive Catalog of Scientific Objectives in Space.

As an example of how the scientific effectiveness index varies with alternative space plans, assume that there are two possible space plans, the first plan smaller than the second. Both plans contribute to each of the five research objectives within the broader national goal of advancing science (the earth, extra-terrestrial life, the solar system, the universe, and space as a laboratory). By assumption, Plan 1 includes: (1) a short-duration earth orbital space station; (2) a two-year earth orbital space station; and (3) a two-man, short-duration lunar orbit space station. On the basis of the value judgment of scientists, Plan 1 warrants, say, 0.52 value array points on the scientific effectiveness index. In addition to the three missions outlined above, Plan 2, again by assumption, includes a three-man, 90-day lunar-orbital laboratory. Although the second plan contributes additional points to each of the five scientific research goals, it adds the largest number to the goal of investigation of the universe beyond the solar system. By including a more sophisticated lunar platform, Plan 2 warrants, say, 0.65 scientific-effectiveness points.[38]

RAND's Method of Measuring National Security Effectiveness

Although national security is enhanced by military strength, economic strength, advanced scientific and technical capabilities, and prestige, an analysis of all of these factors was beyond the scope of the RAND project. Rather, the group focused attention on the contributions that NASA could make to national defense by obtaining information to support military space operations and by developing technology applicable to military space systems. The RAND model proposes to identify NASA activities that provide cost savings to the Department of Defense by eliminating duplicate DOD programs. Classification and measurement of NASA's potential contribution to national security would be made by a joint DOD-NASA long-range space planning group.

An assessment scheme proposed by the RAND team for space missions that have defense applications is: (1) space missions that have an explicitly identified DOD need; (2) those in which there is a potential DOD need; and (3) those in which the DOD need is remote. This broad assessment is suggested because many of the components of alternative space plans have not been sufficiently defined to permit adequate evaluation of their possible contributions to national defense. For example, most of the spacecraft in the RAND model, scheduled for launch after 1972, have not yet been designed.

Cost-Estimating Procedures

To estimate costs, the RAND research team proposes to gather cost data for various subsystems used on NASA programs. The data would be collected on as low a level of aggregation as possible. Because RAND has been developing cost-estimating techniques for NASA for a number of years, some of the cost data are already available. Various combinations of cost information, by subsystem, would permit a cost estimate for alternative space projects. The RAND group believes that detailed information about the costs of various subsystems can be used to allocate the probable costs of any possible space plan at the design, the development, and the production phase of the program. Some of the subsystems separately identified and costed by RAND include: spacecraft structure, re-entry structures, electrical power systems, environmental control systems, and space propulsion systems. Data about the cost of subsystems, together with the value array for scientific worth, would be used to determine the cost-effectiveness of alternative scientific space plans.

The RAND team proposes a different measurement scheme for the cost of defense-oriented space missions. To obtain the fraction of the total cost of a space plan that pertains to a specific contribution to national defense, the RAND research group suggests the following apportionment scheme: (1) For a contribution to operational capability, include the R & D and the flight test costs of the entire spacecraft. (2) For a technical contribution, include the R & D costs of any new subsystem or payload. Do not include flight test costs because the Department of Defense would have to conduct its own flight tests for any spacecraft using the new subsystem or payload. (3) For a contribution to information, include the R & D costs of the payload that provided the information and the R & D costs of other essential support subsystems. (4) To avoid double counting for a space program that makes more than one contribution to defense, add only those cost increments that have not been included.[39]

Application of Space Projects to the National Security Goal

To illustrate how the RAND group measures the space contribution to the national security goal, some arbitrary numbers will be applied to the two possible space plans described earlier. Assume that the total cost for Plan 1 amounted to $35 billion during the period 1970 to 1987, and that the cost of Plan 2 amounted to $49 billion for the same period. Recall that Plan 2 differs from Plan 1 because it includes a three-man, 90-day lunar orbital laboratory. As indicated, there are three possible defense contributions—explicit, potential, and remote. If all three of the contributions are realized, about 32 percent of the cost of Plan 1 would be also savings to the Department of Defense. If only those space missions that satisfy an explicitly identified defense need are realized, about 17 percent of the cost of the space plan would be savings to the Department of Defense. In Plan 1, potential defense applications amount to 7 percent of the program cost, and remote defense applications amount to 8 percent.

For Plan 2, also assume that the realization of all three DOD contributions would amount to 32 percent of the cost of the space plan. Although the percentage allocated to DOD savings might be the same for Plan 1 and Plan 2, the absolute savings are higher for Plan 2. Plan 2 costs about $14 billion more than Plan 1. Because the more expensive space plan includes the three-man, 90-day lunar orbital platform, remote cost savings to DOD would materialize if the defense profile required deep space and lunar orbital missions. If such missions materialized, the remote DOD contribution would amount to about 14 percent of the cost of the space plan, compared with 8 percent for Plan 1. The shorter-term, explicit Defense applications in Plan 2, if realized, would amount to 13 percent of the program cost.[40]

Synthesis of Space Plans: Science, Security, Economic Benefits, and Prestige

Without attempting to measure the effectiveness of economic benefits or the potential contributions to national prestige, the RAND group qualitatively examined the dimensions of these factors in multi-goal planning. The RAND team believes that the greatest direct economic benefits would result from earth-orbit satellites and space station applications. Space plans that emphasize these projects also result in the largest explicit cost savings to the Department of Defense. To the RAND team, this suggests complementarity between the economic and national security goals for space projects that stress earth-orbit satellites and space station applications. In contrast, the RAND group believes that programs with the greatest prestige value are those that include manned interplanetary flights and spacecraft landings. These kinds of programs, according to the RAND group, probably also have relatively high scientific effectiveness levels. Direct economic benefits, however, appear to be smaller and the potential savings to the Department of Defense remote.

NATIONAL ACADEMY OF SCIENCES: SUMMER STUDY OF SPACE APPLICATIONS

In the fall of 1966, NASA asked the National Academy of Sciences to sponsor a study of the peaceful uses of satellite systems to determine future earth-oriented applications. James E. Webb requested the National Academy to obtain: "the considered reactions and recommendations of highly qualified scientists and engineers on the nature and scope of a research and development program believed to be necessary to provide the technology required to exploit these applications."[41]

The National Academy organized technical panels that convened during the summers of 1967 and 1968. Each panel was responsible for the identification of at least one use of a satellite system that would demonstrate foreseeable benefits to that scientific field. The panels were also asked to estimate the costs of the satellite applications. Panel members prepared reports for twelve fields in applied science. The twelve fields were: (1) forestry-agriculture-geography; (2) geology; (3) hydrology; (4) meteorology; (5) oceanography; (6) sensors and data; (7) point-to-point communication, i.e., multi-point collection but single-point distribution of information; (8) systems for remote-sensing information and distribution; (9) broadcasting; (10) point-to-point communications, i.e., fixed-voice communication rather than multi-point broadcasting; (11) navigation and traffic control; and (12) geodesy and cartography. In addition, an economics panel was established to analyze the probable economic costs of the satellite systems proposed by the technical panels.

About one hundred experts met at Woods Hole during each of the summer sessions. Panel members had academic affiliations with more than thirty different colleges, universities, and other nonprofit organizations. Some of the universities included: the University of California, Carnegie Institute of Technology, University of Illinois, Lehigh University, Scripps Institution of Oceanography, and the University of Wisconsin. Other panel members had affiliations with about twenty-five different business and consulting firms. Among these organizations were: Aerospace Corporation, E.I. du Pont de Nemours, General Electric Company, Lockheed Aircraft Corporation, and Raytheon Company. Another group of participants were employees of the federal government.

The Study Group's View of Cost-Benefit Relationships

A number of economists worked closely with members of the various technical panels during the summer sessions. Participating economists, as well as the members of the economic analysis panel, took a conservative view of the applicability of conventional cost-benefit analysis to the rapidly developing fields of satellite and sensor technology. The consensus among the economists at Woods Hole was that formal cost-benefit analysis is not suitable for judging rapidly changing technologies. The study group adopted instead a technique that has been widely used by business for planning and developing new products, processes, and services. That technique is a four-step evaluation process, usually extending over a five- to ten-year period. The procedure calls for project evaluation before each successive commitment is made. The first phase in the sequence is the commitment for basic and exploratory research. Because both the potential costs and benefits are quite conjectural at this phase, judgment usually determines the investment decision. The second stage is development, i.e., preliminary design and limited testing. The third step is the decision whether to undertake market testing programs. For the second and third steps, judgment is reinforced with an increasingly larger number of facts. Compared with the first three stages, the fourth phase requires the greatest financial commitment. That stage, of course, is the operational phase, which might require the design, construction, and operation of a commercial plant. According to the Woods Hole group, if the final commitment is made, the decision-maker usually obtained sufficient data about relevant costs and benefits to make a sound judgment.

The Woods Hole study group believes that earth-oriented applications of space systems that came under review by the panels fit into this sequential decision-making process. According to the Central Review Committee, "Some applications are challenging but of uncertain benefit; some warrant consideration for funding; others meriting support are now in competition with other pressing demands."[42]

Estimated Benefits from Earth-Oriented Applications

Despite the fact that the National Academy of Sciences study group denies the applicability of a formal cost-benefit framework, each of the panels identified areas in which benefits would accrue from the practical use of earth-oriented satellites. Clothing their remarks with caution and numerous caveats, some of the panels attempted to measure the dollar value of possible benefits. Because of the diversity of scientific fields and the speculative nature of the analyses, there was little uniformity among the several panels in the method of benefit analysis. Twelve different case studies materialized. To indicate the flavor of the work accomplished by the scientists meeting at Woods Hole, a brief summary of the findings of several panels is given. It is hoped that the brevity of the synopsis does not do grave injustice to the careful work of the committees.

Forestry-Agriculture-Geography

The principal uses of earth-orbital satellite systems in the fields of forestry, agriculture, and geography would be the provision of synoptic photographs. Information from wide-coverage photographs can be used for land-use maps; crop-yield estimates; area measurements of water, snow, ice, and vegetation; and for surveys of urban areas. The members of this panel believe that conventional methods (on-the-spot surveys made with air photography and ground visits) for predicting changes in water yield, air and water pollution, crop conditions, and the effects of natural disasters are too costly for repeated coverage. The panel members also believe that such information usually cannot be obtained in time for application by potential users.[43]

To obtain a rough approximation of benefits that might accrue to agriculture from satellite data-collection systems, the panel members assumed that the use of information satellite systems would have a value at least equal to the expenditures being made, or planned, for identical data collected by conventional methods. The committee estimated that agricultural benefits to developed countries resulting from the replacement of obsolete data systems and from new data systems might amount to between $15 million to $22.5 million annually. Benefits accruing to undeveloped countries might amount to an additional $3 million to $22.5 million annually. World-wide benefits to agriculture might then range between $18 and $45 million annually. In the combined fields of forestry, range management, recreation, and wildlife preservation, the panel members believe that the annual world-wide benefits from satellite information systems might amount to an additional $26 million to $36 million each year. Of that amount, between $7.5 million and $10.5 million might accrue annually to the United States.[44]

Geology

The use of satellite systems would provide synoptic photographic surveys to help solve geological as well as agricultural problems. These surveys, which locate geological features that are not evident from close observation, ultimately might be used in explorations for minerals, petroleum, and gas. The data provided by satellite systems, in the form of photographic imagery, would be distributed directly to the thousands of users in industry and in government. The geology panel members believe that the recipients of satellite-derived information are already competent in analyzing synoptic photographs. Because of this, additional data-processing would be minimal.

The geological panel thinks that benefits from earth-orbital satellites would result from increases in exploration activity and from markedly higher finding rates. The quantification of the benefits, according to the panel, would be difficult to defend because such estimates necessarily assume a positive relation between collected data and the realization of mineral and oil discoveries. On the basis solely of the potential benefit value from royalties, the panel estimates that total benefits resulting from the use of satellite survey systems probably would be in excess of $2 billion for a twenty-year period, or approximately $100 million annually. World-wide royalty earnings, enjoyed by governments and private parties, were estimated to average about 16 percent on the gas and oil production and about 5 percent on the production of metallic minerals. The acceleration in the rate of discovery and ultimate increase in production resulting from satellite systems was estimated to be 10 percent per annum.

Hydrology

As an applied science, hydrology is concerned principally with agricultural, home, and industrial uses of water. The panel members of the hydrology committee believe that there are large gaps in hydrological theory and in empirical information about hydrologic matters. According to the panel, hydrologists have only limited knowledge about the water discharged by many of the major rivers of the world. Hydrologists do not know, with confidence, the world-wide amounts of annual snowfall, precipitation, evaporation, and soil moisture.[45] This is true particularly of inaccessible mountain areas in the United States, and of almost all areas in underdeveloped countries. Satellite information systems, according to the panel, can provide technical information about these hydrological matters. The panel thinks that it has been too costly to obtain adequate data, on a repetitive and global scale, with conventional observation techniques. The ultimate goal of hydrology, according to the Woods Hole group, is for man eventually to be able to control the ancient scourges of flood and drought, as well as to abate the new and mounting problems of water pollution. The panel contends that the practical benefits of achieving this goal would be enormous but not quantifiable.

The hydrology panel did, however, make an estimate for one of many benefits that might result from improved hydrological data. If superior information increased the design and management efficiency of municipal water supplies by 5 percent, benefits would amount to about $5 million dollars per annum over a 50-year period. The estimate is based on the panel's assumption that the investment for municipal water supplies in the United States will amount to about $5 billion during the next half century.

Meteorology

As a result of the successful operation of the Tiros and Nimbus satellite systems, the use of satellites to observe and forecast weather conditions is one of the best known earth-oriented applications of space technology. Meteorological satellites have provided world-wide weather forecast information about the life cycles of hurricanes, typhoons, cyclones, and other weather disturbances. In addition to improved weather forecasting techniques, meteorological satellites can provide economic benefits to other fields of applied science. The fields include agriculture, forestry, hydrology, navigation, and oceanography. According to the meteorology panel, the present state-of-the-art is such that relatively accurate 1- to 2-day weather forecasts can be made. More sophisticated satellite systems might provide comparably accurate information for 5- to 7-day forecasts.

The meteorological panel gave the following subjective appraisal of the possible benefits of a 5- to 7-day accurate weather forecast: (1) more efficient management in the routing and scheduling of air, highway, and water traffic; (2) reduced spoilage of perishable commodities in transit or at terminal facilities; (3) more effective schedules for on-site filming in the motion picture industry; and (4) improved planning for recreational activities and sporting events.

The panel believes that the greatest economic benefits would accrue to the following industries and sectors of the economy: (1) agriculture might realize savings by reducing unnecessary reseeding, fertilizing, or spraying operations, or by improved timing in the harvest of hay, grain, and fruit crops; (2) the construction industry would benefit from more efficient scheduling of the work forces, and from more effective handling of materials and equipment at construction sites; (3) water management and conservation could realize benefits by more efficient flood control techniques provided by advanced warning of weather turbulence; and (4) the public utilities would derive benefits from more efficient methods of facility repair, maintenance and replacement, and switch-over.[46]

The members of the meteorological panel cited an estimate, made by John A. Russo, of the annual savings to the construction industry from the availabillty of a 5-day weather forecast. The construction industry, of course, is one of the most weather-sensitive industries in the United States. Russo estimates that the total annual savings from a 5-day weather forecast would amount to no less than $250 million in the construction industry. The components of the savings estimate include: wages paid for less than full production; idle or inefficiently used equipment; perishable material loss and unrecoverable overhead items.[47]

Estimated Costs of Satellite Information Systems

In contrast with the benefits analysis, the analysis of the costs of satellite information systems was uniform. As indicated, the analysis of costs was undertaken by a single panel, the economic analysis panel, rather than a dozen different panels. To obtain the maximum amount of comparability in its estimate of the costs of the hypothetical satellite systems, the panel established a two-factor framework for cost analysis. One part of the framework was for the time cycle of satellite systems; the other part was for the functional role played by the principal systems and subsystems required for the operation of the satellites. The time cycle of the costing system included: (1) research, development, engineering, integration, and testing; (2) investment in plant and equipment; and (3) operations and maintenance. The functional division separated those costs arising from space vehicle requirements from those costs arising from processing and distribution of information to the final users in agriculture, industry, and government. For certain kinds of hardware, the panel adopted a standard cost-estimating procedure, i.e., a given cost was established for a Thor-Delta launch, another for an Atlas-Centaur launch; a specific cost was also established for any given spacecraft and for any given sensor. The cost-estimating technique did not take into account changes in costs that might result from inflation, nor did the technique include costs that might be incurred in educating and training personnel in organizations using satellite-obtained information. In addition, the panel did not discount future costs to their present value.

During the 1967 session, most of the cost analysis was based on the assumption that each of the proposed satellite information systems would operate independently. Toward the end of that session, it became clear to panel members that a smaller number of satellite systems could operate and still provide each discipline with its critical information. The members of the economics panel think that the joint use of satellite systems would result in little, if any, sacrifice to the objectives of any one discipline. Logically, the panel developed two approaches for estimating the cost of satellite-obtained information. One is a separate discipline approach; the other is a common-use approach.

Table 3–5 gives information on the estimated cost of developing and operating different satellite systems in five separate applied scientific fields. Cost components are divided by time cycle and by function. If the hypothetical satellite systems were funded, the largest component of cost for the several space systems would be investment for plant and equipment. Compared with investment costs, research and development costs would be relatively minor. In contrast, operation and maintenance costs would be the largest component for the data acquisition, processing, and distribution part of the satellite information systems.

A comparison of Table 3–5 with Table 3–6 shows that the common use of satellites would result in a reduction in cost of about $83 million over a seven-year period. The economics panel believes that data acquisition costs for the common-use approach would be approximately the same as for the separate disciplines approach. Savings from the common-use approach would result from the use of only two principal satellite systems. The first system would include earth-resources satellites and data-collection relay satellites. This system fulfills the requirements recommended by the forestry-agriculture-geography, geology, hydrology, and oceanography panels. Because of some unique requirements, the economics panel proposes a separate system for meteorological uses. Although not the primary source, the data-collection relay satellite also would provide some of the data necessary to satisfy part ($30 million out of $182 million) of the information requirements outlined by the meteorology panel. An additional piece of hardware, required by the

TABLE 3–5: ESTIMATED COST OF THE HYPOTHETICAL EARTH-ORIENTED SATELLITE SYSTEMS PROPOSED BY SUMMER STUDY PANELS: SEPARATE DISCIPLINES APPROACH *(MILLIONS OF DOLLARS)*

	COST OF SPACE SYSTEMS[a]			DATA ACQUISITION COSTS[b]			TOTAL[c]
FIELD OF APPLICATION	RESEARCH AND DEVELOPMENT	INVESTMENT	OPERATION	RESEARCH AND DEVELOPMENT	INVESTMENT	OPERATION	
Forestry-agriculture-geography	$16	$ 39	$ 18	$ 7	$ 6	$ 56	$ 142
Geology[d]	1	18	19	2	0	5	45
Hydrology	19	51	30	19	11	78	208
Meteorology	27	182	65	5	2	160	441
Oceanography	8	71	44	8	8	44	183
Total	$71	$361	$176	$41	$27	$343	$1,019
Percentage distribution	7.0	35.4	17.3	4.1	2.6	33.6	100.0

[a] Includes cost of: Launch operations, spacecraft and sensors, ground support, and management of space systems.

[b] Includes costs for the processing and distributing of data in a form suitable to user groups.

[c] The approximate time-frame for each system is seven years, with about three years required for the research and development phase and four years for the operational phase.

[d] In addition to two satellites, geology requirements call for an aircraft system that provides coverage by radar.

Source: Adapted from National Academy of Sciences-National Research Council, *Summaries of Panel Reports, Useful Applications of Earth-Oriented Satellites* (Washington, D.C.: National Academy of Sciences, 1969), p. 65.

TABLE 3-6: ESTIMATED COST OF THE HYPOTHETICAL EARTH-ORIENTED SATELLITE SYSTEMS PROPOSED BY SUMMER STUDY PANELS: COMMON DISCIPLINES APPROACH *(MILLIONS OF DOLLARS)*

SATELLITE SYSTEMS AND DATA ACQUISITION SYSTEMS	RESEARCH AND DEVELOPMENT	INVESTMENT	OPERATION	TOTAL[a]
Earth-resource satellite[b]	$14	$117	$ 62	$193
Data-collection relay satellite[c]	7	35	32	74
Aircraft radar[d]	0	2	12	14
Meteorology's geosynchronous satellite[e]	27	152	65	244
Total cost of space systems	48	306	171	525
Cost of data acquisition[b]	41	27	343	411
Total cost[f]	$89	$333	$514	$936
Percentage distribution[f]	9.5	35.6	54.9	100.0

[a]The appropriate time-frame for each system is seven years, with about three years required for the research and development phase and four years for the operational phase.

[b]System would be used in the fields of agriculture-forestry-geography, geology, hydrology, and oceanography.

[c]System would be used in the fields of hydrology, oceanography, and meteorology.

[d]The system is the same as the separate-discipline system for geology.

[e]The cost of the system is reduced by $30 million because of a less expensive geosynchronous satellite. The Data-Collection Relay Satellite provides part of the data requirements of the meteorology discipline.

[f]Costs applicable to all disciplines.

Source: Adapted from National Academy of Sciences-National Research Council, *Summaries of Panel Reports, Useful Applications of Earth-Oriented Satellites* (Washington, D.C.: National Academy of Sciences, 1969), p. 66.

geology field and costing about $14 million, would be an aircraft radar photography system that supplements information provided by satellites. The total cost of developing, building, and operating the common-use satellite systems would amount to an estimated $936 million over a seven-year period. As Table 3-6 shows, more than 50 percent of the total estimated cost would be spent for maintenance and operation. Less than 10 percent would be required for research and development.

Recommendations Made by the Space Applications Summer Study Group

First, the central review committee of the Woods Hole summer study group stated that satellites are the newest resource of the nation. According to the committee, the benefits from the practical applications of satellite systems would be far larger than those initially expected by the panel members. In addition, the benefits would be much greater than the estimated costs of achieving the benefits. The committee strongly recommended that NASA develop a space program for the 1970's that emphasizes earth-oriented satellite programs. The committee also urged that the United States commit itself to an annual expenditure of at least $200 to $300 million for satellite applications programs. In implementing an earth-oriented satellite program, the committee suggested that NASA employees work closely with personnel in potential user agencies to insure the development and use of viable and effective earth-oriented satellite information systems.

SUMMARY

In August 1965, President Johnson issued a directive to agency heads in the federal government to use the planning-programming-budgeting technique, i.e., cost-effectiveness analysis, as a management tool. Cost-effectiveness analysis is an attempt to emulate in the public sector the resource allocation mechanism of the price system in the private sector of the economy. For years, economists, especially those working in the fields of public finance and pure welfare theory, have been trying to identify, to measure, and to obtain the optimum allocation of resources.

Conceptually, it is possible to construct cost-effectiveness models for at least four levels in the governmental decision-making process: (1) the allocation of resources between the public and the private sector; (2) the allocation of resources within the public sector; (3) the allocation of resources between programs within a national given goal or federal agency; and (4) the least-cost means of achieving a given program objective. Because of incommensurabilities, it is probably a utopian dream to believe that operationally valid models can be constructed for the first two, if not three, levels in the governmental decision-making process. To achieve an optimum allocation of resources between the private and the public sectors of the economy and also among disparate programs within the public sector, it is necessary to give empirical content to what economists call a "social welfare function." The "social welfare function" is an ivory tower notion that, if derived, would reflect the true and agreed-upon economic, political, social, and ethical desires of all people in our society. The potential use of cost-effectiveness analysis is greatest in the fourth tier of the governmental decision-making process—when it is used as a means to achieve the lowest possible cost of a clearly defined program objective.

In applying the cost-effectiveness technique, analysts encounter and must solve a number of problems in method. The most important are: the selection of discount rates that reflect the opportunity costs of using resources on one project rather than another; the appropriate amortization period; a way to quantify the results of the research and development effort; a way to measure changes in the quality of goods and services provided by the federal government; and a way to allocate the joint costs and joint benefits of multi-purpose plant and equipment.

By the late 1960's, NASA had not applied formal cost-effectiveness analysis to its agency-wide decision-making process. There are three principal reasons for this: (1) The size and the direction of NASA's largest project, the manned lunar landing and exploration program, were a Presidential decision and therefore not a program to be decided by the planning-programming-budgeting method; (2) NASA is a research-oriented agency, and, at best, quantification of the benefits arising from the R & D process is elusive. By definition the results of basic research are both unknowable and unpredictable; and (3) It takes time and funds to assemble a staff that can do good cost-effectiveness analysis. Throughout most of the federal government, the use of the planning-programming-budgeting technique is in its infancy.

Although not in a formal cost-effectiveness analysis framework, the choice of the mode for the manned lunar landing is an example of the use of cost-effectiveness analysis in NASA. Between mid-1961 and mid-1962, study groups in NASA and under contracts with NASA, made comparative cost and effectiveness studies of the direct ascent technique, the earth orbital rendezvous technique, and the lunar orbital rendezvous technique. The third method, as is now well-known, was selected. In 1962, NASA officials estimated that the cost of the lunar rendezvous technique would be between 10 and 20 percent less than the cost of the other modes, and that it would permit a manned lunar landing between a year and a year and a half earlier than either the direct ascent mode or the earth orbital mode.

In an attempt both to comply with the 1965 Presidential directive and to learn more about the applicability of cost-effectiveness analysis for planning space programs, NASA let a number of research contracts for cost-effectiveness studies during the late 1950's. The studies were exploratory in nature. The results of the research have not provided NASA with a meaningful and operational technique for examining the cost-effectiveness of possible space programs. Stanford Research Institute made a comparison of the benefits and the costs of policy options in the management of a manned orbital research laboratory. As a first step in measuring the scientific effectiveness of space programs, the RAND Corporation of Santa Monica, California, constructed what that research group calls a Comprehensive Catalog of Scientific Objectives in Space. The RAND group proposes to give their catalog, which contains 1,033 space objectives in pure science, scientific research, exploration and discovery, empirical content by establishing a value array point system based on the results of questionnaires sent to experts in a number of scientific fields.

The Planning Research Corporation made a study of the economic benefits of space station operations. For five uses of satellite-obtained information (two in agriculture and one in health, conservation, and hydrology), the PRC research group believes that the results of their research yield sufficiently high benefit-cost ratios to show the economic feasibility of satellite information systems.

During the summers of 1967 and 1968, the National Academy of Sciences, on behalf of NASA, sponsored a study on the peaceful uses of earth-oriented satellite systems. About one hundred people from government, the academic world, and industry participated in each summer research session. The research was conducted by panels in 12 different fields of applied science. Although the National Academy of Sciences study group did not believe that the formal cost-effectiveness framework was suitable for judging technologies in a formative state as satellites, the group's research effort led the NAS Central Review Committee to conclude that the benefits from the practical application of satellite systems are far larger than those initially expected by panel members and the benefits are also far larger than the costs of achieving them.

NOTES

1. The distinctions between these concepts made by professionals in the field need not be of concern here.

2. Alain C. Enthoven, "Economic Analysis in the Department of Defense," *American Economic Review,* LIII, 2 (May 1963), pp. 413–14.

3. U.S. President, *Public Papers of the Presidents of the United States,* Book II (Washington, D.C.: Office of the Federal Register, National Archives and Records Service), Lyndon B. Johnson, 1965. Also published in U.S. Congress, Senate, Subcommittee on National Security and International Operations of the Committee on Government Operations, *Planning-Programming-Budgeting,* 90th Cong., 1st Sess., 1967, p. 1.

4. *Ibid.*

5. U.S. President, Weekly Compilation of Presidential Documents, Vol. 3, No. II, March 20, 1967. Also in U.S. Senate, Subcommittee on National Security and International Operations, *op. cit.* p. 6.

6. Historically, the two great names in economic welfare theory are Vilfredo Pareto (Italian, 1848–1923) and A. C. Pigou (British, 1877–1959).

7. U.S. Congress, Senate, Statement of Charles L. Schultze, *Planning-Programming-Budgeting, Hearings,* before the Subcommittee on National Security and International Operations of the Committee on Government Operations, 90th Cong., 1st Sess., August 23, 1967, p. 12.

8. David Novick, "The Origin and History of Program Budgeting," *California Management Review,* Fall, 1968, pp. 7–12.

9. Roland N. McKean, ed., *Efficiency in Government Through Systems Analysis, with Emphasis on Water Resources Development* (New York: John Wiley, 1965), pp. 18–20.

10. Novick, "The Origin and History of Program Budgeting," pp. 8–9.

11. Robert N. Grosse, *Principles of Cost Effectiveness Analysis,* Transcript of materials

presented to the Industrial College of the Armed Forces, Washington, D.C., Publication No. L66-24, September, 1965.

12. For a discussion of such a voting system, see Richard A. Musgrave, *The Theory of Public Finance: A Study in Public Ecônomy* (New York: McGraw-Hill, 1959), Chapter 6, "Budget Determination Through Voting," pp. 116-35.

13. The issue of whether the United States should install a defensive missile system was highly controversial in 1967. It seemed to be settled until Congress reopened the debate shortly after it convened in 1969.

14. Robert L. Banks and Arnold Kotz, "The Program Budget and the Interest Rate for Public Investment," *Public Administration Review*, XXVI, 4 (December, 1966), p. 287.

15. For a discussion of discount rates in PPBS work based on opportunity costs, see U.S. Congress, Senate, Statement of William J. Baumol, *Planning-Programming-Budgeting, Hearings*, before the Subcommittee on National Security and International Operations, *op. cit.*, August, 1967, pp. 150-59. Also see William J. Baumol, "On the Social Rate of Discount," *American Economic Review*, LVIII, 4 (September 17, 1968), pp. 788-802.

16. The report of the Comptroller General was published in U.S. Congress, Subcommittee on Economy in Government of the Joint Economic Committee, *Interest Rate Guidelines for Federal Decision-Making*, 90th Cong., 2d Sess., January 29, 1968, pp. 33-65.

17. One of the few studies on quality change in the provision of government services is that of Werner Z. Hirsch, "Quality of Government Serivces," in *Public Expenditure Decisions in the Urban Community*, ed. by Howard G. Schaller (Washington, D.C.: Resources for the Future, Inc., 1963), pp. 163-79.

18. U.S. Congress, House, Statement of Elmer B. Staats, Deputy Director, Bureau of the Budget, *Federal Research and Development Programs: The Decision-Making Process, Hearings*, before the House Committee on Government Operations, 89th Cong., 2d Sess., January, 1966, pp. 9-14.

19. Thomas C. Schelling, *PPBS and Foreign Affairs* (memorandum prepared at the request of the Subcommittee on National Security and International Operations of the Senate Committee on Government Operations, 90th Cong., 1st Sess., January, 1968).

20. Milton A. Margolis and Stephen A. Barro, "The Space Program," in *Program Budgeting, Program Analysis and the Federal Government*, ed. by David Novick (Cambridge, Mass.: Harvard University Press, 1965), pp. 128-33.

21. U.S. Congress, Senate, Statement of H. G. Rickover, *Selected Comment: Planning-Programming-Budgeting* (prepared by the Subcommittee on National Security and International Operations of the Senate Committee on Government Operations, 90th Cong., 1st Sess., July 26, 1967), pp. 37, 43-44.

22. Alain C. Enthoven, presentation prepared for the Special Subcommittee on the Utilization of Scientific Manpower, Senate Labor and Public Welfare Committee, 89th Cong., 2d Sess., May 17, 1966.

23. Novick, *California Management Review*, p. 7.

24. U.S. Congress, Statement of Henry S. Rowen, *Planning-Programming-Budgeting: Progress and Potentials, Hearings*, before the Subcommittee on Economy in Government of the Joint Economic Committee, 90th Cong., 1st Sess., September, 1967, p. 185.

25. U.S. Congress, Senate, Subcommittee on National Security and International Operations of the Committee on Government Operations, *Memorandum, Planning-Programming-Budgeting: Uses and Abuses of Analyses*, 90th Cong., 2d Sess., April 22, 1968.

26. U.S. Congress, Senate, Testimony of Henry M. Jackson, *Planning-Programming-Budgeting, Hearings*, before the Subcommittee on National Security and International Operations of the Senate Committee on Government Operations, Part I, 90th Cong., 1st Sess., August 23, 1967, p. 38.

27. A reply from NASA to Elmer B. Staats, Comptroller General of the United States

in January 1968. Mr. Staats inserted it into his testimony about the use of discount rates among federal agencies. See Joint Economic Committee, *Hearings, Interest Rate Guidelines for Federal Decision-Making*, p. 11.

28. Robert L. Rosholt, *An Administrative History of NASA*, 1958–1963 (Washington, D.C.: U.S. Government Printing Office, 1966), pp. 212–13.

29. H. L. Nieburg, *In the Name of Science* (Chicago: Quadrangle Books, 1966), p. 29.

30. U.S. Congress, House, Committee on Science and Astronautics, *NASA Lunar Orbit Rendezvous Decision, Hearings*, 87th Cong., 2d Sess., July 12, 1962, p. 10.

31. National Aeronautics and Space Administration, "Lunar Orbit Rendezvous" (News Conference on Apollo Plans held at NASA Headquarters, July 11, 1962).

32. John G. Meitner, "Priority Analysis of Manned Orbital Research Applications," 2 vols. (Menlo Park, Calif.: Stanford Research Institute, September, 1965, SRI Project M-5465). (mimeographed)

33. Allan H. Muir, "A Study of the Economic Benefits and Applications of Space Station Operations" (Washington, D.C.: Planning Research Corporation, 1968). (mimeographed)

34. Income and employment multipliers are explained in footnote 5 in Chapter 6.

35. S. H. Dole, H. G. Campbell, D. Dreyfuss, W. D. Gosch, E. D. Harris, D. E. Lewis, T. M. Parker, J. W. Ranftl, and J. String, *Methodologies for Analyzing the Comparative Effectiveness and Costs of Alternative Space Plans*, 2 vols. (Santa Monica, Calif.: The RAND Corporation, RM 5656-NASA August, 1968). (mimeographed)

36. *Ibid.*, Vol. II, p. 10.

37. The reader should not think that the RAND work is a method for analyzing the actual comparative effectiveness and the actual costs of alternative space projects. The model has not yet (1974) been given empirical content.

38. *Ibid.*, Vol. I, pp. 18–27, and Vol. II, pp. 64–101.

39. *Ibid.*, Vol. II, p. 121.

40. *Ibid.*, Vol. II, pp. 141–45.

41. National Academy of Sciences-National Research Council, *Report of the Central Review Committee, Useful Applications of Earth-Oriented Satellites* (Washington, D.C.: National Academy of Sciences, 1969), p. 1.

42. *Ibid.*, p. 11.

43. National Academy of Sciences-National Research Council, *Summaries of Panel Reports, Useful Applications of Earth-Oriented Satellites* (Washington, D.C.: National Academy of Sciences, 1969), pp. 1–2.

44. National Academy of Sciences-National Research Council, *Space Applications Summer Study: 1967 Interim Report* (Washington, D.C.: National Academy of Sciences, 1968), p. 39.

45. National Academy of Sciences, *Summaries of Panel Reports*, p. 13.

46. *Ibid.*, p. 18.

47. John A. Russo, "The Operational and Economic Impact of Weather on the Construction Industry in the United States," Travelers Research Center, Hartford, Connecticut, 1965, U.S. Weather Bureau, Contract CWB-10948. (mimeographed)

4 The Decision-Making Process: A Case Study of the Use of Systems Analysis in the Office of Manned Space Flight

In recent years, professional schools of business and public administration have increasingly emphasized the importance of systems analysis as a tool of management. Management, of course, is rational decision-making. Managers transform the disorganized resources of people, plant and equipment, money, and time into an operationally effective organization. The managerial process calls for efficient planning, organization, communication, and control.[1]

SYSTEMS ANALYSIS

The notion of a "system" includes a wide range of concepts. One definition of a system given in Webster's Dictionary is "harmonious arrangement: order out of confusion." Another definition is "an interdependent group of items forming a unified whole."[2] As applied to complex military and space programs, systems analysis is a framework used for integrating the various elements of a massive program.

A good description of systems analysis, as it has been applied to the space program, appeared in the July 1, 1968 issue of *Forbes:* "systems analysis involves looking at the world in big chunks; not just at end products, but their uses and their effects on other projects and on the environment—and, in turn, the influences of all these back to the product itself . . . Systems management is the art of tying the whole thing together."[3]

Cost-benefit analysis is one of several systems approaches to the decision-making process. Other systems approaches include: activity analysis, decision theory, game theory, information theory, operations research, and queuing theory.[4] This book is a systems approach to an analysis of the political economy of the space program. Although many different definitions might be applied, systems analysis is neither more nor less than an approach to decision-making that uses clearly identified means to accomplish clearly defined ultimate ends or objectives.

The purpose of this chapter is to examine briefly the dozen or so management tools employed by NASA's Office of Manned Space Flight that focus on objective criteria and milestones that were accomplished for the manned lunar landing in July 1969. To be sure, the same management tools apply to other manned space flights and also to other programs in the Office of Manned Space Flight.[5] For the most part, this chapter leaves open such philosophical issues as whether or not NASA's management policies created a research environment that stimulated creativity.[6] Nor will there be a discussion of systems techniques developed with NASA funding, most of which have never been employed by program offices in NASA. For example, NASA funded a joint study by the Jet Propulsion Laboratory and the General Electric Company for the development of management techniques that would result in an optimum strategy for the exploration of Mars during the 1970's.[7] In the late 1960's, the National Academy of Public Administration was engaged in research on several NASA-wide management studies. The projects included: the NASA management reorganization of 1966–1967; the management of NASA's world-wide data tracking system; the development of a publications clearing house for studies in public administration sponsored by NASA; the development of a management data bank; a study on how technical specialists are converted into managers; and biographical sketches of NASA's program directors and program managers.

THE MANAGEMENT STRUCTURE IN THE OFFICE OF MANNED SPACE FLIGHT

As discussed in the sections on program and site selection in Chapter 2, the Office of Manned Space Flight is the largest of four program offices in NASA.[8] Each of four program directors is an Associate Administrator and is responsible to the NASA Administrator. Table 4–1 gives a simplified view of the management structure in the Office of Manned Space Flight in 1969. As the table shows, the managerial organization in the Office of Manned Space Flight can be viewed as a four-tier structure: the headquarters office, four program offices, three main field centers, and ten principal contractors. Grouped in the first tier of the management structure is the Management Council, as well as the Office of Associate Administration for Manned Space Flight. Membership on the Management Council consists of the directors of each of the three Manned Space Flight Centers (tier three in the Management structure) and the As-

TABLE 4–1: STRUCTURE OF THE MANNED SPACE FLIGHT MANAGEMENT OPERATION

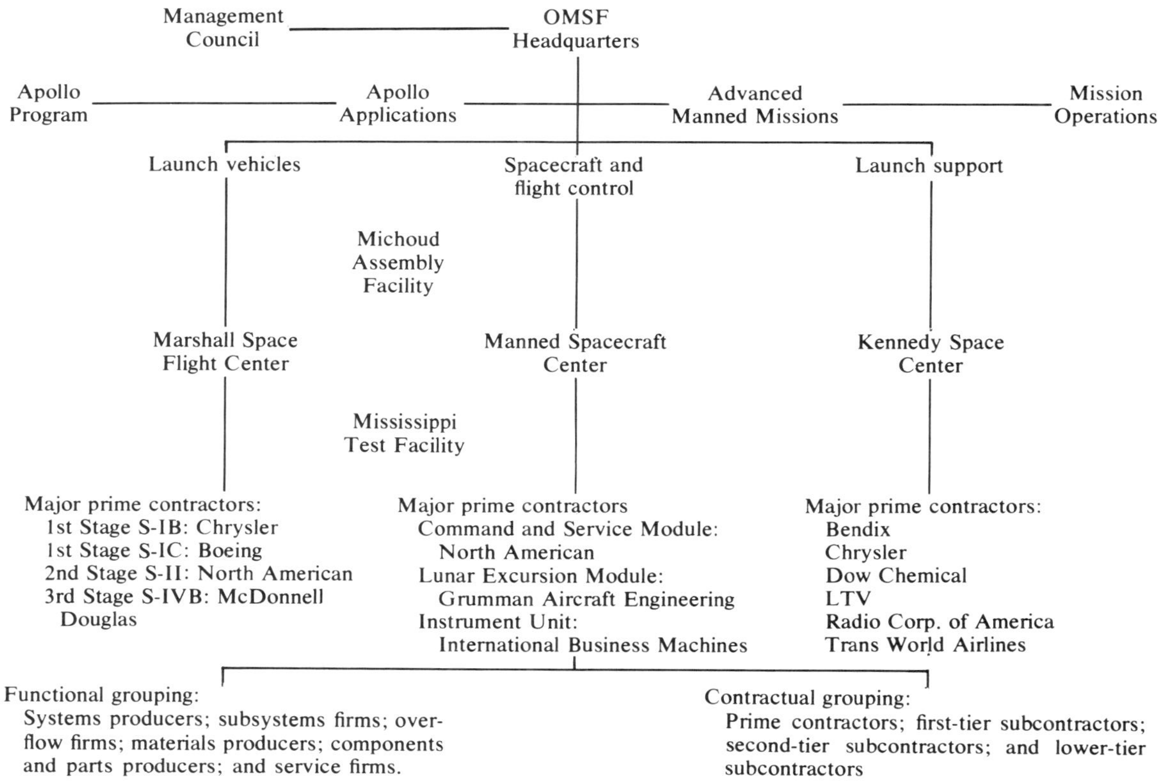

sociate Administrator (MSF). With the accomplishments of Apollo flights throughout 1968 and 1969, along with the comprehensive television and press coverage of the flights, the names of these NASA managers became well known. At that time George E. Mueller was Associate Administrator for Manned Space Flight; Wernher von Braun, Director of the Marshall Space Flight Center; Robert R. Gilruth, Director of the Manned Spacecraft Center; and Kurt H. Debus was Director of the Kennedy Space Center.

The Management Council is what its name implies; it is the in-house executive group that solves problems, approves plans, and makes the "go or no go" decisions for specific flight plans. The Manned Space Flight Management Council is subordinate to and works closely with the NASA-wide Management Council. All of NASA's Associate Administrators belong to the NASA-wide Management Council, which is chaired by the NASA Administrator, Thomas O. Paine. Although not shown in Table 4–1, there are other executive groups that consist of the corporate officials of NASA's principal prime contractors and NASA executives. There are still other executive groups that include notables in the academic and scientific community.

On the second tier of the management structure are the four Manned Space Flight Program offices. Of the four, the Apollo Program Office is the largest and the best known to the public. Samuel C. Phillips (Major General, U.S.A.F.) was Director of the Apollo Program at the time of the first manned lunar landing. Logically, the directors of these program offices are responsible for the successful operation of the component programs in the Manned Space Flight Program (Apollo, Apollo Applications, Advanced Manned Missions, and Mission Operations). Before completion of the programs, Mercury and Gemini were also program offices.

As indicated, the three manned space flight field centers form the third tier of the management structure. The fourth tier is American industry. The grouping and classification possibilities for the industrial complex that provides hardware and services for the Manned Space Flight Program are wide. A grouping of NASA's 93 largest contractors by economic function was discussed in Chapter 2. A grouping by contractual arrangement would be possible, but for most purposes somewhat less useful.

A primary task of each of the three field centers is the technical and administrative management of contracts. Wernher von Braun's Space Flight Center has had the greatest participation in the development, assembly, and test of launch vehicles. That center has worked most closely with the corporate officials at Boeing, Chrysler, McDonnell Douglas, and North American Rockwell. Gilruth and his group at Houston have coordinated with contractors building the spacecraft, and Kurt

Debus's team at Cape Kennedy has had principal responsibility for the management of contracts for launch support activities.

The size of the manned space flight management operation is given in Table 4–2. At the approximate peak of the Apollo Program effort, George E. Mueller had direct and indirect responsibility for the actions of about 300,000 people. Less than 1 percent of those were employed in the NASA headquarters office in Washington; only 6 percent were employed by the federal government; and only 50 men were astronauts.

TABLE 4–2: SIZE OF THE MANNED SPACE FLIGHT MANAGEMENT OPERATION AT APPROXIMATE PEAK OF THE APOLLO PROGRAM EFFORT

ORGANIZATION	MANPOWER JUNE 1966		FUNDING FISCAL 1966 (MILLIONS OF DOLLARS)	
	NUMBER	PERCENT	NUMBER	PERCENT
MSF Headquarters[a]	1,500	.5	32	.9
MSF Field Centers[a]				
Marshall Space Flight Center[b]	7,300	2.7	128	3.6
Manned Spacecraft Center	4,700	1.7	87	2.5
Kennedy Space Center	2,600	1.0	81	2.3
MSF Contractors[c]	258,200	94.1	3,200	90.7
Total	274,300[d]	100.0	3,528	100.0

[a] Includes permanent civil service positions and funds for administrative operations.

[b] Includes operations of the Mississippi Test Facility and the Michoud Assembly Facility.

[c] Includes contracts let for research and development, and for construction of facilities.

[d] Employment on the Manned Space Flight Program peaked during the spring of 1966 with an estimated work force of 300,000.

Source: Office of Manned Space Flight, National Aeronautics and Space Administration.

MANAGEMENT CHARACTERISTICS

The manned space flight management system, as well as NASA in general, has four conspicuous management characteristics: (1) phased-project (program) planning; (2) a project-oriented (program-oriented) organizational structure; (3) small-group management review and control; and (4) concentrated decision-making authority.

Phased-Project Planning

With the issuance of a NASA directive, phased-project planning became a formal NASA policy in November 1965, although in fact the technique had been used since the early 1960's.[9] The Department of Defense adopted phased-project planning in July 1965. Phased-project planning simply means that a program and its components are clearly defined in separate phases. After completion, the successes and failures of a given phase are reviewed and analyzed. If evaluations are favorable, the program proceeds to the next phase. The time lag between phases depends on the nature of problems encountered and the magnitude of corrections that might be required. For example, NASA did not launch a manned space flight for more than a year after the Apollo 204 accident on January 27, 1967. Apollo 8, launched on December 21, 1968, was so successful that the Apollo 9 launch followed about two months later. Each phase of a large program is known as a "major milestone" inside NASA.

The following schedule of major milestones illustrates phased-project planning for the Apollo Program: (1) unmanned earth-orbital Saturn I-B flights; (2) unmanned earth-orbital Saturn V flights; (3) manned earth-orbital Saturn V flights; (4) manned lunar-orbital Saturn V flights; (5) tests of the Lunar Excursion Module in earth orbit; (6) tests of the Lunar Excursion Model in lunar orbit; (7) the manned lunar landing; and (8) lunar explorations. The components, subsystems, and systems that go into the accomplishment of each of these major milestones also are developed and tested on the basis of phased-project planning.

Project-Oriented Structure

A glance at an organization chart of the Space Agency, as well as the organization charts of each of the four headquarters program offices and field centers, quickly shows the program and project orientation of the agency's management structure. The four NASA Associate Administrators are responsible for the management of NASA's four broad programs. As indicated in Chapter 2, they are: Advanced Research and Technology, Manned Space Flight, Space Science and Applications, and Tracking and Data Acquisition. The principal segments of the Manned Space Flight Program, in turn, are separate program offices. See Table 4–1. Each of the manned space flight field centers is responsible for an identifiable segment of the lunar landing objective. Take the Marshall Space Flight Center in Huntsville, Alabama, as an example of project-oriented management at the field centers. During 1968 and 1969, the main specialized functions undertaken at the Marshall Center included the development and procurement of the Saturn I-B launch vehicle, the Saturn V launch vehicle, equipment for Apollo Applications Program, and engines for launch vehicles. Each of these separate operations was a program office at the Marshall Center in 1968 and 1969. On a lower level, there were three project offices in the Engine Program Office—one for the H-I engine, one for the J-2 engine, and one for the F-1 engine.

Small-Group Management Review and Control

Small-group management review and control is the mirror image of the project-oriented organization structure. Small-group management review and control are necessary because the components of sophisticated research and development projects require specialized scientific expertise in many widely diverse fields of technology. With small-group control, managerial tasks are delegated to successively lower tiers. In the jargon used at NASA, small-group management reflects the "work breakdown structure" of a program. Going from a lower to a higher management decision point, a subtask group is responsible to a task force; the task group to those managing a subsystem; subsystem management groups report to those responsible for a system; the systems management groups report to project managers; and project managers to program management groups.[10] Most of the management tools discussed in this chapter show how management review is delegated to small control groups.

Centralized Decision-Making Authority

It is not paradoxical to say that managerial tasks are delegated to small specialized groups and at the same time to say that the decision-making authority in the Office of Manned Space Flight, and in NASA in general, is centralized. The difference is in the kinds of decisions that are made at various levels within the organization. In his book on government contracting, Danhof strongly argues that the growth of contractual research and development has resulted in the concentration of responsibility for specific and specialized decisions at high administrative levels in the executive departments. Danhof contends that concentrated decision-making results from the fact that the execution of proposed programs is necessarily influenced by experience and knowledge gained from the successful performance of preceding programs. It is simply a matter of a small number of able people knowing the numerous important alternatives that must be considered before decisions can be made. Concentrated decision-making in such science-oriented agencies as the Atomic Energy Commission, National Institutes of Health, the National Science Foundation, NASA, and the Department of Defense is subject to review for feasibility and desirability by technically qualified staffs and advisory councils within the given executive agency and also by the Executive Office of the President. Additional review comes from Congressional committees.[11]

Information Flows

An essential ingredient in an effective management operation is the flow of information. The transmission of information can be formal or informal. The principal forms of information exchange in NASA, as in other organizations, consist of personal exchange, formal periodic reviews and meetings, and written reports.

In 1964, the Office of Manned Space Flight established a data management system to organize and to control data gathered and disseminated about the Apollo Program. The functions of the data management system include: (1) identification of data users, i.e., NASA and contractor personnel, and their requirements; (2) acquisition and storage of data; and (3) the distribution of information to users. The kinds of information that are readily available from the system include plans, specifications, drawings, schedules, and technical manuals. A data review and documentation analysis are made for every large contract by NASA procurement officers. The evaluation process is concurrent with both contract negotiation and actual contract performance. Because contractors also evaluate information requirements, they frequently propose additions, deletions, and changes in the data-acquisition process. Acquired items ultimately appear on what is called a "document index." The index lists documents and other kinds of information that might be essential for the management of programs and their various components. The information system is, in effect, an input for other management tools used for the administration of the Manned Space Flight Program.

Program Development Plans

The apex of the information pyramid in the Office of Manned Space Flight is a document called the Program Development Plan.[12] NASA published an *Apollo Program Development Plan in May 1966,* which was designated for internal use only. The purpose of that document was to identify program requirements, responsibilities, tasks, resources, and the time frame of major projects required to accomplish the objectives of the Apollo Program. The Apollo Program, of course, is the third of three step-by-step programs planned to achieve the manned lunar landing.

The Mercury Program established the fact that man can perform effectively in the environment of orbital flight. That program was the foundation for the development of a manned space flight technology. The Gemini Program was designed to improve operational proficiency in space. During that program, old techniques were perfected and new techniques were developed. The most important new technique was the rendezvous and docking technique. As indicated in Chapter 3, the manned lunar landing required a lunar-orbital docking of the Lunar Excursion Module and the Command and Service Module for return to earth. The main objective of the Apollo Program in 1969 was the manned lunar landing, and the objective for the early 1970's investigations of different sites on the moon's surface.

THE ENDS-MEANS RELATION IN THE SYSTEMS PROCESS

The best way to order the management systems used in the Office of Manned Space Flight is in an "ends-means" framework. An "end" is an objective to be accomplished and a "means" is the way in which the objective is accomplished.

Any structure or system can be visualized as a series of ends-means relations.[13] For example, a strong national defense is a given national objective (End A). A means to that end is an effective space program (Means to End A). Depending on the point of view and the policy context, an effective space program is another objective (End B). A viable civilian space program is a means to an effective space capability (Means to End B). As an executive department, the civilian Space Agency, and its successful operation, is still another end in itself (End C). The Manned Space Flight Program is a means to the accomplishment of that end (Means to End C). Yet, the accomplishment of a manned lunar landing is also a national objective (End D). The Mercury, Gemini, and Apollo Programs are means to that end (Means to End D). The information given in Table 4–3 carries the ends-means relation into a narrower context and into the management setting of NASA's Office of Manned Space Flight.

TABLE 4–3: ENDS-MEANS RELATIONS IN THE MANAGEMENT OF THE MANNED SPACE FLIGHT PROGRAM

END	MEANS
Quality	Configuration management Experiment Packages System Safety system Panel Review Board
Time	Schedule system (PERT) Logistics system
Cost	Budget review process Production cost model "All-up" system Work Packages System

The management decisions that have governed the Manned Space Flight Program and other large research and development programs generally consider balancing system performance against program schedule and against program cost.[14] It is the task of management to predict the effect of alternative courses of action on these crucial factors and to make the correct decisions. For this analysis, quality, time, and cost are ends. Each, of course, can also be viewed as a means to the higher-order end of the manned lunar landing.

As used here, quality is the technical performance and the reliability of a given space system; the time and cost objectives have conventional definitions. For the manned lunar landing, of course, the time objective was to be achieved during the decade of the 1960's. The principal management tools used in the Office of Manned Space Flight are also listed in Table 4–3. The tools include systems to manage hardware configuration, experiments, safety, schedule, logistics, budget review, and production costs. To be sure, not all of the tools described in this chapter are actually identified in NASA's management manuals as "systems."

The different systems used to accomplish the objectives of technical performance, time, and cost overlap. For example, failure to detect hardware defects can cause program delays that can increase costs by staggering amounts. The systems (i.e., the means displayed in Table 4–3) are grouped as closely as possible, according to the primary role played by each in providing information about technical performance, time, and cost.

Research groups have been able to compare, and to compute the error margins between the predicted and the actual values for the time and cost variables. Because the performance criteria usually have a multitude of qualitative aspects, it has been more difficult to give empirical content to the degree to which original promises were above or below actual performance.[15]

TECHNICAL PERFORMANCE

The technical performance criteria for NASA's Manned Space Flight Program differ from those of some other major research and development projects because the Manned Space Flight Program requires hardware to be "man-rated." This means the hardware must approach the zero defect point as closely as possible.

In contrast with other Department of Defense R & D programs involving human life, the NASA program is not enshrouded in secrecy. Because it is daring and spectacular, the program has captured the imagination of the public and of the press. Many Americans believe that they are an integral part of the space program. Unlike military and civilian test pilots, or even aquanauts, the astronauts and their families are in the public image. They are national heroes and champions of international good will. The Apollo accident that took the lives of Virgil I. Grissom, Edward H. White, and Roger B. Chaffee on January 27, 1967 was much more than a personal tragedy to three families. The accident affected the entire nation. Because of this national sentiment about manned space flights, the assurance of technical quality becomes an awesome responsibility for the top managers of the program.[16]

In the early days of the Manned Space Flight Program (1959), three imperative technical matters had to be investigated and solved before man could go into space. According to the then Administrator, T. Keith Glennan:

> The problems known to exist include (1) high-energy radiation, both primary and cosmic ray and the newer plasma type discovered in the IGY satellite series; (2)

man's ability to withstand long periods of loneliness and strain while subjected to the strange environment of which weightlessness is the factor least evaluated; and (3) reentry into the atmosphere and safe landing. The reliability of the launching rocket must be increased before a manned capsule is used as payload. Once these basic questions have been answered, then we can place a manned vehicle in orbit around the earth.[17]

During the following ten years, these and other technical problems were resolved. By 1969, the problem of loneliness in space might have been transformed into a quest for privacy in space. The seemingly well-rehearsed television performances by astronauts from outer space have delighted millions of people.

In the spring of 1968 shortly before the speed-up in the manned space flight launch schedule, certain technical problems were being solved. Many of these resulted from modifications in hardware made after the Apollo 204 accident. Some of the technical problems included: (1) developing materials to be used on the spacecraft that would be fire-resistant; (2) minimizing increases in the weight of the Command and Service Module that arose from technical changes incorporated after the fire; (3) investigation of the capabilities and limitations of the Apollo spacecraft for early launch abort contingency; (4) stability of the ascent engine used on the Lunar Excursion Module; and (5) insuring that some 600 computers performed correctly.[18] The almost flawless success of the Apollo missions during 1968 and 1969 provides ample proof that these problems were successfully resolved.

Configuration Management

Configuration management is a system used to assure that technical design and hardware specifications are accurate and complete for each space mission. It is a means to control any changes that are required in specifications and drawings, hardware, and contracts. Configuration management is not unique to NASA; it is used throughout the Department of Defense as a tool to control the large quantities of equipment that ultimately become complex weapons systems.[19]

In the Office of Manned Space Flight, configuration control is implemented by Configuration Management Offices and by Configuration Control Boards. The four program offices (i.e., Apollo, Apollo Applications, etc.), the three field centers, and the principal contractors all have Configuration Management Offices and Configuration Control Boards. In addition, there are inter-center panels that perform coordination tasks. Configuration management operates through a series of formal technical reviews and inspections at checkpoints in the design, production, test, and launch of space hardware. Table 4–4 presents a simplified view of the process.

The first phase in the process is the definition and design of hardware.

TABLE 4-4: THE PROCESS OF CONFIGURATION MANAGEMENT FOR A MANNED SPACE FLIGHT MISSION

PROGRAM PHASE	REVIEW
Definition and design of hardware items	Preliminary Design Review
Factory design of hardware items	Critical Design Review
Factory manufacture of hardware items	First Article Configuration Inspection and Design Certification Review
Site testing of hardware items	Flight Readiness Review
Launch mission	

The specifications for hardware or other end items are prepared jointly by NASA personnel at the field centers and by contractor representatives. The requirements established in these specifications define all of the technical aspects about the quality and the performance of the item to be produced. The first major engineering review of any given end item, piece of hardware, or system is called the Preliminary Design Review (PDR). The purpose of the PDR is to review the basic design of the item and also to approve the technical documentation that will accompany the item through the procurement and production process. The Critical Design Review, which is the second review, provides the engineering specifications that are to be incorporated in the first article manufactured.

As Table 4-4 shows, the third review is called the First Article Configuration Inspection (FACI). The First Article Configuration Inspection is an examination of a selected (earliest possible) end item that has been manufactured. The criteria for examination are based on the approved technical specifications and the accompanying engineering documents. The First Article Configuration Inspection is acceptance testing. If defects are identified, the inspection is repeated several times, if necessary, to insure that the deficiencies have been corrected.

There are two additional phases in the configuration management process: the Design Certification Review (DCR) and the Flight Readiness Review (FRR). The DCR, conducted by the Associate Administrator for Manned Space Flight, certifies that the design of the complete space vehicle is flightworthy and safe for man. Some of the documents evaluated at the Design Certification Review include: the results of development tests; the results of qualification tests; evaluations of test failures; and statements signed by NASA and contractor personnel certifying that the hardware is reliable for manned flight. That kind of information is provided for each subsystem, the launch vehicle, and the spacecraft.

The Flight Readiness Review is the final review, and consists of two parts: a review conducted by the program director and one conducted by the mission director. Satisfactory completion of the Flight Readiness Review initiates the actual launch mission period. The purpose of the review by the program director is to insure that the hardware used on the space vehicle and the launch complex are ready to begin the mission. The result of a favorable judgment made by the mission director is the deployment of necessary manpower and equipment throughout the world in support of the mission. Some of the factors evaluated at the Flight Readiness Review include: the preparedness of the manned space flight tracking network; flight control capabilities; medical reports; recovery plans; the preparedness of flight crews; and public information communications networks. The Flight Readiness Review is, in effect, the "go or no go" decision point for any given space flight.

Experiments System

The Manned Space Flight Experiments Program is designed to cover a wide range of objectives in the fields of space sciences, technology, and operations. The process of defining and developing space experiments requires the help of almost all program offices in NASA, a number of other government agencies, and many groups from the scientific and industrial community.

Manned space flight experiments can be put into three major groups: (1) science and applications experiments, (2) operational experiments, and (3) technology experiments. Science experiments are designed to advance knowledge of the laws, the composition, and the history of the universe. Applications experiments are planned to gain better understanding of terrestrial phenomena and to apply this knowledge to exploitation of the earth's resources. Operational and technology experiments seek to establish a quantitative understanding of the influence of the space environment on man. Operational experiments are also designed to improve the capability of man and spacecraft to operate outside the earth's atmosphere. Together, the three groups of experiments include a broad array of scientific and technical disciplines. Some of the scientific disciplines are: astronomy, biomedicine, bioscience, communications and navigation, meteorology, natural resources, physical science, and solar physics.

Organizationally, the Office of Space Science and Applications is responsible for science and applications experiments; the Office of Advanced Research and Technology for technology experiments; and the Office of Manned Space Flight is responsible for operational experiments. Each organization develops its own basic experiments plan. These plans collectively comprise the total Manned Space Flight Experiments Program. The planning and execution of the program are carried out on an experiment-by-experiment basis. The typical life cycle of an experiment from conception to the return of experimental data can be divided into five distinct steps: (1) pre-definition phase, (2) definition, (3) development, (4) operations, and (5) data analysis. Management review occurs throughout the entire life cycle of an experiment.

Pre-definition Phase

The primary purpose of the feasibility phase is to identify potential experiments and to determine the scientific and technical merits of an array of possible experiments. The pre-definition phase usually takes about one year. Both NASA scientists and other experts, under grant or contract, participate in the identification of meritorious experiments. Experiment selection tasks can be managed either by a center or by a headquarters office. During this phase, working groups prepare information about possible experiments. Such information might include: descriptions of equipment that will be required; statements about possible constraints; descriptions of astronaut participation; and statements about data requirements. To warrant progress to the next stage, an identified experiment must usually be accompanied by a preliminary development plan that provides information about costs and schedules.

Definition Phase

The objective of the definition phase is to plan the development and the operation of the experiment. Experiment definition activities consist of engineering studies, laboratory tests, and simulated flight tests. These studies are designed to define the technical characteristics of experiments. The process results in a comprehensive outline of the experiment, along with specifications for the development of required hardware. Improved information about expected costs and schedules is provided in what is called a "hardware development plan."

Development Phase

The development phase is concerned with the design, engineering, fabrication, ground testing, and flight qualification of hardware required for the experiment. During this step, an operational profile is supplied by the principal investigator. That profile includes refined and complete information about schedules and astronaut participation.

Operations and Data Analyses Phases

The operations phase, of course, is the conduct of the experiment in flight and the return of the experiment data for analysis. The final phase in the life cycle of an experiment is reached when the data are analyzed, and actual results are compared with those postulated in the experiment objective.

Management evaluation and review of an experiment are made by the Manned Space Flight Experiments Board. The board coordinates experiment programs to be conducted on manned space flights. The board has four principal functions: (1) it recommends the approval or disapproval of experiments; (2) it recommends assignment of experiments to specific flight programs; (3) it makes recommendations about the relative priorities of experiments scheduled for specific missions; and (4) it reviews the status of approved experiments. The Associate Administrator for Manned Space Flight is chairman of the board. In late 1969, other members of the Experiments Board were the Deputy Director for Strategic and Space Systems of the Department of Defense; the Associate Administrator for Space Science and Applications, NASA; the Associate Administrator for Advanced Research and Technology, NASA; and the Director of Space Headquarters, United States Air Force.

Program offices in NASA submit experiment proposals for consideration by the Manned Space Flight Experiments Board after the definition phase of the process. If an experiment is approved and assigned to a flight program, the Experiments Board usually requests the originating program office to continue to manage the experiment. As in most NASA procedures for procurement, authority for development of flight equipment for the experiment is delegated to a field center.

Safety Systems

Throughout the program's life, the principal feature of the safety system for manned space flight has been the provision of maximum redundancy in components and hardware. The matter of safety, of course, became a major national concern after the Apollo 204 accident. As a result, the Office of Manned Space Flight established its own safety office in August, 1967. Jerome Lederer, an international expert on safety matters, was named head of that office. Previously, Lederer had been director of the Guggenheim Aviation Safety Center.[20]

From the Manned Space Flight Safety Office, which is subordinate to the NASA-wide safety office, extends a network of safety offices. Each program office and each manned space flight field center has a safety office. Safety offices at the centers are responsible for implementing the safety plans developed for assigned programs. The centers are also responsible for directing the safety-related tasks of each prime contractor and for assuring that lower-tier contractors, suppliers, and vendors comply with safety standards. Contracts usually include clauses on safety requirements to be followed by contractors. Under a contract for technical integration and evaluation of contractor effort on the Apollo Program, the Boeing Corporation has the responsibility of coordinating safety programs among contractors and between contractors and field centers.

The Manned Space Flight Safety System is designed to resolve three different kinds of problems: (1) crew and flight safety; (2) community and public safety; and (3) industrial safety.

Crew and Flight Safety

For the most part, identification and control of potential hazards associated with hardware are engineering functions. Safety engineering analyses are conducted throughout the design and development of individual components for each flight mission. Another safety review comes after components and subsystems are joined. Intersystem reliability checks are also made after the launch vehicle and spacecraft have been assembled. The technical integration and evaluation function of the Boeing Company provides the final step in safety evaluation. Safety engineers with that company conduct a comprehensive analysis of the total space package, its launch requirements, and ground support equipment. The Headquarters Manned Space Flight Safety Office, in conjunction with the safety organizations at the field centers, makes periodic safety surveys to insure that NASA's safety standards are met. During the first six months of 1968, for example, survey teams evaluated the astronaut lunar landing training vehicle program, the vehicle assembly building, astronaut activity at the Manned Spacecraft Center in Houston, and engine test activities conducted at the Marshall Space Flight Center in Huntsville.[21] During 1969, the Advanced Manned Missions Program Office undertook safety studies that were primarily concerned with in-space rescue missions. One of the studies examined alternative methods that would provide escape and rescue devices for an orbiting manned space station or laboratory. Another study investigated hazards associated with space walks and explorations outside of the space capsule.[22]

Community and Public Safety

NASA's public safety program is designed to protect the life and property of people who might be exposed to space-related hazards. Criteria established for public safety standards were an important factor in the selection of field centers. As indicated in Chapter 2, the government obtained easements from the owners of over 125,000 acres surrounding the Mississippi Test Facility to provide an acoustical buffer zone to protect the public against any adverse effects from vehicle test activities. Easement rights to the buffer zone permit nonresidential use of the land for farming, grazing, and lumbering. Criteria have been established to prevent prohibitively high noise levels at the Mississippi Test Facility. Safety criteria for launch operations at Cape Kennedy provide against the random re-entry on land of space vehicle components that have been launched.

Industrial Safety

The manned space flight industrial safety program is more mundane and is similar to plans of other safety-conscious organizations. The purpose of industrial safety activities is to provide a safe working environment for civil service and contractor personnel. NASA's industrial safety program encompasses such usual safety activities as: the development and distribution of educational material about safety, the development of accident investigation and reporting procedures, the analysis of injury statistics and employment compensation data, sponsorship of safety contests and campaigns, and the establishment of guidelines for the well-being of physically handicapped employees.

The Panel Review Board and Intercenter Coordination Panels

In 1961, the Office of Manned Space Flight established a system of intercenter coordination panels to make recommendations for a solution of technical problems that arise in the development of launch vehicles, spacecraft, supporting facilities, and other equipment. The panels are working groups composed primarily of scientists and engineers. Each panel is responsible to a group called the Panel Review Board. The system is designed to facilitate the exchange of technical information between the three manned space flight field centers.

As organized in the late 1960's, the system consists of the Panel Review Board, an executive secretariat, eight panels, and twenty-three subpanels. In January 1968, about 340 people participated in one way or another in this intercenter coordination system. The Panel Review Board is composed of eight members. The members are technical personnel from the headquarters office of Manned Space Flight and from each of the three field centers. The Apollo Program Director serves as chairman of the Panel Review Board. Although meetings of the Panel Review Board are called whenever necessary, a meeting must be held at least once every three months. To achieve the maximum amount of technical competence, nonmember experts frequently attend Panel Review Board meetings.

The executive secretariat performs administrative and housekeeping tasks and insures that documents from the Panel Review Board are acted upon by relevant offices. One of the roles of the secretariat is to keep informed about work performed by each panel and subpanel and, thereby, to prevent overlap and duplication of effort among the various panels.

When established, each of the panels was designed to cope with one of the eight principal technical problems that confronted the three manned space flight field centers. The missions of each of the eight panels are listed below.

Crew Safety Panel

The crew safety panel investigates matters relating to space vehicle emergency conditions and crew safety before launch and after launch until the spacecraft separates from the launch vehicle.

Electrical Panel

This panel is responsible for the efficient operation of computer support, other electrical support equipment, and emergency detection systems.

Flight Evaluation Panel

The purpose of this panel is to define and to coordinate the post-flight evaluation process. It also investigates the causes of any malfunctions during flight.

Flight Operations Panel

This panel establishes mission control requirements for launch vehicles. It also monitors flight mission control activities.

Flight Mechanics Panel

The responsibility of this panel is to coordinate such flight mechanics matters as: guidance, navigation, control, trajectories, and range safety.

Instrumentation and Communications Panel

This panel is responsible for the efficient operation of instrumentation and communication systems used on spacecraft and on the ground support equipment.

Launch Operations Panel

This panel resolves problems that can occur at the launch site. The problems include the operational compatibility between the spacecraft and ground support equipment and facilities, emergency egress for the astronauts, and the efficient operation of power supplies.

Mechanical Panel

The mechanical panel investigates matters associated with the handling, scheduling, and transportation of hardware between the centers and other facilities. It pays particular attention to problems that might arise from the handling of explosives and other hazardous materials.

Recommendations made by the panels go to the Panel Review Board. Unless challenged by a center director or vetoed by either the NASA Administrator or the Associate Administrator for Manned Space Flight, a technical decision made by the Panel Review Board is accepted by the three centers and by the headquarters office.

TIME VARIANCE

Managerial interest in effective scheduling is as old as management itself. One of the early techniques developed for scientific management was the Gantt Progress Chart, devised by Henry L. Gantt during World War I. Most textbooks on management cite the Gantt Chart as the origin of the Program Evaluation and Review Technique, or PERT, currently in wide use to solve scheduling problems. Working for the Army Bureau of Ordnance, Gantt developed a structured production plan for the World War I munitions program. Time was the principal common denominator by which the progress of programs then undertaken by Army Ordnance was assessed.[23] The relatively low cost and greater availability of high-speed computers during the last decade or so have provided an effective means to change the simple Gantt Chart into complex scheduling control schemes for use in industry and in government.

Schedule Slippage

The research of Marshall and Meckling, and of Peck and Scherer,[24] provided the first publicized empirical information about the magnitude of schedule slippage on military research and development programs carried out during the 1950's. For ten weapons systems development programs, Marshall and Meckling found that the time variance factor was about 1.50 for actual, compared with predicted, schedule. This means that, on the average, the ten weapons systems were actually available "50 percent later" than the original estimate. Analyzing twelve different military systems, Peck and Scherer found the schedule slippage factor to be 1.36.[25] Less well known is another set of findings by Peck and Scherer. Their research showed that schedule slippage is not solely the plight of government procurement. Peck and Scherer estimate that the time variance factor for five advanced commercial development projects undertaken during the 1950's was 1.40—slightly higher than that in their weapons sample. They also cite other evidence (for the construction industry, the field of atomic energy, and for commercial research and development) which indicates that private industry also is plagued by schedule slippages.[26] Both research teams attribute most of the time slippages to technological uncertainty.

Using schedule projections made by NASA in early 1959, Archie L. Wood finds that the time slippage factor on Project Mercury was 2.25. He contends that the first manned orbital flight was about 21 months later than planned. The principal causes for the delay, according to Wood, were the lack of a clearly defined program at the inception of NASA and of the Mercury Project, and the necessary tendency of NASA to decide almost all trade-off matters with astronaut safety as the paramount criterion, i.e., schedule was much less important than technical reliability.[27]

Program Evaluation and Review Technique

It seems well established that the program evaluation and review technique, PERT, was first used on a large government research and development project in 1958. At that time, the Special Projects Office of the United States Navy introduced PERT to control scheduling problems encountered on the Polaris Project. It is suggested that those with an understanding of PERT skip to the discussion of the use of the technique in the Office of Manned Space Flight. The next few pages, on the general usefulness of the program evaluation and review technique, are for those who have been working outside of the management field.

The program evaluation and review technique is a network model.[28] The network establishes the sequential relation between tasks and subtasks that must be completed to accomplish a given project. Probability estimates are made for the completion time of each task and subtask. The model identifies those tasks that require much time and those tasks that require little time. The activities that require the longest time, of course, determine how long the entire project will take. Depending on the kind of project and also the degree to which tasks and subtasks are divided, the PERT network model can be simple or it can be intricate. The intricate and complicated models require the use of computers.

An extension of the original PERT concept is the PERT/COST model that provides information about both possible schedule over-runs and possible cost over-runs. Although not operationally identical with the concept, the manned space flight production cost model approximates the PERT/COST technique.

Optimism and Pessimism

As a managerial tool, PERT is like any other kind of model. Its usefulness for schedule control depends on the quality of the estimates made about the completion time for each task and subtask. Either unrealistically optimistic or pessimistic biases can be built into any PERT model. Unwarranted optimism, of course, would result in schedule over-runs. Virtually the only way to reduce or to minimize unduly optimistic or pessimistic estimates is to insure that those supplying the data inputs into the PERT model are experts with ample experience. Some PERT models give three probability distributions for major activities to be completed: one for an "optimistic" schedule, with all tasks going right the first time; another for the "most likely" or "normal schedule"; and a "pessimistic schedule," based on the maximum expected time an activity might take, i.e., when unusually bad luck is encountered.[29]

Improved Communication

Several research groups have found that the use of PERT provided an unexpected benefit to its users in the form of improved communications between groups working on different facets of a given project. Improved communications came about because people were forced, through the PERT technique, to become familiar with all of the important phases of a given project.[30]

When To Use PERT

It has been argued that federal procurement regulations do not specify adequately when PERT models should be used by contractors.[31] In an attempt to clarify the controversy, Jack C. Hayya sent questionnaires to a group of government representatives and industry officials working on 27 projects using PERT, and on 34 projects using the PERT/COST technique. One of Hayya's findings was that most of those surveyed, both in industry and in government, believed that PERT should be used for cost-reimbursement contracts. This is in contrast to fixed-cost contracts, when contractors can estimate uncertainties with much higher degrees of accuracy. Among those surveyed, Hayya found only a few who favored the use of PERT on contracts let for basic research, because it was considered impossible to define a meaningful network of activities because "ex ante" the results of basic research are unknown. This contrasts sharply with the consensus of those surveyed that PERT should be required on applied research and development projects.[32]

PERT and Schedule Over-Run

In a study of the use of PERT on 36 large government research and development projects initiated between 1957 and 1963, researchers with the Sloan School of Management found a marked difference in adherence to schedule between those projects using PERT and those not using the technique. About two-thirds of the projects using PERT were "non-over-run" projects. This compares with about one-third for those projects that did not use PERT as a scheduling tool.[33]

The Cost of PERT

In his questionnaire, Hayya asked those using PERT about its cost. His purpose was to prove or disprove the belief held by many in the aerospace industry that PERT was an excessively costly tool for management. Information provided to Hayya by an aerospace company[34] showed that the cost of PERT amounted to about 3 percent of the cost of contracts under $100,000 and to 1 percent of the cost of those in excess of $100 million. The memorandum from the aerospace company indicated that the PERT/COST technique cost was about twice as much as simple PERT. Hayya believes that the aerospace industry overestimates the cost of PERT because the results of his questionnaires showed that the cost of using PERT amounted to about .3 of 1 percent (rather than 1 percent) of the cost of contracts in excess of $100 million.[35] Hayya also believes that the cost of using PERT will decline as firms gain experience in the use of the technique.

The Scheduling System Used in the Office of Manned Space Flight

The use of the program review and evaluation technique for scheduling events in the Manned Space Flight Program is an integral part of phased-project planning in NASA. As its name implies, PERT evolved as a means of solving single project-oriented rather than multi-project problems. The solution of many multi-project problems usually is a matter of arriving at an optimum resource allocation between competing projects. One of the major criticisms of PERT is that it contributes little or nothing to the solution of the multi-project allocation problem.[36] Cost-effectiveness analysis, of course, is a multi-project–oriented management tool.

In the spring of 1964, NASA gave Congress an estimate of the major launch milestones that would have to be completed successfully during the next five years if the manned lunar landing were to be accomplished by 1970. Viewed broadly, that schedule of annual flight milestones became the framework for NASA's scheduling system for manned space flights. Table 4–5 lists the forecast date on each flight and also the date that the objective was actually accomplished. NASA, of course, did meet the schedule objectives outlined to Congress in 1964. Armstrong and Aldrin set foot on the moon on July 21, 1969, five months before the end of the decade of the sixties. The Gemini Program was also achieved with a few months to spare. As the table shows, however, there was a long schedule slippage as a result of the Apollo 204 accident. That slippage

TABLE 4–5: MANNED SPACE FLIGHT SCHEDULE OBJECTIVES, 1964–1969

FORECAST DATE OF ACCOMPLISHMENT[a]	MILESTONE	DATE ACTUALLY ACCOMPLISHED
1964	(1) 1st Gemini unmanned flight	1964
1965	(2) 1st Gemini manned flight	1965
1966	(3) 1st Gemini rendezvous flight	1965
1967	(4) Gemini operations and completion	1966
1966	(5) 1st Apollo Saturn I-B unmanned flight	1966
1967	(6) 1st Apollo Saturn I-B manned flight	1968
1967	(7) 1st Apollo Saturn V unmanned flight	1967
1968	(8) 1st Apollo Saturn V manned flight	1968
1969[b]	(9) Apollo operations and manned lunar landing	1969[c]

[a] Forecast made in 1964 after the reorganization of NASA and the Office of Manned Space Flight.

[b] As a national goal, to be accomplished by December 31, 1969.

[c] Accomplished on July 21, 1969.

was not made up until the end of 1968, when Apollo 8, using the Saturn V (Milestone 8), followed Apollo 7, using the Saturn I-B (Milestone 6), by a few months rather than by the original flight plan of approximately one year.

Schedule Control Levels

Each of the schedule objectives presented to Congress in 1964 for the Apollo Program became part of the master launch schedule for the 1966 through 1969 period. Other launches, of course, are also included in the master schedule. Viewed within the structure of the PERT system, the master launch schedule is the culminating point. As Table 4–6 shows, it is the first of four major schedule and review levels. Responsibility for the master launch schedule rests with the Office of the Deputy Administrator of NASA; the Associate Administrator for Manned Space Flight; and for Apollo, with the Apollo Program Director. The master launch schedule provides information about the launches included in each series. For example, the Apollo launches using the Saturn I-B vehicles were a distinct launch series, as are those using the Saturn V vehicle. Apollo Flights 8 through 11 were in the same launch series. Each used the Saturn V launch vehicle. Some of the information inputs required to develop the master launch schedule include the availability and readiness of vehicles, spacecraft, launch support facilities, mission control activities, and tracking systems.

Table 4–6 shows that the PERT system is a functional breakdown of the structure of work required for any given launch or any given launch series. The number of activities in the PERT network broadens markedly at each successively lower level. With the exception of starting activities, each activity has one or more predecessor events. Usually, any given activity can be completed only after all predecessor activities have been

TABLE 4–6: THE STRUCTURE OF THE SCHEDULING SYSTEM USED FOR THE MANNED SPACE FLIGHT PROGRAM

INFORMATION NETWORKS	SCHEDULE OBJECTIVES AND THEIR COMPONENTS
Headquarters	Level I: Launch schedules, and hardware deliveries required to support launches.
The centers	Level II: Principal project components, such as the Saturn V launch vehicle and the spacecraft.
Contractors	Level III: Primary systems used on projects, such as the various vehicle stages, the Lunar Excursion Module, and the Command and Service Module.
Subcontractors	Level IV: Subordinate components of systems, such as the environmental control system and earth landing system.

completed. As used for the Manned Space Flight Program, Level II activities are project events. Schedule integration provides information about the delivery time for the principal pieces of hardware used for launches. Not all schedule information controlled at this tier, however, pertains to critical hardware. For example, completion of astronaut training and space medicine requirements are also integrated into the scheduling system at the second tier. For the most part, schedule estimates for activities performed at Level III and Level IV in the PERT structure consist of information about primary systems, subsystems, and components required before the delivery of major hardware items is possible.

The left side of the table shows that subcontractors, contractors, and field centers provide PERT status reports. As new contracts are let, the assigned activities of new contractors are integrated into the PERT network. Similarly, as projects are completed, the activities are deleted from the network. Simply conceived, the manned space flight schedule system integrates operating reports that go from prime contractors to field centers, and then to the relevant Manned Space Flight Program office. What NASA calls milestones become the objects of the primary scheduling control activity, as well as the base for supporting events. To repeat, milestones are events of major significance in the Manned Space Flight Program.

Because of the size of the entire system, a summary PERT network is maintained for the Office of Manned Space Flight. That network integrates the activities of the Apollo Program, the Apollo Applications Program, and the Advanced Manned Missions Program. The purpose of the summary PERT network is to provide an integrated picture of the entire Manned Space Flight Program.

In addition to the launch schedule system and the summary PERT network, there is a schedule system for the construction of facilities. That system was more important in the mid-1960's than in the late 1960's because NASA's spending for the construction of facilities peaked during fiscal year 1963. More than 40,000 different activities were included in the PERT network that was used to control the installation of plant and equipment required for Launch Complex 39 at the Kennedy Space Flight Center.[37]

Operation of the System

The core of the PERT system consists of continuous computer checks that compare expected with actual progress. Any difference between expected and actual progress is called either positive or negative slack. Positive slack means that an activity is ahead of schedule, whereas negative slack connotes the opposite. If the slack value is equal to zero, actual progress is equal to expected progress. Activities in the network that consistently show much positive slack generally require less management attention than activities with a high degree of negative slack, because they do not result in program delays. Negative variance directs managerial attention to potential or actual problems. One function of the manned space flight configuration control system is to review and to make corrective recommendations for strategic activities that are shown by PERT to have much negative slack. Depending upon their severity, schedule slippage problems are resolved by middle or top management in NASA.

The Logistics System

A discussion of logistics planning for the Manned Space Flight Program must begin with the site selection process. In the earliest phases of the program, it was recognized that massive rockets, thousands of tons of fuel, and all kinds of heavy equipment would have to be handled on a fairly routine basis. As indicated in Chapter 2, one of the most important criteria for the selection of Houston, the Michoud Assembly Facility, and the Mississippi Test Facility was the availability of year-round navigable water for barge transport of large and cumbersome space vehicles. For scheduling purposes, it was deemed important to have the launch vehicle test site within a maximum barge trip of two weeks from both the Marshall Space Flight Center in Huntsville and the Atlantic Missile Range at what was then Cape Canaveral.

Although efficient transportation of heavy equipment is still an important logistics matter, the primary concern of the current Manned Space Flight Program's logistic system is maintenance. As used here, the maintenance function is broadly conceived and includes the provision of components, parts, and other equipment required to implement configuration change. The maintenance function also includes the provision of spare parts for launch vehicles, spacecraft, ground support equipment, and other facilities. Apart from providing spare parts for use on space flights, the logistics system used for the space program faces the usual kinds of inventory control problems. They include the funding of spare parts, selection and allocation of spare parts, storage and use of spares, repair and overhaul of spare parts, and the disposal of used and superfluous spare parts.

COST VARIANCE

Because of the close relationship between schedules and costs in the production of weapons and space systems, much of the available information about the ability of contractors to adhere to original cost estimates comes from the same data sources as information about adherence to schedule. The results of several empirical studies show that the cost variance factor is much higher than the time variance factor. Marshall and

Meckling found that the mean cost variance factor on the contracts they examined was between 2.4 and 3.2.[38] For the twelve weapons systems contracts analyzed by Peck and Scherer, the factor averaged 3.2.[39] A cost factor of 3.2 means that actual costs averaged about 220 percent higher than the original estimate. The data used by Archie Wood showed that the budget over-run factor for the manned space flight Mercury Project was between 1.92 and 2.20. Wood attributes about one-half of the cost over-run on the Mercury Project to increases in overhead that resulted from schedule slippages. Expansion of the tracking system was another important cause for the cost over-run. According to Wood, flight failures accounted for only 10 percent of Mercury's budget over-run.[40] Peck and Scherer believe that business firms and other organizations conducting the research and development for weapons systems have been fairly successful in meeting quality standards but not in adhering to their initial cost and schedule standards.[41]

Cost Over-Runs on the Manned Lunar Landing Program

Aggregate data on the magnitude of the cost over-run for the manned lunar landing project between 1964 and 1969 can be obtained from information provided by NASA to Congress for the annual authorization hearings. As Table 4–7 shows, NASA estimated in March, 1964, that the research and development component of the manned lunar landing project would cost $14.8 billion. Although it is not shown in the table, NASA also estimated that costs for tracking and data acquisition, facilities, and installation operations would amount to an additional $4.7 billion. Together, R & D and other costs would amount to $19.5 billion. At the hearings conducted during the following year, NASA estimated that the research and development component would total $17.6 billion. By January, 1969, estimated cost had risen to about $19 billion. Cost estimates for tracking, facilities, and other operations amounted to an additional $4.9 billion, bringing the cost estimate to $23.9 billion by January, 1969. In current dollars, the cost over-run on the research and development component was 1.28. The total cost variance was slightly less – 1.23. This means that the actual cost of the manned lunar landing was about 25 percent higher than estimates made in 1964.[42]

Not all of the cost over-run factor for the manned lunar landing program can be attributed to uncertainties in the research and development process or to optimism on the part of NASA officials when they presented cost estimates to Congress during the early phases of the program. Like consumers, business, or other federal agencies, NASA suffered from the impact of inflation following the rapid build-up for the Vietnam War in mid-1965. The constant 1964 dollar cost of the program gives a conserva-

TABLE 4-7: CHANGES IN ESTIMATED COST OF RESEARCH AND DEVELOPMENT FOR THE MANNED LUNAR LANDING, 1964–1969[a] *(BILLIONS OF DOLLARS)*

PROJECT COMPONENTS	MARCH 1964	MARCH 1965	MARCH 1966	MAY 1967[b]	JANUARY 1968	JANUARY 1969
Apollo spacecraft: Current dollars	5.05	6.45	6.64	7.13	7.46	7.94
Saturn launch vehicles: Current dollars	7.70	8.88	8.94	8.89	8.83	8.77
Engine development: Current dollars	1.19	1.11	1.05	.89	.85	.85
Operations support: Current dollars	.86	1.15	1.08	1.22	1.57	1.43
Total research and development cost:						
Current dollars	14.80	17.59	17.71	18.13	18.71	18.99
Constant 1964 dollars[c]	14.80	17.11	16.77	16.79	16.71	16.37
Implicit price deflator[d]	100.0	102.8	105.6	108.0	112.0	116.0
Cost over-run: 1964 = 1.00						
Current dollars	1.00	1.19	1.20	1.23	1.26	1.28
Constant 1964 dollars[c]	1.00	1.16	1.14	1.14	1.13	1.10

[a] Estimate given to Congress after the reorganization of NASA and the Office of Manned Space Flight.

[b] Estimate made after modifications caused by the Apollo 204 accident.

[c] Deflated by the implicit price deflator for federal purchases of goods and services.

[d] Compiled by Office of Business Economics, U.S. Department of Commerce. Deflator is for closest calender year quarter.

Source: Files, Office of Manned Space Flight, NASA. The cost estimates for each date were made by the Office of Manned Space Flight for various Senate authorization hearings.

tive estimate of the degree to which inflation affected the cost of the manned lunar landing project. To get the constant dollar series, the annual cost of the program was deflated by the implicit price deflator for all federal purchases of goods and services. As the table shows, the index for the cost of goods and services bought by the federal government rose by 16 percent between March 1964 and January 1969.

At least two-thirds of the cost over-run factor for the research and development component of the manned lunar landing project can be attributed to inflation. If it were available, an index for federal research and development purchases would probably give a factor for inflation in excess of the index used. Using NASA's cost estimates, the variance factor is somewhat smaller when tracking, facilities, and operations costs are also included. In constant 1964 dollars, the January, 1969, cost estimate of $23.9 billion drops to $20.6 billion. Taking inflation into account, the cost variance factor would be 1.06. To be sure, a cost over-run factor of that magnitude, i.e., 6 percent, on a multi-billion dollar project is more significant than a somewhat higher cost variance factor on a multi-million dollar project.[43]

The Budget Review Process

In the ends-means structure of the management decision-making process in the Office of Manned Space Flight, four principal systems (i.e., means) are used to accomplish the objective (i.e., end) of cost minimization and the optimum allocation of resources. They are the budget review process, the production cost model, the work packages system, and the "all up" concept. Of these, the budget review process is the only system that NASA has in common with all other federal agencies and departments.

The annual budget review is the exercise of justifying programs, their objectives, and their resource requirements to the Bureau of the Budget, to the President, and to the Congress. From start to finish the budget review process takes about eighteen months in the Office of Manned Space Flight. It begins in about January of the year preceding the start of a given fiscal year. For example, preparation of the budget for fiscal year 1971 (July 1, 1970 through June 30, 1971) began in January 1969.

The Preliminaries

The typical budget cycle begins after the Administrator of NASA issues agency-wide budget guidelines. During the next five months, each program office and each field center prepares a detailed, but preliminary, estimate of its anticipated resource requirements. In NASA, these estimates are called Program Operating Plans. After review and coordination within the Office of Manned Space Flight and within other NASA offices, a preliminary budget plan goes to the Bureau of the Budget for review. In early summer the Bureau of the Budget considers the request made by NASA and those made by other federal agencies. On the basis of the preliminary information supplied by his Executive Departments, the President establishes budget guidelines.

The President's Budget

The President usually sets his guidelines during July, about six months before his annual budget request to Congress. Preparation of the final budget is based on Presidential directives. It takes about three months for NASA to prepare its final budget request. Sometime in September or October, officials in NASA Headquarters review the estimates and determine the final budget position that the agency will attempt to maintain. NASA's budget document is again sent to the Bureau of the Budget for review. During October and November NASA and other federal agencies defend their positions and negotiate with the Bureau of the Budget. Any unresolved issues are taken to the President for decision sometime in late November or early December. The entire Presidential budget goes to the Government Printing Office in December and to Congress in January.

Authorization

NASA's authorization hearings are before the House Committee on Science and Astronautics and the Senate Committee on Aeronautical and Space Sciences. At these hearings, which usually take place between February and April, NASA defends the President's budget request. Depending on launch activities, NASA's top management team is an especially mobile group at this time—the usual flight pattern being between Cape Kennedy, Houston, Huntsville, and Washington.

After Congressional committee and subcommittee hearings, each chamber submits an authorization bill for floor action. It is during April or May, and sometimes as late as October or November, that the Senate and House Conference Committee meets to negotiate and to resolve any differences between the House and Senate bills. After the committees reach agreement, the President signs the authorization act.

Appropriation

Concurrent with NASA's hearings before the Senate and House Committees that oversee the space program are hearings before the Independent Offices Subcommittee of the House Committee on Appropriations. Hearings before the House Appropriations Subcommittee usually take place a month or so earlier than those before the Senate Independent Offices Subcommittee. If the budget cycle functions on schedule, the

House and Senate Appropriations Committees come to agreement so that floor action can be taken by each body in time for the President to sign the appropriation act into law by June 30. In recent years, however, Congress frequently has not enacted appropriations legislation before the beginning of the new fiscal year. When this occurs, Congress must pass a temporary appropriations act, by joint House-Senate resolution, to enable federal agencies to pay their bills. Also, at times, NASA and other agencies are forced to make program adjustments to align their spending with Congressional appropriations. As Chapter 2 indicated, that happened when Congress forced NASA to absorb its share of the proposed $6 billion decrease in all federal spending during fiscal 1969. For that fiscal year, NASA requested $4,370 million; the Congress authorized $4,013 million; and appropriated $3,995 million. It was not until October of 1968 that NASA submitted its revised operating plan to the Congress.[44]

Production Cost Model

Since its inception, NASA has supported a considerable amount of research for computer simulation models. Most of these models can be classed as resource analysis models, designed to develop cost-estimating relationships (CER's). The modeling is undertaken to find functional relationships between cost, performance, and other technical characteristics of specific pieces of space hardware. The models are designed for long-range planning. The objective, of course, is to improve the Space Agency's ability to forecast the research and development costs and the operational costs of proposed programs. Much of the model-building has been done by the RAND Corporation and by the Jet Propulsion Laboratory. The Department of Defense has also been constructing cost-estimating models for its long-range planning. Broadly conceived, these models are the cost adjunct to the models being developed in the field of technological forecasting.[45]

Use of a Short-Run Cost Model

Although long-range planning models are important, the Office of Manned Space Flight has made much greater use of short-run cost-reporting and estimating techniques.[46] Currently, the cost-estimating technique or system is called the Manned Space Flight Cost Study. The model provides a means to assess changes in cost that can result from the addition or deletion of hardware from a given launch schedule. It also indicates changes in cost that can result from schedule slippages. Because the cost data are in fine detail, the model can be used to estimate the cost of variations in configuration. For example, the model would enable NASA to estimate the cost of an intermediate vehicle—one between the size of the Saturn I-B and the Saturn V. The estimate would be based on a re-arrangement of the required hardware components and their costs. The model is also useful for management information about hardware items that have long-lead times in the procurement process. With long-lead procurement, management must frequently make decisions about hardware that may require subsequent revision.[47]

The Data

Unlike the long-range cost models, the manned space flight cost model emphasizes possible changes in production rather than research and development costs. It predicts changes in production costs in response to various hardware and schedule options that could be pursued on the Manned Space Flight Program. The body of data used for the model

TABLE 4-8: STRUCTURE OF THE MANNED SPACE FLIGHT COST STUDY, 1967

HARDWARE STAGE	HARDWARE CONFIGURATION	CONTRACTORS	NUMBER OF PLANTS
Launch vehicles:			
SI-B stage	Saturn I-B	Chrysler	2
SI-C stage	Saturn V	Boeing	3
S-II stage	Saturn V	North American Rockwell	3[a]
S IV-B stage	Saturn I-B and Saturn V	McDonnell Douglas	2
Instrument unit	Saturn I-B and Saturn V	IBM and Bendix	4
Engines:			
H-1	Saturn I-B	North American Rockwell	2
F-1	Saturn V	North American Rockwell	3[b]
J-2	Saturn I-B and Saturn V	North American Rockwell	3[b]
Spacecraft:			
Command and Service Module	CSM	North American Rockwell	2[a]
Lunar Excursion Module	LEM	Grumman	1
Guidance and navigation system	CSM and LEM	General Motors	2
Systems support for launch vehicle and spacecraft	Saturn I-B and Saturn V	Boeing and Chrysler	3

[a] Includes one plant that is common to the production of the Command and Service Module and to the S-II Stage of Saturn V.

[b] Common facilities for the production of the F-1 and the J-2 engines.

Source: Office of Manned Space Flight, National Aeronautics and Space Administration.

comes from annual surveys of principal contractors. Contractors are asked by the Office of Manned Space Flight to estimate the effect on cost of changes in the rate of production of the specific pieces of hardware produced at their plants. Contractors are also asked to forecast the outlook for new business at their plants and to assess any possible effects on overhead costs that might result from changes in the level of business at the plant. NASA personnel visit the plants of contractors to obtain, or to validate, the cost estimates. After evaluation of cost estimates made by contractors, sometimes NASA personnel change the estimates. With the decrease in space spending during the late 1960's, NASA personnel at times found that a contractor's estimate of future cost reflected a procurement position or level of operation that the contractor would have liked to maintain rather than the most likely level of space business.

In 1967, the production cost model included information obtained from 26 different plants engaged in work on the Manned Space Flight Program.[48] Table 4–8 shows the contractors that provided information for the model in 1967. The table also gives a simplified view of the structure of the model.

Kinds of Costs

Although the model uses a complex cost structure, it includes essentially two kinds of costs: nonrecurring costs and recurring costs. Nonrecurring costs are one-time charges against a program, usually the start-up costs for the development of a given hardware item. Recurring costs, of course, are those incurred to sustain a program or a piece of hardware throughout its life. Because nonrecurring costs are unique, the model necessarily emphasizes changes in those costs that recur through the production process.

The Rate Effect. The model subdivides recurring costs into those that vary and those that remain relatively fixed when the number of units produced changes.[49] As used in the cost model, a relatively fixed recurring cost might be personnel costs at a test site. For example, if the number of tests conducted by a crew that normally handles two tests per year increases to five, the manpower cost might double. In the model, the relatively fixed costs are expressed in the computer as an average rate per year. If a hardware component cost $10,000 over a two-year period, the model would calculate the cost at $5,000 per year. As used in the model, the relatively fixed costs produce a "rate effect," and do not reflect minor or short-lived fluctuations in the level of business at the plant.

The Learning Curve Effect. The variable costs are expressed in terms of the learning curve. Conventionally, the slope of the learning curve is the percentage reduction in cost that occurs as the cumulative quantity of production is doubled. Thus, on an 80 percent slope, the cost of unit two is .8 times the cost of the first unit; the cost of unit four is .8 times the cost of the second unit produced; and so on. On a 100 percent slope, there would be no reduction in the cost of successive units produced. Each contractor providing information for the model defines the learning curve function for his equipment. The function is based on changes in the costs of production actually experienced by the contractor. The learning curve effect provides a means to show changes in cost that result from possible changes in schedule, i.e., procurement lead-time. For example, if a piece of hardware is modeled for a three-year lead-time, the computer divides that cost (defined by the relevant learning curve effect) by twelve to get the unit cost for each quarter during the three years. Assuming a learning curve effect of less than 100 percent, a four-year lead-time for the same piece of hardware would yield a different unit cost for each quarter. In the model, this is called the "learning curve effect."[50]

The Rate and Learning Curve Effects at Reporting Plants. For each plant and for each hardware configuration, the computer sums the several cost components to get the total cost, i.e., the unit cost based on assumptions about the "rate effect" plus the unit cost that results from the "learning curve effect." The information is supplied for each quarter of a fiscal year. For the production of the SI-C Stage of the Saturn V, at the three reporting plants of the Boeing Company, there were four rate effects and three learning curve effects for the combined hardware components. For the components of the Lunar Excursion Module, Grumman reported fifteen different rate effects and seventeen different learning curve effects. On the H-1 engine, North American Rockwell reported four different rate effects and one learning curve effect. For the 26 plants combined, there were 66 rate effects and 46 learning curve effects included in the production cost model in 1967.

In the model, the rate effect, determined by the relatively fixed recurring costs, has a larger impact on changes in cost than the learning curve effect, determined by the variable recurring costs. This results from the fact that in the model the relatively fixed costs comprise a larger percentage of recurring unit costs than the variable costs.

Assumptions and Limitations

Like all models, the production cost model is based on certain assumptions. To some extent, these assumptions become built-in shortcomings. One assumption of the model is that follow-on pieces of hardware are produced in approximately the same way as earlier units. To be sure, this does not mean that a given unit is a "Chinese copy" of the preceding unit. The model provides some latitude for change orders, but that freedom does not extend to production by different contractors under different procurement conditions. A second shortcoming of the model is that it cannot estimate future costs that will be unique or peculiar to a given mission. Third, the model is not constructed to reveal possible changes in nonrecurring costs that might result from schedule change.

A final weakness of the model is that all relationships are tied to the delivery of hardware. For example, if a contractor has produced the last piece of hardware for a given set of flights, which have a launch schedule extending three years beyond the date of hardware completion, the model cannot estimate hardware maintenance or spare parts costs that might arise during the three-year flight schedule.

An ancillary benefit of the Manned Space Flight Cost Study, however, has been the creation of a data bank for cost information. Since 1964, cost data have been collected by means of a special and standardized survey technique. The existence of this data bank means that consistent cost data are available for the principal hardware components used on the Manned Space Flight Program for a relatively long period. The data will be of much value for any future cost-effectiveness modeling that NASA may undertake.

The "All-Up" Flight Testing System

Shortly after being appointed Associate Administrator for Manned Space Flight, George E. Mueller recommended that NASA change its flight verification test system. In 1963, the Office of Manned Space Flight adopted what is called the "All-Up System." With all-up testing, the complete space vehicle, which includes all live stages of the launch vehicle and all components of the spacecraft, is tested. This contrasts with what was called a "step-by-step" test procedure, which had been used before 1963. For example, before 1963 the Apollo Saturn V series called for 15 space flights, with the fifteenth vehicle being the most probable one for the manned lunar landing. As is well-known, each of the Saturn V vehicles has three live or burning stages. Step-by-step testing would have required the first Saturn V vehicle to be launched with only the first stage (S-IC) live; the second (S-II) and the third (SIV-B) burning stages would have been launched as "dummies." After successful test of the first burning stage, the second Saturn V vehicle would have been launched with both the first and second stages live; the third stage of the vehicle would have been launched as a dummy.

The step-by-step procedure would have continued until the entire launch vehicle and the major components of the spacecraft had been flight tested. Informal and internal NASA staff discussion papers indicate that with step-by-step testing, the first manned flight in the Saturn V series probably would have begun with the seventh Saturn V vehicle. This means that six unrecoverable Saturn V vehicles would have been launched before a man was placed aboard the fully-assembled space vehicle. With the step-by-step approach, manned testing in the vicinity of the moon probably would have taken place on the ninth Saturn V flight. As one result of the all-up test system, the July 21, 1969, actual manned lunar landing used the sixth Saturn V vehicle rather than the fifteenth, which would have been the most probable vehicle under the step-by-step approach.

The estimated cost for the hardware used on the Apollo 11 mission is about $280 million. That cost estimate includes the three stages of the Saturn V vehicle, as well as the Command and Service Module and the Lunar Excursion Module. Although it is impossible to be certain because the step-by-step testing approach was abandoned, use of the all-up test method must have resulted in a savings of hundreds of millions of dollars for hardware used on the Manned Space Flight Program.

Work Packages System

Of the management systems discussed in this chapter, the Work Packages System is the newest. On a trial basis, it was first used in the Office of Manned Space Flight during the summer of 1968. The system is a planning tool for the use and allocation of manpower. Because of the timing of its inception, the system must have been devised, in part, to insure effective compliance with personnel ceilings placed upon NASA by Congress.

The system contains employment and workload information about all civil servants, military personnel, and employees (except temporary employees) of principal support service contractors. The work forces of support service contractors are included in the system because the personnel of these contractors work so closely with NASA employees—at times in the same offices. In 1968, 46 different support service contractors supplied information to the Work Packages System. The system does not extend to contractors and subcontractors working on production contracts. The vast majority of NASA's industrial labor force, of course, works on production rather than support services contracts. If it operates effectively, the procurement process puts the onus on production contractors to insure the efficient use of the industrial labor force working on government projects.

The purpose of the Work Packages System is fourfold. First, the system provides information about the manpower required to accomplish any given mission or milestone. Second, the system forces line managers to justify, by documentation, their manpower requirements. In the third place, the system integrates manpower resource requirements with the budget cycle. The system calls for manpower reports, actual and forecast estimates, during a time frame of three fiscal years. Reports, made in June and December, include descriptions of significant changes in the Manned Space Flight Program. Finally, the system provides a uniform data base for the preparation of manpower reports. The number of civil servants and military personnel reported is based on an actual headcount. Data for contractor personnel are expressed in terms of personnel equivalents. Any overtime work of civil servants and employees of contractors is also reported in personnel equivalents. Members of Congress, personnel in the Bureau of the Budget, and others continually request information about manpower requirements for the space program.

A Work Package

As used by the Office of Manned Space Flight, a work package is a document that delineates major mission responsibilities by project and by function. Within the Apollo Program in 1969, nine different work packages were identified by project. The projects included such distinct mission objectives as: engine development, launch instrumentation, mission control, the Saturn V, and spacecraft. There were fifteen different project work packages for the Apollo Applications Program in 1968, although it was smaller than the Apollo Program. By project, some of the work packages for the Apollo Applications Program were: experiment definition, experiment development, launch vehicle modification, the Saturn I workshop, and the space station. The division of work by function is less exotic. Some of the work packages, divided by function, included: administrative services, data processing, legal services, photographic services, programming and budgeting, and reliability and quality assurance control. By project and by function, there were a total of 77 different work packages in the system in 1968. Table 4–9 displays a simplified matrix of the work package structure.

TABLE 4–9: MANPOWER INFORMATION AVAILABLE FROM THE WORK PACKAGES SYSTEM

STRUCTURE OF THE WORK PACKAGES SYSTEM	MANPOWER RESOURCE
Organization: Headquarters, Marshall, Kennedy, etc.	Employer: civil service, support contractor, and military
Program: Apollo, Apollo Applications, Lunar Excursion, etc.	Skill characteristic: scientists and engineers, technical, wage board, administrative, and clerical.
Project: Engines, experiments, spacecraft space station, etc.	Overtime: civil service and support contractors
Function: Data processing, finance, legal, safety, etc.	

Kinds of Skills

Using NASA and civil service occupational classes, the Work Packages System calls for the reporting of manpower by skill. A manpower grouping by skill is made for each work package. The system divides skill characteristics into five groups—scientists and engineers, technical personnel, wage board employees (i.e., blue collar workers), professional administrative personnel, and clerical workers. The numbers and activities of military personnel are reported separately.

As the information in Table 4–9 indicates, the Work Packages System provides the means to coordinate manpower requirements between the field centers and between the program offices. For example, data available from the system show the number of those working on the Apollo Applications Program at the various centers and at headquartes. The system subdivides that information by project, by function, by employer, and by skill. As designed, the system is a management tool. As a tool for economic analysis, however, it is limited. This is true because, at most, it includes between 10 and 15 percent of those actually working on the Manned Space Flight Program. Extending the system to include information about manpower resources of production contractors and subcontractors, however, could not be justified in terms of cost-effectiveness. Data are available about the structure of NASA's industrial work force, although they are not divided into as many small packages. The data are collected by NASA, in cooperation with the Department of Defense. That body of data, which is called the joint NASA/DOD Economic Information System, is described in the next chapter.

SUMMARY

The Office of Manned Space Flight, NASA, employs what is commonly called systems analysis by those working in the fields of business and public administration. Systems management is the art of tying complex management problems together. In the late 1960's, the structure to be tied together in the Office of Manned Space Flight consisted of a headquarters office; four program offices (Apollo, Apollo Applications, Advanced Manned Missions, and Mission Operations); three field centers (Marshall in Huntsville, Manned Spacecraft in Houston, and Kennedy in Brevard County, Florida); about a dozen principal prime contractors; hundreds of other prime contractors; and thousands of lower-tier subcontractors. At the approximate peak of the Apollo Program in early 1966, about 300,000 people were working on various phases of the nation's Manned Space Flight Program. Only 6 percent of that entire work force were employees of the federal government; the other 94 percent were employees of NASA prime contractors and subcontractors.

In the late 1960's, the management technique used in the Office of Manned Space Flight, as well as in NASA generally, had four distinctive characteristics: (1) phased-project planning; (2) a project-oriented organizational structure; (3) small-group management review and control; and (4) concentrated decision-making authority.

To accomplish its objectives, the principal one being the manned lunar landing, the Office of Manned Space Flight used at least ten different management systems. It is convenient to view these management schemes in an ends-means relationship. An "end" is simply an objective to be accomplished and a "means" is a way to do it. The most important management decisions made in the Office of Manned Space Flight, or in any large-scale research and development project office, pertain to weighting the relative importance of technical performance against program schedule and program cost. Although superior technical performance, efficient scheduling, and the minimization of costs are each a "means" to the defined objective of the manned lunar landing, the three factors can and should also be viewed as specific "ends" to be achieved. The ten management systems used in the Office of Manned Space Flight then become the "means" to accomplish the separate three objectives of superior technical performance, efficient scheduling, and the minimization of costs. To be sure, the ten management systems employed to meet the objectives of quality, time, and cost overlap. For example, on a project as large as the Apollo Program, hardware defects (quality) can result in program delays (time) and can cause costs to rise by staggering amounts.

The four main systems used on the Apollo Program and on other programs in the Office of Manned Space Flight to attain superior technical performance included: (1) a configuration management system that assured hardware design and manufacture was accurate and complete for each space mission; (2) an experiment packages system designed to maximize the technical and scientific payoff of experiments conducted on manned space flights; (3) a safety system; and (4) a panel review board system for inter-center cooperation and coordination on technical matters. The program review and evaluation technique (PERT) and a logistics system were the main tools used to insure that schedule objectives were met. Finally, the Office of Manned Space Flight had four management systems to meet the minimum cost objective: (1) a budget review system; (2) a production cost model that provided information about the recurring costs at 26 different plants of large contractors; (3) an "all-up" flight verification test system that eliminated the need to test each of the three stages of the Saturn V launch vehicle and the principal components of the Command and Service Module and the Lunar Excursion Module with a separate launch for each new component tested; and (4) a

Work Packages System that provided information on manpower requirements by project and by function.

The successful manned lunar landing in July, 1969, almost six months ahead of the national goal-oriented time schedule announced by President Kennedy in 1961 and within 6 percent of a cost estimate made in early 1964, is sufficient evidence to prove the managerial excellence underlying the largest single R & D project undertaken by the United States.

NOTES

1. Richard A. Johnson, Fremont E. Kast, and James E. Rozenzweig, *The Theory and Management of Systems* (New York: McGraw-Hill, 1967).

2. *Webster's Seventh New Collegiate Dictionary* (Springfield, Mass.: G. & C. Merriam Company, 1967).

3. "The Systems Approach," in "The Case for Space," *Forbes*, July 1, 1968, p. 80.

4. For a concise description of each of these approaches to systems analysis, see Guy Black, "Methods of Systems Analysis," in *The Application of Systems Analysis to Government Operations* (New York: Frederick A. Praeger, 1968), Chap. 3, pp. 24–36. Also see Ole C. Nord and David W. Packer, "Management Science Approaches," in *Management Science: A Primer for Managers* (Sudbury, Mass.: Management Science Associates, 1964), Part II, pp. 51–107.

5. For a discussion of the management of the entire Space Agency, see the eminently readable book written by the former Administrator of NASA, James E. Webb, *Space Age Management: The Large-Scale Approach* (New York: McGraw-Hill, 1969).

6. For a discussion of such issues, see Albert F. Siepert, "Creating the Management Climate for Effective Research in Government Laboratories," and also Norman Kaplan, "Organization: Will it Choke or Promote the Growth of Science?" Both papers were published in *The Management of Scientists*, ed. by Karl Hill (Boston: Beacon Press, 1965).

7. National Aeronautics and Space Administration, Office of Technology Utilization, *Applications of Systems Analysis Models: A Survey* (Cambridge, Mass.: ABT Associates, 1968).

8. Much of the information used in this chapter was assembled by Eugene R. Spangler with TRW, Incorporated. Each of the Manned Space Flight management systems described summarizes existing NASA instructions, directives, or other publications.

9. National Aeronautics and Space Administration, *Phased Project Planning*, NASA Issuance NPD 7121.1 (Washington, D.C.: NASA, November 29, 1965).

10. Peter K. Hatt, "Management Systems Planning on New Flight Programs," in *The Management of Aerospace Programs*, Vol. 12, ed. by Walter L. Johnson (Washington, D.C.: American Astronautical Society, 1967), pp. 53–75.

11. Clarence Danhof, *Government Contracting and Technical Change* (Washington, D.C.: Brookings Institution, 1968), pp. 127–28.

12. Samuel C. Phillips, "Management Scheme for Apollo," in *The Management of Aerospace Programs*, Vol. 12, ed. by Walter L. Johnson (Washington, D.C.: American Astronautical Society, 1967), pp. 193–211.

13. For an excellent discussion of the ends-means structure in economic policy, see Donald S. Watson, *Economic Policy: Business and Government* (Boston: Houghton Mifflin, 1960), pp. 86–145.

14. Samuel C. Phillips, speech presented to the Electronics Industries Association, Systems Effectiveness Conference, Washington, D.C., October 19, 1965.

15. A. W. Marshall and W. H. Meckling, "Predictability of Costs, Time and Success of

Development," in *The Rate and Direction of Inventive Activity: Economic and Social Factors* (Princeton: Princeton University Press, 1962), pp. 461–75. Merton J. Peck and Frederic M. Scherer, "The Unique Environment of Uncertainty," in *The Weapons Acquisition Process: An Economic Analysis* (Cambridge, Mass.: Graduate School of Business Administration, Harvard University, 1962), pp. 19–25.

16. The documents from the Congressional investigation of the Apollo 204 accident are readily available but voluminous. A relatively concise description of the salient factors appears in U.S. Congress, Senate, Committee on Aeronautical and Space Sciences, *Report: Apollo 204 Accident*, 90th Cong., 2d Sess., 1968. Also, NASA submitted a two-volume report to Congress on December 29, 1967 about the actions taken by the Apollo Program Office in response to the Apollo 204 Review Board Report.

17. U.S. Congress, House, Select Committee on Astronautics and Space Exploration, *Report: The Next Ten Years in Space, 1959–69*, 86th Cong., 1st Sess., 1959, p. 120. For a discussion of the man-rating problems and their solutions for Project Mercury, see Loyd S. Swenson, Jr., James M. Grimewood, and Charles C. Alexander, "Man-rating the Machines," in *This New Ocean: A History of Project Mercury*, NASA Historical Series, SP-4201 (Washington, D.C.: U.S. Government Printing Office, 1966), Chap. 3, pp. 167–221.

18. U.S. Congress, Senate, Testimony of George E. Mueller, *NASA Authorization for Fiscal Year 1969, Hearings*, before the Senate Committee on Aeronautical and Space Sciences, Part I, 90th Cong., 2d Sess., 1968.

19. For a discussion of the use of configuration management on the Army's Pershing Missile system, see Charles Borum, "Found: A Practical Configuration Management System," *Defense Industry Bulletin*, 2, 5 (May, 1966), pp. 2–4. Also see William Seith, "Configuration Management in the Navy," *Defense Industry Bulletin*, 3, 4 (April, 1967), pp. 4–8.

20. Lederer was also a former Vice President and Technical Director of the Flight Safety Foundation, Inc., and had served as Director of the Safety Bureau of the Civil Aeronautics Board. See testimony of George E. Mueller, *Hearings, NASA Authorization for Fiscal Year 1969*, Part I, pp. 150–56.

21. *Ibid.*, pp. 156–57.

22. *Ibid.*, p. 265.

23. Robert W. Miller, *Schedule, Cost and Profit Control with PERT: A Comprehensive Guide for Program Management* (New York: McGraw-Hill, 1963), pp. 5–6.

24. Marshall and Meckling, *The Rate and Direction of Inventive Activity*, p. 473; Peck and Scherer, *The Weapons Acquisition Process*, pp. 431–50.

25. Peck and Scherer, *ibid.*

26. *Ibid.*, pp. 433–34.

27. Archie L. Wood, *The Analysis of Large-Scale R & D Programs: A Case Study of Project Mercury* (Cambridge, Mass.: Sloan School of Management, Massachusetts Institute of Technology, 1965), pp. 10, 11. Report No. 115-65. (mimeographed)

28. Since the late 1950's much has been written about the use of PERT and variations of the model to solve management problems. An excellent article on how to interpret the vast literature about PERT appeared in the March–April, 1964, issue of the *Harvard Business Review*. In that article, Arch R. Dooley divides articles, monographs, and books into three groups: (1) the pioneer contributions for managers who require only an acquaintance with the technique; (2) the literature for those who wish deeper penetration; and (3) the literature for those who must put the technique in practical use [Arch R. Dooley, "Interpretations of PERT," *Harvard Business Review* (March–April, 1964), pp. 161–68].

29. Miller, *Schedule, Cost and Project Control with PERT*, pp. 40–42.

30. M. W. Hilton, "The Use of PERT in Industry" (Unpublished master's thesis submitted to the Sloan School of Management, Massachusetts Institute of Technology, June, 1966), p. 28. Also, Wyckman D. Seelig and Irwin M. Rubin, "The Effects of PERT on

R & D Organizations" (Cambridge, Mass.: Sloan School of Management, Massachusetts Institute of Technology, 1966), pp. 2, 6–7. (mimeographed)

31. Jack C. Hayya, *PERT and Procurement Policy* (Los Angeles: Graduate School of Business Administration, University of California, 1966), pp. 3–5.

32. *Ibid.*, pp. 9–31.

33. Seelig and Rubin, "The Effects of PERT on R & D Organizations," pp. 10–11.

34. The information was an informal interdepartmental memorandum (of an unidentified aerospace firm) that summarized the results of conversations with personnel from Hughes, Convair, the United States Navy, and the United States Air Force. Hayya, *PERT and Procurement Policy*, pp. 41–43.

35. *Ibid.*, pp. 44–45.

36. Robert W. Miller, "Multiproject Problem," in *Schedule, Cost and Profit Control with PERT*, pp. 77–78.

37. "PERT Aids Saturn—Apollo Site Selection," *Technology Week*, August 15, 1966, p. 40.

38. Marshall and Meckling, *The Rate and Direction of Inventive Activity*, p. 473.

39. Peck and Scherer, *The Weapons Acquisition Process*, p. 427.

40. Wood, *The Analysis of Large-Scale R & D Programs*, pp. 11–19.

41. Peck and Scherer, *The Weapons Acquisition Process*, p. 430.

42. The cost estimates used for the manned lunar landing project are not comparable with those made by Marshall and Meckling and by Peck and Scherer for weapons systems. Cost over-run estimates made by those researchers were based on the difference between the estimated and actual cost of a group of sampled contracts. The manned lunar landing cost estimates are comparable with the data used by Wood to estimate the cost variance factor of the Mercury Project. Wood also used data supplied by NASA to Congress during authorization hearings.

43. An official estimate made by the Office of Manned Space Flight, NASA, for the cost of the first manned lunar landing is slightly less than the estimate made in January 1969. The January 1969 estimate included costs for the Apollo Program through December 1969. Because the actual manned lunar landing was in July, NASA made a cost adjustment—shifting the cost of post-Apollo 11 launch vehicles and spacecraft into fiscal 1970 and fiscal 1971 cost calculations. With the cost adjustment, the total R & D cost of the first manned lunar landing drops from $17.6 billion to $16.9 billion; the tracking, facilities, and operations cost component drops from $4.9 billion to $4.4 billion; and the total estimated cost drops from $23.9 billion to $21.3 billion. If inflation is taken into account, the total cost would amount to about $18.0 billion in 1964 constant dollars, and would be almost 8 percent below the March 1964 estimate of $19.5 billion for the first manned lunar landing. Another official estimate, published in the January 25, 1970, *Washington Post*, citing a press conference with Thomas O. Paine, NASA Administrator, puts the current dollar cost at $19.4 billion rather than $21.3 billion. The difference between the estimates results from disagreement within the Space Agency on the correct allocation of joint costs between Apollo 11 and subsequent manned lunar missions.

44. U.S. Congress, Senate, *NASA's Proposed Operating Plan for Fiscal Year 1969, Hearings*, before the Senate Committee on Aeronautical and Space Sciences, 90th Cong., 2d Sess., October, 1968.

45. James R. Bright (ed.), *Technological Forecasting for Industry and Government* (Englewood Cliffs, N.J.: Prentice-Hall, 1968). See especially the article by Howard A. Wells, "Technological Forecasting in Product Planning, with Emphasis on Systems Development," pp. 311–20.

46. F. T. Rosenberg and A. T. Diamond, "Manned Space Flight Cost Models and Their Applications," Cost Analysis Branch, Manned Space Flight Program Control, NASA, May 17, 1965. (mimeographed)

47. Ronald M. Konkel and Frank T. Rosenberg, both of the Office of Manned Space Flight, NASA, provided much of the information about the Manned Space Flight Production Cost Model.

48. As used here, the concept of a plant is somewhat different from the conventional notion of a plant in economics. In the model, the plant is a level of aggregation used for purposes of allocating overhead. Some of the cost aggregations cover activities at more than one location.

49. The distinction between variable and relatively fixed costs used here should not be confused with the distinction between variable and fixed costs used in economics. At best, the notion of relatively fixed recurring costs can be likened to a constant marginal cost function over a wide range of the production process. That marginal cost function might float up or down as a result of indivisibilities, and again be constant over a wide range of the production process.

50. For a discussion of the difference between the rate effect and the learning curve effect, see DeMarquis D. Wyatt, "Cost Models for Complex Programs," in *The Management of Aerospace Programs,* Vol. 12, ed. by Walter L. Johnson (Washington, D.C.: American Astronautical Society, 1967), pp. 77–93.

5 Measures of the Impact of the Space Program

NASA, perhaps more than any other science mission-oriented federal agency, has done much to try to understand the social and economic ramifications of its objectives. The explanations for this must be multifarious. The Space Act explicitly directs NASA to undertake studies "of the potential benefits to be gained from . . . and the problems involved in the utilization of aeronautical and space activities. . . ." The Space Act also directs the Administrator to "provide for the widest practicable and appropriate dissemination of information concerning its activities and the results thereof."[1]

The mandates of the Space Act are not the sole explanation for NASA's support of research in the social sciences. Interest in the social and economic effects of the space program, which had a budget that once approached $6 billion, necessarily became the concern of NASA officials. Also it is obvious that NASA courts the intellectual community for support of the space program. It is also politically expedient to have the answers to social and economic questions that will inevitably be asked by Congressmen and by state and local politicians.

Speaking at the annual meeting of the American Statistical Association in September 1965, Murray Weidenbaum expressed the sentiments of many economists, particularly those working in fields of public policy. To Weidenbaum and to others, information about space and defense spending is necessary for substantive reasons. Such information is not a statistical luxury.[2] It is necessary for an analysis of regional growth, the structure and growth of industries, the government's possible contribution to monopoly and economic control, public policies to offset shifts in the level and product-mix of government procurement, and efficiency in the allocation of resources within the public sector.

This chapter contains a review of the stock of information that NASA, other federal agencies, and other organizations have developed to permit analysis of the social effects of the space program. The emphasis will be on economic information systems.

SOCIAL AND ECONOMIC ISSUES

The evolution of the social and economic issues that have caused concern within NASA can be discerned from an examination of the kinds of grants and contracts let for research in the social sciences. The social research studies funded by NASA reflect not only changes in the program structure of NASA, but also changes in the research fads of the academic community.

Table 5–1 gives information on the kinds of research in the social sciences sponsored by NASA. As is indicated by the nine fields of study, which include education, general economics, international studies, legal studies, management studies, public policy, regional studies, technology utilization, and urban studies, the definition of social science used here is broad. With the exception of the multi-disciplinary contracts, each of the areas of research listed in the table is in one of the "soft" sciences. In its early years, NASA concentrated most of its research studies on public policy issues and also on legal studies. The principal concern of these early research projects was with the peaceful uses of space, especially uses related to communication satellites. The American Bar Association Foundation received a contract from NASA to do a literature search on national and international laws applicable to satellite operations. United Research, Incorporated did a project on the legal regulatory aspects of communication satellites. The RAND Corporation of Santa Monica, California, received NASA's first economics research project, under one large contract and a number of follow-on contracts. RAND's dual assignment was (1) to study economic and international policy issues arising from communication satellites, and (2) to investigate the economic aspects of meteorological satellites.

As the space program expanded, regional studies grew in importance. During the 1962–1964 period, NASA funded two main kinds of regional studies. One group was concerned with increasing state and regional participation in the space program, including the use of space technology for state and regional economic growth. The second group of regional studies was designed to determine the economic impacts of the space program on specific localities. The Department of Economics, Washington University, St. Louis, studied the effects of the Mercury and Gemini Programs on the economic structure of the St. Louis area. The McDonnell

TABLE 5-1: GRANTS AND CONTRACTS LET BY NASA FOR RESEARCH IN THE SOCIAL SCIENCES, FISCAL YEARS 1959[a]–1968[b]

FIELD OF STUDY	1959–1961[a]	NUMBER OF CONTRACTS 1962–1964	1965–1968[b]	TOTAL AMOUNT ($THOUSANDS)	PERCENT OF TOTAL	AVERAGE SIZE OF GRANT CONTRACT ($THOUSANDS)
Technology utilization	1	30	88	$16,441	47.3	$138
Public policy	6	9	17	3,645	10.5	114
Legal studies	2	1	1	114	0.4	28
Education	1	17	15	875	2.5	27
General economics	1	5	16	4,695	13.5	213
Regional studies	0	18	7	2,151	6.2	86
Management studies	0	2	20	1,973	5.7	90
International studies	0	1	5	1,696	4.9	283
Urban studies	0	0	5	292	0.8	58
Multi-disciplinary[c]	0	13	84	2,850	8.2	29
Total number	11	96	258	[d]	100.0	[d]
Total amount ($Thousands)	$859	$8,225	$25,648	$34,732	–	$ 95

[a] October 1, 1958 through June 30, 1961.
[b] Includes only the first six months of fiscal 1968.
[c] Includes 10 percent of amount for multi-disciplinary research.
[d] Not applicable.

Source: Compiled from data in the appendices of the National Aeronautics and Space Administration, *Semi-Annual Reports to Congress.*

Aircraft Company (now McDonnell Douglas), located in St. Louis, was one of the largest contractors for the Mercury and Gemini Programs. That company was the systems firm for the production of the Mercury and Gemini space capsules. The University of Florida, Florida State University, and the East-Central Florida Planning Commission all received NASA funds to study the social, political, and economic effects of the space program in the communities surrounding Cape Kennedy. The Department of Regional Sciences at the University of Pennsylvania received a large grant to construct a regional input-output model for Philadelphia. NASA's principal interest in that study was to learn more about the economic effects of its rapidly expanding program in electronics research. Mississippi State College received NASA money to explore the economic contribution of the Mississippi Test Facility to the economy of Southwestern Mississippi, as did the University of Boulder for the Boulder area. Ball Brothers, Incorporated, located in Boulder, was engaged in much of the research and engineering work on NASA's orbiting observatories.

NASA has also exerted a great effort in its attempt to promote the transfer of technology to American industry. Almost one-half of NASA's funds for research in the "soft sciences" has been for technology transfer or technology utilization projects. Some may object to including research projects on technology utilization and technology transfer in this discussion of research in the social sciences sponsored by NASA. However, with the rapid growth of government spending for research and development since the end of World War II, the process of technology transfer and the various effects of technology utilization have become increasingly important to social scientists. For example, economists are seeking to learn how the process of technology transfer affects economic development and economic growth; psychologists want to know how technology affects the individual; and sociologists are studying the ways in which technology influences society.

NASA's first grant for exploring technology utilization went to the University of Utah in the amount of $15,000. NASA-funded studies for technology utilization have been wide in content. Technology utilization grants and contracts have included projects on identification, evaluation, measurement, stimulation of use, trade fairs, conferences, abstracting and indexing, and information exchanges. The recipients of NASA money for technology utilization work have been as diverse as the projects. The names of a few research institutions are: the Smithsonian Institution, Armour Research Foundation, Southern Research Institute, Stanford Research Institute, National Academy of Sciences, Aerospace Medical Association, Library of Congress, Lovelace (Medical) Foundation, and the American Meteorological Society. In addition, a dozen or more different universities have worked on the matter of technology utilization. The American Institute of Aeronautics and Astronautics has received by far the largest amount of money—over $3 million by the end of fiscal 1968. That organization performs a scientific and technical information service for NASA that includes the abstracting, indexing, and publishing of aerospace literature.

NASA funds for education studies have been made available mostly to small teachers colleges. The $875,000 shown for these studies does not include the money NASA has spent on its Predoctoral Traineeship Program.[3] Most of the research in education studies has been for the development of elementary and secondary teaching materials about the space sciences that can be used in classroom instruction. Grants and contracts to the dozen or so different state teachers colleges have typically been small—many for less than $5,000. From time to time, the National Science Teachers Association has received money from NASA to develop and publish books to enrich the science programs of elementary and secondary schools.

NASA has also awarded research grants and contracts for multi-disciplinary research. Part of the research money is for research in the social sciences. For example, a NASA project at the Massachusetts Institute of Technology calls for "multidisciplinary research in space related physical, engineering, social, and life sciences." NASA's Semi-Annual Reports give a brief description of the kind of research to be included in the multi-disciplinary projects. Only those that appear to incorporate the social sciences are included in the tabulation shown in Table 5–1.

Without careful examination of the work statements in contracts, or of the resulting contractor reports, it is difficult to know the proportion of multi-disciplinary contracts that were actually for research in the social sciences. Again using the multi-disciplinary contract to the Massachusetts Institute of Technology as an example, that contract has been a continuing one, with several supplements. A literature search revealed that the Massachusetts Institute of Technology had submitted 377 different unclassified reports to NASA under the terms of the contract as of September 1, 1968. Between fiscal year 1962 and 1968, NASA awarded almost 100 contracts for multi-disciplinary research. For purposes of this analysis, the cost of examining contract work statements and contractor reports far exceeds the benefits that would accrue from knowing the relative importance of the social research incorporated in NASA multi-disciplinary research awards. NASA's Office of University Affairs estimates that about 10 percent of all research on multi-disciplinary projects was for research in the social sciences. The $2,850,000 estimate for multi-disciplinary research displayed in Table 5–1 is 10 percent of the total amount of awards for multi-disciplinary research contracts that included the social sciences.[4]

The twenty contracts NASA let for management studies between fiscal year 1965 and 1968 reflect the current zeal for using the systems approach for solving managerial problems. During the preceding six years, NASA let only two contracts for management studies. Another interest of the Space Agency has been the effects of the space program on urban affairs. NASA's interest in funding urban studies is at least twofold. First, the growing national awareness of the magnitude and complexity of the urban problem has given rise to the idea of using the technological knowledge developed for the space program in the solution of some urban problems. Such urban problems include transportation systems, pollution control, minimum acceptable noise levels, safety systems, and perhaps even crime detection and control. Another NASA interest in urban studies is more immediate and focuses on the rapid decrease (after mid-1966) of its contractor work force. That interest is in the possibility of using the scientists, engineers, and managers separating from the space program to

TABLE 5-2: GRANTS AND CONTRACTS LET BY NASA FOR RESEARCH IN THE SOCIAL SCIENCES AS A PERCENTAGE OF AWARDS TO NONPROFIT ORGANIZATIONS AND TOTAL AWARDS

FISCAL YEAR	SOCIAL SCIENCE RESEARCH ($THOUSANDS)	SOCIAL SCIENCE RESEARCH AS A PERCENTAGE OF AWARDS TO NONPROFITS[a]	SOCIAL SCIENCE RESEARCH AS A PERCENTAGE OF TOTAL NASA AWARDS
1959–1960[b]	362	1.61	0.07
1961[c]	497	2.03	0.07
1962	793	1.58	0.05
1963	3,983	3.90	0.12
1964	3,449	2.43	0.08
1965	5,385	3.27	0.10
1966	5,738	3.22	0.11
1967	10,214	5.92	0.22
1968[d]	4,311	7.87	0.20
Average	34,732[e]	6.36	0.13

[a] Excludes awards let to the Jet Propulsion Laboratory.

[b] Procurement awards during the period October 1, 1958 to June 30, 1960. Grants and contracts for social research for the period October 1, 1958 to March 30, 1960.

[c] Procurement awards during fiscal 1961. Grants and contracts for social research for the period April 1, 1960 to June 30, 1961.

[d] First six months of fiscal 1968.

[e] Total rather than average.

Sources: Compiled from data in the appendices of the National Aeronautics and Space Administration, *Semi-Annual Reports to Congress*, and from *Annual and Semi-Annual Procurement Reports*.

work on the solution of urban problems. For a small number of people, urban problem study groups are a form of offset employment for the decreasing employment level in the aerospace industry.

How much has NASA spent for research in the social sciences? The figure for a ten-year period is about $35 million. To the universities and research organizations receiving the funds, the $35 million has meant much. In terms of NASA's total expenditures during the last decade, spending for social studies has not been large—about .1 of 1 percent of the agency's ten-year budget. Table 5-2 shows the relative importance of the social sciences in NASA's total budget and also NASA's funds to nonprofit research organizations.

NASA'S VIEW OF THE BENEFITS IT DERIVES FROM ITS RESEARCH IN THE SOCIAL SCIENCES

From time to time, NASA establishes task forces to evaluate its programs. These task forces may or may not consist solely of NASA personnel. In 1968, the Assistant Administrator for University Affairs established a task force to assess NASA university programs; the task force consisted of NASA personnel and four outside consultants. As part of its report, the group evaluated the usefulness of NASA-financed social science research to the recipients of the funds and to NASA. For the most part, the evaluation was applicable to social science research projects sponsored under NASA's Sustaining University Program. Initiated in fiscal 1962, the Sustaining University Program was designed to help universities develop their educational and research capabilities. The report of the task force is not applicable to all research conducted by universities for NASA. The results of scientific research undertaken by universities have made important contributions to the success of NASA programs. This is true especially of many of NASA's unmanned space projects.

In regard to research in the social sciences sponsored under the Sustaining University Program, NASA believes that it has helped universities more than it has been helped by universities. Although it puts some of the fault on the academic research community, NASA shoulders most of the blame itself.

The report of the task force states that universities have not taken advantage of the opportunities offered by the Space Agency. NASA finds little evidence of the development of a university capability to respond as an institution; it finds little evidence of the development of a university capable of engaging in successful multi-disciplinary research; and it finds little evidence that technology transfer has been accelerated as a result of research expenditures under the Sustaining University Program. Viewed in terms of productivity, as measured by publications and by faculty and student participation, NASA's program, it believes, has been successful.

Yet NASA believes that few, it any, management decisions within the agency have been influenced by the results of the social science research it has funded. The report of the task group attributes this to the lack of an in-house mechanism for utilizing research results. Those on the task force believe that NASA-funded research in the social sciences has had no centralized direction, and that there has been little participation by the NASA Field Research Centers where many management decisions are made. Elaborating on the lack of NASA use of the results of its social science research, the report of the task force states that NASA has no technique for identifying those within the agency who can put the information to use. Further, when research results are transmitted from the research group to the NASA policy-maker, there is little communication because of the difference in the outlook of the researcher and the civil servant. The task force proposes that a "research-oriented management science group" be established to solve this communications problem.[5]

The Practicability of NASA's Using the Results of Its Research

The NASA task force to assess university programs judges itself too harshly, and somewhat naively. Certainly on a need-to-know basis, the Administrator, his Associate Administrators, his Assistant Administrators, and others along the top management line are broadly aware of the contents of much research in the social sciences funded by NASA. When required, those at the decision-making level at NASA use the phone and memoranda systems approach to find out the specifics of the results of research projects.

Both at higher and lower management levels in NASA, and in other government agencies, use of research results can be a Dr. Jekyll and Mr. Hyde situation. Dr. Jekyll supports and sponsors research work because he is anxious to learn ways to improve his organization. But when project results come in, Mr. Hyde finds that he has no use for the finished product.[6] That is to say, after commissioning and fostering research, it is frequently the sponsoring client or the contract monitor himself who is directly responsible for the non-utilization of the results of research.[7] An important reason for this rests in the fact that much research is not funded until after the problems become publicized and pressing issues. Because of lags in the research process, the issues are frequently settled before the fruit of the research is available. New and different problems, of course, displace those prevailing earlier.

Another reason for not using, or at least not publicizing, the results of sponsored research is that the conclusions can be unfavorable and negate a long-standing agency position. This possible cause for the non-use of research results is as applicable to opposing groups in business firms as it is to agencies in the federal government. The Office of Supersonic Transport of the Federal Aviation Agency found itself in this situation after getting the results of research undertaken by three independent research groups. The FAA was later criticized for keeping the results of that research confidential because the conclusions, so it was argued by several members of Congress, about the relative costs compared with the benefits that would accrue from the development of the supersonic transport were so critical in the decision to undertake production of the airplane.[8] In addition, most research projects contain analyses of subordinate matters or side issues, as well as the principal research problem. Disapproval and disagreement with secondary findings can cause the entire research project to be discarded. The disapproval of secondary findings can be the result of personal objection on the part of a contract monitor, but it can also reflect superior insight by the contract monitor about adverse public reaction to the research findings.

Another reason that project initiators are frequently those who throttle the use of research results relates simply to difficulties associated with innovation in a bureaucratic environment. Finally, the results of some research projects can be of such poor quality that there is no use for them.

In May 1969, the National Science Foundation published the report of its Special Commission on the Social Sciences on how to improve the nation's use of the social sciences. In addition to the lack of communication between social scientists and policy-makers and the outright failure of many social scientists to meet the demand for providing usable results, the Commission believes that there are five main obstacles to the use of the social sciences: (1) the lack of clients or agencies to act upon new knowledge in the social sciences; (2) the inability of many social scientists to translate theoretical concepts into a form useful for practical application by policy-makers; (3) the tendency of people and organizations to reject social science knowledge that threatens to destroy existing views; (4) the failure to implement the application of the results of social research frequently because it is too costly; and (5) the inability to implement much research in the social sciences because it incites too much controversy in the political arena.[9]

FEDERAL AGENCIES CONCERNED WITH THE ECONOMIC IMPACT OF FEDERAL PROCUREMENT

The National Aeronautics and Space Administration is only one of many federal agencies concerned with the socio-economic effects of federal procurement. Table 5–3 lists twelve different government departments that are directly or indirectly concerned with the economic effects of changes in the level and product-mix of federal procurement.

TABLE 5-3: FEDERAL AGENCIES CONCERNED WITH THE ECONOMIC IMPACT OF FEDERAL PROCUREMENT

AGENCY	
Arms Control and Disarmament Agency	Sponsors research; disseminates information
Atomic Energy Commission	Sponsors research; provides economic adjustment assistance
Department of Commerce	
Office of Business Economics	Data collection and analysis, including a national input-output model
Bureau of Census	Data collection and special surveys
Economic Development Administration	Sponsors research; data collection and analysis
Business and Defense Services Administration	Data collection, with emphasis on strategic materials and labor
Department of Defense	Sponsors research; provides economic adjustment assistance; data collection and analysis
Department of Labor	
Manpower Administration	Sponsors research; data collection and analysis
Bureau of Labor Statistics	Data collection and analysis, including an employment input-output model
Federal-State Unemployment Insurance System	Provides economic adjustment assistance
General Services Administration	Provides economic adjustment assistance
Department of Housing and Urban Development	Provides economic adjustment assistance
National Aeronautics and Space Administration	Sponsors research; data collection and analysis
National Science Foundation	Sponsors research; data collection and analysis
Office of Emergency Preparedness	Analytical and planning studies on resource preparedness in emergencies
Small Business Administration	Data collection; provides economic adjustment assistance
Veterans Administration	Provides economic adjustment assistance.

The interest of each agency is also indicated in the table. Some of the agencies sponsor contract research; others collect, analyze, and disseminate information; still others have the legal or administrative responsibility to provide economic adjustment assistance to people, firms, and communities.

A number of government agencies that have large in-house staffs engaged in research in the social sciences or that sponsor contract research in the social sciences have been excluded from the list because the missions of the agencies are outside the areas of defense, space, and exotic technology procurement. For example, the Bureau of Economic Research in the Department of Agriculture is one of the strongest, largest, and oldest social science research organizations in the federal government. The Department of Health, Education and Welfare, the Department of Transportation, and the Office of Economic Opportunity are other agencies that support or conduct special mission-oriented research.

Not unexpectedly, a principal concern of the Arms Control and Disarmament Agency is that of the ability of American industry and communities to adjust to shifts in defense spending. That agency's funds, most of which have gone for contract research, averaged about $10 million annually between 1966 and 1968.[10] The Arms Control and Disarmament Agency also operates as a coordinating body for other federal efforts designed to increase national understanding of arms reduction, measures of war risk, and arms inspection and verification.

The Department of Defense, and the Atomic Energy Commission on a smaller scale, collects data about the distribution of their prime contract awards. DOD has undertaken much in-house and contract research in the area of cost-effectiveness analysis. Because much of the analysis is security-classified, the results of these studies are not usually available. Unlike NASA, the Department of Defense does not collect information about the economic effects of subcontracting. To compensate for that information gap, DOD has let contracts to such organizations as Stanford Research Institute, Research Analysis Corporation, and CEIR, to study subcontracting patterns in industries and in geographic regions. The Office of Economic Adjustment in the Department of Defense (established in 1961) advises localities faced with dislocation caused by changes in the levels of defense spending about the exploitation of commercial activities for economic growth. The Atomic Energy Commission has an Office of Economic Impact and Conversion that provides similar services to communities faced with the closing of AEC facilities. The Department of Defense also has a computer facility in Dayton, Ohio (DOD Centralized Referral Activity) that registers and refers displaced workers to jobs available throughout the entire nation.

As Table 5–3 shows, four offices in the Department of Commerce, in one way or another, participate in the social research process for defense- and space-related matters. The Office of Business Economics in the Department of Commerce has the principal function of gathering information about the components of the gross national product. As part of this work, that office has developed an inter-industry input-output model that provides a tool for analysis of the industrial effects of changes in federal spending. The model covers about 400 different industries. In cooperation with the Office of Business Economics, the Bureau of Labor Statistics of the Department of Labor has developed an input-output model that can be used to measure changes in employment resulting from shifts in defense spending.

The Economic Development Administration in the Department of Commerce has sponsored research on the impact of federal procurement on regional economic development. The Economic Development Administration also has the authority to designate communities faced with economic dislocations caused by government actions as redevelopment areas. That organization then can provide economic assistance by means of matching grants, technical advice, and low-interest loans to develop commercial facilities and industrial plants. The Business and Defense Services Administration in the Department of Commerce collects information on the use, production, and capacity to produce strategic products and materials. The Bureau of the Census, of course, conducts many special surveys for the United States government. One of these surveys is of "Shipments of Defense-Oriented Industries." This annual survey provides information about government procurement, by industry and by geographic area.

The Office of Emergency Preparedness has an Economic Impact Evaluation Group engaged in econometric model building to permit evaluation of changing economic and resource conditions under possible emergency situations. The General Services Administration conducts research on economic problems associated with the stockpiling of strategic materials. The General Services Administration also has the authority to donate federal property to communities affected by shifts in government procurement. In addition, that agency can sell or lease surplus property and facilities for industrial use.

Because of their statutory missions, the Department of Housing and Urban Development, the Veterans Administration, and the Small Business Administration are concerned with the economic aspects of defense and space procurement. Both the Department of Housing and Urban Development and the Veterans Administration have the authority to initiate mortgage forbearance agreements to prevent the collapse of housing markets in localities affected by reduced federal spending. The Housing Act of 1954 also provides for assistance in urban planning as a result of economic dislocation. Under the Act, grants can be made to localities for studies on economic development. The Small Business Administration can assist small businesses through its lending programs, and it can provide technical assistance to firms for product and market development. The Small Business Administration is also developing a data bank with information about small businesses engaged in defense activities, as well as information about the impact of defense contracting on small businesses.[11]

SOURCES OF DATA FOR MEASURING ECONOMIC EFFECTS OF DEFENSE AND SPACE SPENDING

A list of non-NASA—originated information systems and sources of data for measuring the economic effects of space and defense spending is presented in Table 5–4. The organization compiling the data, the frequency of publication, and a brief description about the information are also given. Each of the publications is issued on a recurring basis; no occasional sources are listed. Of the eighteen publications listed, all but the last three come from federal government sources. Nearly every one of the data systems listed in the table, like almost all sources of information, has gaps or shortcomings of one kind or another. To be sure, merely collecting all of the information given in each of the sources will not give a picture of the economic effects of either the defense program or the space program. In fact, indiscriminate use of the documents can distort rather than reflect the effects of federal spending.[12] Also the data from one source need not agree or be consistent with data from another source.

There are three important information gaps that make impact analysis of space and defense spending far from simple. The first is the lack of uniform and continuing data about first- and lower-tier subcontract awards made by the Department of Defense. Some prime contractors let 50 percent or more of their procurement awards out to subcontractors. The net contract position of companies is also determined by the inflow of subcontracts. The economic dependence of firms, industries, and geographic locations is determined by both the inflow and outflow of subcontract awards. Although NASA has gathered much more information about subcontracting patterns than the Department of Defense, it is virtually impossible to discern the net contract position of firms or geographic locations at a point in time, because the NASA data do not reflect lags in the procurement process.

A second large data gap is the lack of consistent information about current income and employment in many locations affected by federal procurement. For example, Item 10 on the list of publications displayed in Table 5–4 is a Bureau of Labor Statistics data series on employment and earnings for states and areas. That series does not provide informa-

TABLE 5–4: NON-NASA INFORMATION SYSTEMS AND SOURCES OF DATA FOR MEASURING THE ECONOMIC IMPACT OF SPACE AND DEFENSE SPENDING

ORGANIZATION COMPILING DATA	TITLE OF PUBLICATION	FREQUENCY OF PUBLICATION	COMMENT
Council of Economic Advisers for the United States Congress Joint Economic Committee	*Economic Indicators*	Monthly	National and industry impact: components of GNP, employment and unemployment, output in manufacturing; sales and inventories; new orders; wholesale consumer prices.
Council of Economic Advisers for the President	*Economic Report of the President*	Annual	National impact: report duplicates statistics published by other agencies; highlights government economic policies directed toward national goals; federal, state, and local receipts and expenditures data.
Bureau of the Budget for the President	*The Budget of the United States*	Annual	National impact: contains the President's annual budget message; executive request for the allocation of federal funds among agencies and among programs within agencies.
Office of Business Economics, Department of Commerce	*Survey of Current Business, Business Statistics, Supplement to the Survey of Current Business*	Monthly Bi-Annual	National, industry, and geographic impact: GNP data; employment by industry; labor turnover rates; orders, sales, and backlog of orders data for the aerospace vehicles industry; income data for cities and states.

Bureau of the Census	*Shipments of Defense-Oriented Industries in Large Standard Metropolitan Statistical Areas (MA-175)*	Annual	Industry and geographic impact: a supplement to the Annual Survey of Manufacturers. The shipment data are gathered to measure the impact of government procurement in metropolitan areas.
Bureau of the Census	*County Business Patterns*	Bi-Annual	Geographic impact: income, population, and industry employment data by county.
Board of Governors, Federal Reserve System	*Federal Reserve Bulletin*	Monthly	National impact: GNP data; rates of plant capacity utilization; investment growth in plant and equipment.
Department of Labor	*Manpower Report of the President*	Annual	National, industry, and geographic impact: data on manpower requirements, resources, utilization and training; usually includes a section on impact of federal procurement and on R&D manpower use and requirements; some forecast data.
Bureau of Labor Statistics, Department of Labor	*Employment and Earnings and Monthly Report on the Labor Force*		National, industry, and geographic impact: data on unemployment; hours worked; earnings; labor turnover rates; unemployment insurance. Many of the data are available by industry.

ORGANIZATION COMPILING DATA	TITLE OF PUBLICATION	FREQUENCY OF PUBLICATION	COMMENT
Bureau of Labor Statistics, Department of Labor	*Employment and Earnings Statistics for the United States 1909–1967*	Annual	National and geographic impact: includes same information as the item above, with industry emphasis: an historical series.
Bureau of Labor Statistics, Department of Labor	*Employment and Earnings Statistics for States and Areas 1939–1967*	Annual	National and geographic impact: includes same information as the last two items above; with geographic emphasis; an historical series.
National Science Foundation	*Federal Funds for Research Development, and Other Scientific Activities*	Annual	National impact: levels and trends in government agency R & D expenditures; expenditures for technical information; and spending for R & D plant and equipment.
National Science Foundation	*Research and Development in Industry*	Annual	National and industry impact: R & D funds by industry; employment of scientists and engineers; geographic distribution of R & D funds.
Department of Defense	*Selected Department of Defense Economic Indicators*	Monthly	National impact: military prime contract awards by kind of procurement; obligations outstanding; net expenditures; and manpower strength.

Department of Defense	*Various Prime Contracts Awards Data*	Semi-Annual and Annual	National, industry, and geographic: separate reports for largest R & D contractors; state distribution of awards; and state distribution of awards by kind of procurement.
Aerospace Industries Association of America	*Aerospace Facts and Figures*	Annual	National impact: summarizes existing data series; aircraft production; missile production, manpower use; international transactions; provides an historical series.
Aerospace Industries Association of America	*Aerospace News* and also *Economic Data Branch Memoranda*	At least monthly	National, industry, and geographic impact: estimates and forecasts of manpower utilization based on its own survey of companies and government contractors; also summarizes existing data series.
Sales Management, Incorporated	*Survey of Buying Power* June issues	Annual	Geographic impact: employment, population, income, and retail sales data for selected cities and all counties; estimates based on United States Bureau of the Census benchmark data.

tion about employment in the areas surrounding Cape Kennedy or the Mississippi Test Facility. Many of the data published by the Department of Commerce and the Department of Labor include only the standard metropolitan statistical areas in the United States.

The third major statistical gap is the lack of consistent information about the dollar value of production and level of employment in nonmanufacturing companies and in the nonmanufacturing facilities of manufacturing companies. Nonmanufacturing firms are the defense and space contractors that provide services, such as research, development, and test work. The Bureau of the Census (Shipments of Defense-Oriented Industries) estimates the value of work performed and the level of employment at these establishments, but the estimates are low. For example, the 1966 Census data show no NASA contractor employment in Mississippi and about 8,600 in Huntsville. Because many NASA contractors at the Mississippi Test Facility are not engaged in manufacturing activities, their work forces are not included in that Bureau of the Census survey. A special NASA survey showed that there were more than 20,000 NASA contractor employees working in Huntsville in 1966.

Although the computer has done much to improve the availability of current data during the last decade, some of the data series have as much as a two-year lag built into them. The National Science Foundation (Items 12 and 13 in Table 5–4) published its report on *Research and Development in Industry during 1966* in 1968. In addition, many of the data are available only for a high level of aggregation. The Federal Reserve System data (Item 7) on investment in plant and equipment and rates of capacity utilization are not broken down by major industry groups.

Another data problem results from the fact that monthly and quarterly information series can be quite erratic. Also, on a monthly or quarterly basis, sets of data can move in opposite directions. For example, the Bureau of Census collects data on new orders, sales, and back orders for aerospace vehicles (Item 4). An index compiled from the data on new orders for aerospace vehicles is extremely erratic—alternating by quarter by as much as 60.0 percentage points from a 1965 base. The index was 104.5 for the second quarter of 1966; 163.8 in the third quarter; 106.5 in the fourth quarter on 1966; 93.6 in the first quarter of 1967; and 126.1 in the second quarter of 1967. The erratic nature of some statistical series makes accurate short-term impact analysis and forecasts difficult.

ECONOMIC INFORMATION COLLECTED BY NASA

Three of the economic data systems compiled by the National Aeronautics and Space Administration warrant special attention in this chapter. Together, they form much of the data base for the analyses in the following chapters on the economic performance of the space program. The first information package is what NASA calls "SCAG"—Status of Contracts and Grants. Originally designed for procurement management, the data comprise the nucleus of NASA's semi-annual procurement reports. The second set of data come from NASA's so-called "Postcard Reporting System." The postcard system provides information about subcontracting patterns. The joint National Aeronautics and Space Administration and Department of Defense "Economic Information System" is the third data source. The Economic Information System provides a wealth of data about employment at the plants of government contractors.

Status of Contracts and Grants

NASA established a formal and systematic procurement reporting system on July 1, 1960. The foreword to the first procurement report stated that the system was designed to provide the statistical data required for the effective management of the NASA procurement program. More than thirty items of information are provided on each individually reportable procurement. Under this system, reports are required for procurement actions of $10,000 and above on research and development prime contracts and grants to nonprofit organizations, and for actions of $25,000 and above on all other prime contract awards.

SCAG is structured to provide information to essentially four groups: (1) procurement management; (2) program management; (3) budget preparation; and (4) Congress. In addition, the data are readily available for public use.

For procurement management, information about prime contract awards helps to furnish answers to three questions. The first question relates to the way in which NASA conducts its procurement activities. The reports contain information about the extent of advertised versus negotiated awards; contract pricing provisions; fees; small business set-asides; and the use of source evaluation boards. The second question concerns the identity of the recipients of NASA prime contract awards. SCAG gives information about the distribution of procurement awards to large business; small business; educational and nonprofit organizations; to the Jet Propulsion Laboratory; to other government agencies; and to individuals. The reporting system identifies prime contract awards by city and by state and answers a third question asked by procurement managers—where does NASA let its awards?

In addition, SCAG separates contract awards by NASA field installation and by program source. Data on obligations, costs, and expenditures are available by program offices for each contract. The computer runs also provide a concise description of each prime contract.

For budget preparation, the SCAG system provides financial data (obligations, costs, expenditures, and unliquidated obligations) for each contract. Target cost estimates and work descriptions are also available. Data collected for Congressional inquiries are intended to provide answers to the five questions most frequently asked by Senators and Congressmen: (1) what is the extent of competition? (2) what is the extent of small business participation? (3) what is the extent of university participation? (4) what are the sizes and numbers of awards to depressed areas? and (5) what is the geographical distribution of NASA's procurement awards? In addition to a statistical breakdown of awards by city, county, and state, NASA can provide prime contract awards data by Congressional district.

As indicated earlier in this chapter, the most serious weakness of the Status of Contracts and Grants Information system is that it provides data only about prime contract awards, whereas, to repeat, business firms, industries, and localities are affected by the inflow and outflow of subcontract awards.

The Postcard Reporting System

To mitigate this weakness, NASA established a subcontractor reporting system in August 1962. At its inception, the main purpose of the system was to provide data on the geographic distribution of first- and second-tier subcontracts.[13] As originally planned, the postcard system included only NASA's top twelve prime contractors. Since then, coverage has been expanded and it is mandatory for contractors receiving prime awards in excess of $500,000 to report. Qualifying contractors provide information about each first-tier subcontract in excess of $10,000. NASA also requires some first-tier subcontractors to participate in the program. Each first-tier subcontractor receiving a NASA subcontract in excess of $50,000 must provide information about each second-tier contract in excess of $10,000. During fiscal year 1967, NASA obtained data from 104 major prime contractors. These prime contractors, along with qualifying first-tier subcontractors, supplied information about 1,800 different first- and second-tier contracts.

As the name of the system suggests, not much information is required of NASA contractors, and this fact, no doubt, increases the rate of response. The first- and lower-tier subcontract report gives information about place of performance; whether the recipient is a small business; the amount of the subcontract award; whether the award is a new subcontract or a modification of an existing award; and whether the work effort is for research and development.

With the establishment of the postcard system, researchers have available for the first time information about subcontract procurement patterns. Previously, only conjectures could be made about the impact of subcontracts. Despite its merits, the postcard system also has its gaps and shortcomings. The most important gap results from lags in the procurement process. Subcontract awards data cover obligation figures rather than actual expenditures. This means that it is impossible to know when work is actually performed. It also means that it is impossible to know the net subcontract position of a given firm or state at a point in time. This is particularly true for companies providing large and sophisticated subsystems. Another weakness is that the system tacitly assumes that prime contractors with awards of less than $500,000 let relatively few subcontracts. But to be manageable, data systems must have cut-off points. The third shortcoming of the postcard system results from NASA's broadening of the reporting base, which caused a discontinuity in the statistical series. For all practical purposes, the data cannot be used to analyze shifts in subcontract procurement patterns before June 1964.

The Economic Information System

In May 1966, the Bureau of the Budget approved the establishment of semi-annual contractor reporting to the joint National Aeronautics and Space Administration and Department of Defense Economic Information System (EIS). The principal purpose of the system is to provide current data on employment patterns at the plants of companies performing work for NASA and for the Department of Defense. In their request for information and accompanying set of instructions, NASA and the Department of Defense state that data supplied by contractors will not influence program and procurement acquisition decisions. Designed to be used by federal, state, and local agencies, the data system indicates possible adverse economic effects resulting from shifts in defense and space procurement. The Economic Information System also provides a continuing and consistent source of data which can be used by the Secretary of Defense and the Administrator of NASA to assess the impact of federal expenditures on employment in individual plants, on individual contractors, on particular industries, and on specific geographic areas.

The body of data in the June 1966 report covered information about the structure of employment at 422 different plants. A year later, contractors provided information about employment at 453 different plants.[14] Total employment at these plants was approximately 1.4 million in June 1966. These plants accounted for about 2 percent of total wage and salary employment in the United States. Employment at these plants, of course, comprised a much larger percentage of total employment resulting from NASA and DOD procurement. In June 1966, employment at plants reporting to the Economic Information System was slightly more than one-half of total NASA contractor employment. For the Department of Defense, EIS plants comprised about one-third of the estimated total amount of defense employment in industry in June, 1966.[15]

Contractors furnish two different sets of data to the Economic Information System. The first set of data is a plant-wide economic report. For this, contractors report employment at each individual plant by source of business – Department of Defense, NASA, all other government employment, and commercial employment. Contractors also report on the skill-mix of plant employment – scientists and engineers versus all other employees. Also, contractors report whether employment originates from prime contracts or whether it comes from subcontract work. The second set of data comprises a report about employment on each individual project, i.e., major weapons and space systems and also large subsystems. For the Economic Information System, a major system or large subsystem is defined as a project employing at least 150 workers.

As a source of data for economic impact analysis, the Economic Information System has at least four major advantages over NASA's Status of Grants and Contracts Report or over the postcard system. First, the data account for a large proportion of those working in industry on defense and space programs. The data cover virtually all of the largest contractor plants that are heavily dependent on government contracts for employment stability. Second, the EIS data reveal the impact of subcontract procurement patterns at a point in time. Third, the EIS data give the sources of employment demand at reporting plants. This means that it is possible to measure the effects of shifting resource requirements between federal agencies and also the effects of shifts between commercial and government employment demand. The fourth advantage is that the data collected for the system provide advance indication of changes in the structure of employment at contractor plants. The system requests contractors to forecast employment levels that might result from future business.

SOME AGGREGATE MEASURES OF THE ECONOMIC IMPACT OF THE SPACE PROGRAM

The space program, like any other large government program, has generated many changes that have permeated the economic structure of communities, industries, and the nation. Before analyzing particular facets about the economic performance of the space program, it is useful to look at some of the program's aggregate impacts for perspective.

Although the space program consists of a multitude of different projects, in the eyes of the public the space program has tended to become synonymous with the manned lunar landing. The Manned Space Flight Program is the largest single program ever undertaken by the United States government. At the end of the first decade of its existence, cumulative spending by the Space Agency amounted to about $31 billion. About two-thirds of that sum was for the manned lunar landing. For comparison, the cumulative cost of the Manned Space Flight Program, in real or constant dollars, has been about four times as large as the total cost of the Manhatten Project of the 1940's.[16]

The Space Program in Relation to the GNP and to Total Employment

Despite its absolute magnitude, the space program never amounted to as much as 1 percent of the gross national product. As Table 2–1 showed, spending for space reached a peak during fiscal 1966. Because of long-lead procurement, that peak was reached well before the manned lunar landing was actually accomplished. Between fiscal 1966 and fiscal 1968, spending for space dropped by more than one billion dollars. As a component of the gross national product, the space program accounted for less in 1968 than it did in 1964.

The aggregate effect of the space program on total employment is even smaller than that on the GNP. In part, this results from the fact that NASA employs a large number of skilled technicians, scientists, and engineers. Because skilled technicians, scientists, and engineers are paid more than semi-skilled workers, a given expenditure produces fewer jobs. Employment on the space program reached a peak of about 400,000 during fiscal 1966. At that time, space employment amounted to one-half of 1 percent of total civilian employment. Although large, the number of people employed by the space program was not nearly as great as the number employed by the largest domestic corporation. At the end of 1966, General Motors employed about 745,000 workers. As Table 5–5 shows, NASA's civil service work force is about one-tenth the size of its contractor labor force. Gerhard Colm, of the National Planning Association, estimated that the space program absorbed about 5 percent of the nation's pool of scientific manpower at its employment peak.[17]

Between the advent of the Manned Space Flight Program and early

TABLE 5–5: NASA EMPLOYMENT IN RELATION TO TOTAL FEDERAL CIVILIAN EMPLOYMENT

	JUNE 1966		JUNE 1968	
	NUMBER (THOUSANDS)	PERCENT	NUMBER (THOUSANDS)	PERCENT
Total civilian employment	74,000	100.00	77,300	100.00
Federal civil service	2,400	3.24	2,800	3.62
NASA civil service	36	.05	33	.04
NASA contractor	360	.49	210	.27
Total NASA	396	.54	243	.31

Sources: U.S. Department of Labor, Bureau of Labor Statistics, National Aeronautics and Space Administration, and Joint NASA/DOD Economic Information System.

1966, combined NASA civil service and contractor employment rose at an average monthly rate of about 5,000 workers. The largest increase in NASA's work force took place between 1962 and 1963, when employment rose by more than 9,000 each month. The data displayed in Table 5–5 clearly reveal that employment trends on the space program have reversed. The 40 percent decrease in employment between 1966 and 1968 means that during the two-year period about 6,000 people separated from the space program each month.

It is impossible to compute meaningful aggregate income and employment multipliers for government spending on space. The multiplier effect shows that primary increases in income and employment, originating from a change in business investment spending or government spending, can generate secondary changes in income and employment that result in total changes in income and employment in excess of the primary changes.[18] The Department of Regional Science at the University of Pennsylvania estimates that the national employment multiplier was about 2.3 in the early 1960's. It would not be accurate to apply that nation-wide multiplier and say that in 1966 NASA spending generated 396,000 new direct jobs and 515,000 new indirect jobs via the multiplier effect. This is true for several reasons. In the first place, spending for space is concentrated in a relatively small number of geographic locations. Because the localities where space employment became an important source of employment had markedly diverse economic structures and growth patterns before NASA, application of an aggregate multiplier has little or no meaning. Furthermore, space-generated income and employment have had different effects in a relatively small number of industries. Unfortunately, comprehensive statistics have not been developed to permit input-output analysis of the industrial structure underlying NASA's expenditures.[19]

Finally, the timing of federal spending for space has been such that it is doubtful whether or not the space program has, in aggregate, created any new jobs. This does not mean that the space program has not influenced the employment skill-mix and the geographic distribution of federally generated employment. Rather, it means that in the early 1960's employment generated by the space program served as offset employment to a reduction in employment by the Department of Defense. During the rapid building of NASA's civil service and contractor work force between fiscal 1962 and fiscal 1965, procurement of missile and space systems by the Department of Defense dropped by about one-third.[20] Since mid-1966 the reverse has occurred; resources that had been employed on the space program have been diverted into the defense sector of the economy. NASA's role in stabilizing employment in the aerospace industry is described in Chapter 7.

Employment Effects of the Space Program: 1965 to 1968

The data displayed in Table 5–6 paint a broad picture of the geographic and industrial employment patterns of the space program between December 1965 and June 1968. The data show the geographic regions and the industries that have benefited most from direct employment generated by NASA procurement. They also show those areas and industries that are most vulnerable to the adverse economic effects that would accompany a continued decrease in space employment that might be coincident with a drop in defense employment resulting from changes in military requirements.

TABLE 5–6: CHANGE IN NASA EMPLOYMENT AT THE PLANTS OF A SELECTED GROUP OF NASA CONTRACTORS, DECEMBER 1965–JUNE 1968

	DECEMBER 1965	JUNE 1966	JUNE 1967	JUNE 1968	DISTRIBUTION IN DECEMBER 1965
Total	100.0	93.4	64.7	53.6	100.0
Region					
Gulf Coast[a]	100.0	100.9	73.6	66.3	20.1
New England and Middle Atlantic[b]	100.0	91.4	72.1	41.9	17.4
Pacific Coast[c]	100.0	92.4	62.3	55.8	51.2
Central and other[d]	100.0	87.4	47.9	38.3	11.3
Industry					
Guided missiles	100.0	93.8	65.6	65.2	38.1
Aircraft and parts	100.0	86.1	62.9	37.7	40.4
Electronic equipment	100.0	107.1	52.1	49.6	12.6
Business and engineering services	100.0	108.0	88.9	84.7	7.2
Instruments and computers	100.0	93.7	74.6	67.7	1.7

[a] Alabama, Florida, Louisiana, Mississippi, and Texas.

[b] Connecticut, Maine, Massachusetts, New Hampshire, Rhode Island, Vermont, Delaware, District of Columbia, Maryland, New Jersey, New York, Pennsylvania, Virginia, and West Virginia.

[c] California, Oregon, and Washington.

[d] All other states.

Source: Joint NASA/DOD/Economic Information System.

The data come from information supplied to the Economic Information System by a selected group of NASA contractors. The selection was based on consistency of reported information. The data include only those plants that had at least 10 workers employed on NASA projects in June 1966, and only those contractors that submitted reports to the Economic Information System in each of the five reporting periods between December 1965 and June 1968. Seventy-four different contractors, operating

134 plants, met the criteria. Combined employment at these plants totaled 168,000, or almost 50 percent of NASA's June 1966 contractor work force of 360,000 people.

For a broad sweep, the geographic and industry distribution of NASA employment shown in the table will suffice. In reality, NASA employment does not fit so neatly into only five industry groups. In subsequent chapters the employment distributions will be refined. For each of the areas and industries, actual employment in December 1965 is the base and is equal to 100 percent. Employment in the other periods is given as a percentage of the base year. The index numbers show, at a glance, the rate at which employment on space projects at the plants of NASA's largest contractors dropped between December 1965 and June 1968. The last column in the table shows the relative importance of NASA employment to four geographic areas and to five industries. Contractors with plants on the Pacific and Gulf Coasts account for about 70 percent of NASA employment. This reflects the structure of the aircraft industry, with plants heavily concentrated in southern California, before the creation of NASA. It also mirrors the establishment of NASA's manned space flight centers in the five Gulf Coast states. To support the work at these centers, a number of NASA contractors established new plants and other facilities in Alabama, Florida, Louisiana, Mississippi, and Texas.

Although the industry distribution given in Table 5–6 refers only to the 134 plants of NASA contractors reporting to the joint NASA-DOD Economic Information System, the data agree with information on the change in total NASA contractor employment. Because of the regional concentration of employment, the largest number of contractor employees separating from the space program have been those working in the Far West. The greatest decrease, to date, occurred during fiscal 1967.

The data show that employee retention has not been geographically uniform. Subcontractors and some smaller primes who have plants located in New England, the Middle Atlantic, and the Central States, were the first to witness the sharpest relative decrease in employment on their space projects. The plants of the contractors in those locations have been supplying major space systems firms with electronics and aircraft subsystems, components, and parts. Least affected by the cutback in NASA procurement have been contractors engaged in test, launch, and mission operations for the Apollo launch series. The plants of most of these contractors are located at the Kennedy Space Center in Brevard County, Florida, and at the Manned Spacecraft Center in Houston, Texas. This contrasts with a much sharper decrease in employment at the Mississippi Test Facility in Hancock County and at the Michoud Assembly Facility in New Orleans.

SUMMARY

Since its inception, NASA has done much in its attempt to understand the socio-economic effects of its actions and programs. Although small compared with total expenditures, about one-tenth of one percent of a ten-year budget, about $35 million was spent by NASA for research in the social sciences between 1958 and 1968. The content of research projects funded by NASA during the period reflects both the evolution of the space program and the social research fads of the academic community. During the late 1950's and early 1960's, NASA concentrated its research effort on public policy and legal issues. With the opening and growth of NASA's manned space flight field centers, as well as the growing concentration of space spending in such areas as Boston, Boulder, and St. Louis, NASA shifted its research in the social sciences to regional studies. NASA hoped that regional research studies would show two main things: (1) how states and localities could increase their participation in the space program, and (2) how the space program was changing the economic and social structure of communities heavily committed to NASA projects. By the late 1960's, NASA had begun to finance more research on the cost-effectiveness of possible space projects; on the application of systems analysis to space-related management problems; on the effects of the space program on urban affairs; and on the transferability of space systems management techniques to the solution of urban problems.

Although it is difficult to give empirical content to the extent to which NASA management decisions have been influenced by the results of social research projects, NASA's top managers are certainly aware of the contents of many research projects. As discussed in a report of the National Science Foundation on improving the nation's use of the social sciences, NASA and other government agencies, as well as business firms, encounter impediments to the implementation of social research findings. Non-utilization of research results can result from a number of causes: (1) many social scientists are unable to translate a theoretical concept into a form useful for policy-makers; (2) lags in the research process can make findings obsolete because issues are frequently resolved before research results become available; (3) research results can negate or threaten to destroy agency positions; (4) the application of research results might simply be too costly to implement; and (5) the application of research results can create unwanted controversy in the political arena.

About a dozen different federal agencies have some responsibility for different aspects of changes in the level and product-mix of federal procurement. The agencies include: the Arms Control and Disarmament

Agency, the Atomic Energy Commission, the Department of Commerce, the Department of Defense, the Department of Labor, the General Services Administration, the Department of Housing and Urban Development, the National Aeronautics and Space Administration, and the Veterans Administration. The concern of these agencies includes (1) sponsoring research projects; (2) data collection and analysis; and (3) the provision of economic adjustment assistance to individuals, businesses, and communities faced with adverse economic conditions resulting from changes in federal procurement. Some of the non-NASA publications and sources of data for measuring the economic impact of space and defense spending are: *The Economic Report of the President, The Budget of the United States, Shipments of Defense-Oriented Industries in Large Standard Metropolitan Statistical Areas, County Business Patterns, Employment and Earnings Statistics for States and Areas, Research and Development in Industry; Selected Department of Defense Economic Indicators;* and *Aerospace Facts and Figures.*

Among the many different data systems compiled by NASA, three were designed mainly to provide economic information. The oldest of these is the information package NASA calls the Status of Contracts and Grants, or "SCAG." For prime contract awards and grants SCAG provides data to answer five questions: (1) What is the extent of competition in procurement awards? (2) What is the extent of small business participation? (3) What is the extent of university participation? (4) What are the sizes and numbers of awards to depressed areas? and (5) What is the geographic distribution of NASA's procurement awards?

To compensate for the most serious weakness of SCAG, which is the lack of information about subcontracting patterns, NASA established what it calls the Postcard Reporting System. By the late 1960's, the Postcard Reporting System could provide information about 1,800 different first- and second-tier subcontracts. No other federal agency has as much information as NASA about the effects of the inflow and outflow of subcontract awards on business firms, industries, and localities.

NASA's third statistical series is the result of a joint undertaking by the Space Agency and the Department of Defense. Called the Economic Information System, that data source provides information about the structure and skill-mix of employment at the more than 450 different plants of the largest contractors engaged in work on weapons and space systems. Contractors provide information about employment on commercial projects, as well as on employment arising from government prime and subcontracts.

Data from the Economic Information System show that employment on the space program reached a peak of about 400,000 during fiscal 1966. At that time, NASA civil service and contractor employment amounted to about one-half of 1 percent of total national employment. The most rapid build-up on the space program occurred between 1962 and 1963, when employment rose by more than 9,000 each month. Employment on the space program reached its maximum more than two years before the manned lunar landing because of the long-lead time for the production of major hardware components. Between 1966 and 1968, about 6,000 people separated each month from work on the space program. Employment on the space program in June 1965 was about 50 percent below the December 1965 level. The sharpest reductions occurred in those plants of contractors located in New England, the Middle Atlantic, and the Central states. Most contractors located in those areas had been firms supplying major space systems with electronics and other space subsystems, components, and parts.

NOTES

1. Section 101 (c)(4) and Section 203 (a)(3) of the Space Act.

2. Murray L. Weidenbaum, "Measures of the Impact of Defense and Space Programs," a paper presented to the annual meeting of the American Statistical Association, Philadelphia, September 9, 1965.

3. Since 1962, NASA has spent about $100 million to support 5,400 graduate students for full-time academic studies in space-related disciplines. See U.S. National Aeronautics and Space Administration, *A Study of NASA University Programs*, NASA SP-185, 1968, pp. 48–52.

4. NASA reported to the National Science Foundation that it spent $6,584,000 for social science research in fiscal 1966. Assuming that 10 percent of NASA contracts and grants for multi-disciplinary research was for the social sciences, NASA contracts and grants information provided in the agency's Semi-Annual Reports to Congress shows that the agency spent $5,738,000 during fiscal 1966 for social science research. See U.S. Congress, House, Committee on Government Operations, Staff Report, *The Use of Social Research in Federal Domestic Programs*, Part I, 90th Cong., 1st Sess., April, 1967, p. 5.

5. U.S. National Aeronautics and Space Administration, *A Study of NASA University Programs*, pp. 2–4 and 22–34.

6. Adapted from the remarks of Theodore Suranyi-Unger, Jr., "The Role of Private Research in Federal Statistical Programs," Eleventh Annual Meeting of the Federal Statistics Users Conference, Washington, D.C., October 19–20, 1967.

7. Charles Y. Glock, "Applied Social Research: Some Conditions Affecting Its Utilization," in *Case Studies in Bringing Behavioral Science into Use*, Part I (Stanford, Calif.: Institute for Communications Research, Stanford University, 1961), p. 7.

8. Fred L. Zimmerman, *Wall Street Journal*, January 9, 1967; reprinted in the U.S. Congress, House, Committee on Government Operations, *The Use of Social Research in Federal Domestic Programs*, Part IV, pp. 599–603.

9. National Science Foundation, *Knowledge in Action: Improving the Nation's Use of the Social Sciences* (Washington, D.C.: U.S. Government Printing Office, 1969), pp. 15–20.

10. The Arms Control and Disarmament Agency publishes a quarterly bibliography, with abstracts and annotations of the current literature on the subject of disarmament. For a listing of projects sponsored by that agency see, U.S. Arms Control and Disarmament

Agency, *Research Sponsored by ACDA on the Economic Impact of Defense and Disarmament in the United States*, June 30, 1968. (mimeographed)

11. For a more detailed discussion of the research sponsored by these agencies and for descriptions of particular research projects, see Committee on the Economic Impact of Defense and Disarmament, Gardner Ackley, Chairman, *Status of Projects and Proposals for Information and Research Relating to the Economic Impact of Defense and Disarmament* (Executive Office Building, Washington, D.C., September 22, 1967). (mimeographed)

12. An issue of *Business Week* featured a lively written article on the defects of Department of Defense documents used by economists to trace the effects of spending for the Vietnam war on the economy. See "The Fairy Tale Figures on Defense," *Business Week*, October 29, 1966.

13. NASA let grants to two institutions for an evaluation of its subcontractor reporting system, Each of the reports is thorough. Research on one covers a $4\frac{1}{2}$-year span. John T. Burgess, *An Economic Analysis of Research and Development Procurement: The NASA Subcontract Program* (St. Louis, Mo.: Department of Economics, Washington University, July 1967). Working Paper 6708. Marvin Hoffenberg, *Analysis of NASA Postcard Subcontract Data* (Los Angeles, Calif.: Institute of Government and Public Affairs, University of California, December, 1966). (mimeographed)

14. Roger F. Riefler and Paul B. Downing, "Regional Effect of Defense Effort on Employment," *Montly Labor Review*, 91, 97 (July 1968), pp. 1–8.

15. The Department of Defense estimates that its procurement generated 2.8 million industrial jobs in June 1966. Vernon M. Buehler, "Economic Impact of Defense Programs," U.S. Bureau of the Budget, *Statistical Reporter* (July, 1967), p. 2. Contractor employment on NASA programs was 360,000 at that time.

16. Data on the cumulative costs of the Manhattan Project are published in Richard G. Hewlett and Oscar E. Anderson, *The New World 1939–1946, A History of the United States Atomic Energy Commission* (University Park, Pa.: Pennsylvania State University Press, 1962), pp. 723–24. The implicit price deflator for federal purchases of goods and services was used to put AEC and NASA purchases in real terms.

17. Gerhard Colm, *The Impact of the United States Civilian Space Program on the U.S. Economy*, a Report by the National Planning Association, Washington, D.C., July, 1966. (mimeographed)

18. Footnote 5 in Chapter 6 gives a brief description of the multiplier effect.

19. NASA let a contract to the Department of Economics at Indiana University in the summer of 1968 to fit NASA's expenditures into the U.S. Department of Commerce's recently completed 1963 national input-output table.

20. U.S. Department of Defense, Office of the Secretary, *Military Prime Contract Awards and Subcontract Payments of Commitments* (Washington, D.C.: Department of Defense, July, 1965–June, 1966).

6 Performance: The Local Impact of the Space Program

Historically, the aerospace industry has had a large number of firms located in California, Massachusetts, and New York. Because of the existing structure of the industry, a large number of NASA's contractors are also located in the technology-based complexes in those states. As indicated in Chapter 2, the space program encouraged the development of several new technology-based complexes in the South. This chapter analyzes the economic effects of the space program on the communities surrounding the five manned space flight field centers. Although the facilities are government-owned and contractor-operated, the Michoud Assembly Facility in New Orleans and the Mississippi Test Facility in Hancock County are called space centers for convenience. Because so many NASA contractors perform work in plants located in Los Angeles, part of the chapter examines the impact of the space program on that metropolitan area.

The main purpose of this chapter is to indicate relationships between the size, structure, and growth of communities surrounding major space installations and the magnitude of the impact of the space program on those communities. The findings have wider applications; they can provide insight into the matter of changing resource allocation resulting from other government programs in other communities.[1] Another purpose of the chapter is to project the future contribution of the space program to the economies of the six localities that have been most affected by NASA.

HISTORICAL PERSPECTIVE

Although communities in almost every state have in one way or another participated in the space program, the impact on them has not been uniform. With the exception of the marked similarities between the Cape Kennedy and Huntsville[2] areas, this is true even of the effects of the program on the localities surrounding the space centers and Los Angeles. The employment impact of the space program can be put in terms of a spectrum. Table 6–1 gives NASA employment as a percentage of total employment for the six geographic locations in June 1966, shortly after the manpower peak of the space program. At one end of the spectrum is employment at the Mississippi Test Facility. Space employment, civil service and contractor, comprised about 63 percent of total employment in Hancock County, Mississippi. At the other end of the spectrum is employment at the Manned Space Flight Center in Houston. Even at its peak, space employment was less than 2 percent of the Houston labor force.

Hidden behind the percentages is the more diverse role played by the space program on the economies of these communities. Because of NASA, high rates of employment growth were sustained in Brevard County, Florida and in Huntsville, Alabama. After decades of virtually no growth, employment in Hancock County doubled after the Mississippi

TABLE 6–1: NASA EMPLOYMENT AND TOTAL EMPLOYMENT, JUNE 1966, IN SELECTED AREAS IN WHICH NASA EMPLOYMENT IS CONCENTRATED

LOCATION	TOTAL EMPLOYMENT	NASA[a] EMPLOYMENT	NASA EMPLOYMENT AS A PERCENTAGE OF TOTAL EMPLOYMENT
Kennedy Space Center, Brevard County, Fla.	96,400	20,000	20.7
Marshall Space Flight Center, Huntsville, Ala.	80,700	23,700	29.4
Mississippi Test Facility, Hancock County, Miss.	8,300	5,200	62.7
Manned Spacecraft Center, Houston, Tex.	681,400	11,400	1.7
Michoud Assembly Facility, New Orleans, La.	393,500	10,800	2.7
Various contractor plants, Los Angeles-Long Beach, Calif.	2,600,000	150,000[b]	5.8

[a] Includes NASA civil service and NASA contractor employment.

[b] Estimated.

Sources: National Aeronautics and Space Administration, U.S. Department of Labor, Florida State Employment Service, and Mississippi State Employment Commission.

Test Facility began its operations. Although space employment is small compared with total employment in New Orleans, it was sufficiently important to reverse a five-year employment slump in that area. In Los Angeles, the principal role played by the manpower build-up on the space program was that of a stabilizing force in the manufacturing sector of the economy. In Houston, it is correct to state that the space program has been more a point of civic pride than a quantifiable increment in the city's dynamic economy. But even in Houston, NASA brought to the area a new kind of research and development complex.

The percentages displayed in Table 6–1, which show the relative importance of space employment, are based only on direct employment created by NASA and by NASA contractors. The data do not take into account secondary changes in the service industries of these localities via multiplier effects.

Comparative Rates of Employment Growth

The Office of Business Economics, United States Department of Commerce, prepared a series that provides information about employment growth, or stagnation, for all counties in the United States between 1940 and 1960.[3] The data provide a basis for comparing employment growth in those counties most heavily committed to the space program with employment change witnessed by all other counties in the United States. On the basis of changes in the level, the industry structure, and the regional growth of employment, the Commerce Department series assigns a rank of between 1 and 8 to each county in the country. The octant scale goes from strong employment gain (1) to strong employment loss (8). Of the six counties, the relative employment growth in Houston (Harris County) and Los Angeles warranted the rank of 1 for each of the areas during the 1950 to 1960 period. This means that during the decade of the fifties the employment growth and industrial development of Houston and Los Angeles were at least as great as those of the most rapidly growing counties in the country.

In contrast, Hancock County, assigned a rank of 6, suffered a moderate employment loss during the period. Employment change in Pearl River County, Mississippi was one of strong loss (8). Part of the buffer zone surrounding the Mississippi Test Facility is located in Pearl River County. As measured by employment, the economy of Orleans Parish, Louisiana also was relatively stagnant between 1950 and 1960. On the octant scale, Orleans Parish ranked 5. Compared with other counties in the country, both Brevard County and Huntsville (Madison County) enjoyed moderate employment gains during the decade of the fifties. Each of the counties received a rank of 3.

THE CAPE KENNEDY AND HUNTSVILLE AREAS

Because of the similarities, in terms of both the pre-NASA economic environment and the effects of the space program, it is convenient as well as useful to discuss the changes that have taken place in the communities surrounding the Kennedy Space Center and the Marshall Space Flight Center together. As indicated in Chapter 2, the John F. Kennedy Space Center was established in 1961 as the launch site for the manned lunar landing program. The Center, which occupies about 88,000 acres in Brevard County, Florida is almost midway between Jacksonville and Miami.

After negotiations with the Department of Defense, approximately 5,000 professional, administrative, and skilled specialists were transferred from the Army's Ballistic Missile Command to the George C. Marshall Space Flight Center of the National Aeronautics and Space Administration in July 1960. As Chapter 2 also indicated, research and development for the large launch vehicles used on manned space flights was assigned as the main mission of the Marshall Space Center in Huntsville, Alabama.

Employment Growth Patterns

Growth in the level of employment and changes in the composition of employment reflect, in part, economic growth. Before World War II, the economy of Brevard County was dominated by minor agricultural activities, such as citrus production, cattle raising, fishing, and truck crop farming. As late as 1950, about 15 percent of the workers in Brevard County were dependent on agriculture for employment. The economy of Huntsville also was dependent upon agriculture, as well as on a depressed textile industry. Since that time, the economies of these areas have become dependent on the federal government.

During the 1950's the Department of Defense was the main source of employment growth in these areas. In 1947, the World War II Banana River Naval Air Station was reactivated, and in 1950, became Patrick Air Force Base, Florida. It now serves as headquarters for DOD's missile launch area known as the Cape Kennedy Air Force Station. In 1949, the U.S. Army Ordnance Corps entered the missile era when it reactivated Redstone Arsenal in Huntsville, Alabama as a center for rocket research and development. The impact of the employment buildup by the Department of Defense in these localities can be seen from the data in Table 6–2. During the 1950's the average annual rate of employment growth in Huntsville was double the growth rate witnessed during the 1940's (Items 8 and 9 in the table). In Brevard County, the rate rose to a high 20 percent per annum during the 1950's. That compares with a growth rate of about 4.0 percent in the preceding decade.[4] There is little doubt that NASA was responsible for employment growth after 1960. Civil service and NASA contractor employment amounted to about 20,000 at the Kennedy Space Center and to about 18,600 at the Marshall Space Flight Center in June 1966 (Item 2). As the data show, employment on Department of Defense programs in these areas was relatively stable between 1960 and 1966 (Item 3). One way to look at the impact of federally generated employment is to estimate what employment growth might have been without the Department of Defense and without NASA. To make such an estimate, it is necessary to assume that there would not have been compensating employment growth from alternative sources. Items 5 through 7 in Table 6–2 present the results of an estimate of what employment in Brevard County and Huntsville might have been in the absence of DOD and NASA civil service and contractor employment. For example, without the direct employment buildup on the space pro-

TABLE 6–2: CONTRIBUTION OF THE FEDERAL GOVERNMENT TO EMPLOYMENT GROWTH IN HUNTSVILLE, ALABAMA, AND BREVARD COUNTY, FLORIDA, 1960–1969

ITEM	CHANGE IN EMPLOYMENT	BREVARD COUNTY, FLORIDA	HUNTSVILLE, ALABAMA
1	Increase in total employment 1960 to 1966	54,100	38,600
2	Increase in NASA employment 1960 to 1966	20,000	18,600[a]
3	Increase in DOD employment 1960 to 1966	1,100	1,700
4	Actual average annual rate of growth: 1960 to 1966	15.0	11.5
5	Estimated growth rate without NASA direct: 1960 to 1966	10.4	6.7
6	Estimated growth rate without NASA direct and induced: 1960 to 1966[b]	3.5	2.6
7	Estimated growth rate without NASA and DOD direct and induced: 1960–1966[b]	2.7	1.8
8	Actual average annual rate of growth: 1940's	4.0	4.6
9	Actual average annual rate of growth: 1950's	20.0	9.2
10	Actual average annual rate of growth: 1966 to 1967	+1.7	−3.6
11	Actual average annual rate of growth: 1967 to 1968	+2.7	−2.4
12	Actual average annual rate of growth: 1968 to 1969[c]	−1.0	−0.7

[a] Excludes 5,300 transferred from the Department of Defense.

[b] Estimate based on an employment multiplier of 2.22 for Brevard County and of 1.66 for Huntsville, Alabama.

[c] Change between March 1968 and March 1969.

Sources: Data for the period between 1940 and 1966, Mary A. Holman and Ronald M. Konkel, "Manned Space Flight and Employment in the South," *Monthly Labor Review* (March, 1968), p. 32. Data for the 1967 to 1968 period, U.S. Department of Labor, Bureau of Labor Statistics, and the Florida State Employment Commission.

gram, the average annual rate of employment growth would have been about half of the actual rate in Huntsville between 1960 and 1966, and about two-thirds of the actual rate in Brevard County.

Simple employment multipliers were derived to estimate the indirect or induced change in employment in the service industries.[5] The coefficient for the multiplier was 2.22 for Brevard County and 1.66 for Huntsville.[6] The service industries, in general, have grown less rapidly in Alabama than in Florida.[7] Almost 80 percent of employment growth in each of the areas between 1960 and 1966 including induced employment resulted from the rapid buildup of the space program. To be more specific, that growth resulted almost entirely from NASA's Manned Space Flight Program. Total Department of Defense and NASA employment, direct and indirect, accounted for virtually all employment growth in these areas between 1960 and 1966.

The last three rows in the table show what happened to employment in the two counties after 1966. Between mid-1966 and early 1969, NASA civil service and NASA contractor employment at the Marshall Space Flight Center dropped from about 23,700 to about 12,700. The largest decrease occurred between 1966 and 1968, when about 8,800 employees separated from NASA projects located in the Huntsville area. As the data show, total employment in Huntsville dropped by 3.6 percent in 1967, by 2.4 percent in 1968, but by less than 1.0 percent in 1969. Although many had hoped that employment on projects for the anti-ballistic missile system would absorb a large number of those separating from space projects in Huntsville, defense employment remained stable, at approximately 22,000 between 1966 and 1968. A relatively small number of former NASA employees, of course, did shift to defense projects in the Huntsville area. A larger transfer, however, took place within the Space Agency. Many NASA employees located in Huntsville in mid-1966 transferred to NASA projects located in the Cape Kennedy area. NASA employment at the Cape continued to rise until mid-1968. At that time, NASA employment reached a peak of about 23,500 people. Manpower requirements for the manned lunar launch operations continued to rise after requirements for the research and development phase of the program were reduced. Total employment in Brevard County did not begin to decline until early 1969.

Despite the decrease in NASA employment in each of these areas, the rate of unemployment has not risen by much. This results from the fact that those working in the aerospace industry tend to be mobile—shifting from one geographic location to another in response to changes in employment demand. To be sure, this is not necessarily true for those working in the service sectors of the economies of these areas. In Huntsville, the average annual rate of unemployment was 3.5 percent in 1968, compared with 2.8 percent in 1966, and 5.0 percent in 1960. Between March 1968 and March 1969 the rate of unemployment in Brevard County rose from 2.6 percent to 3.2 percent. In 1966, the average annual rate in Brevard County was a low 1.9 percent; it was 3.6 percent in 1960.[8]

Change in the Structure of Employment

Marked changes in the structure of employment at Cape Kennedy and in Huntsville mirror the role played by the federal government in these areas. In Brevard County, employment in the manufacturing sector rose from 5.0 percent to 20.0 percent of the labor force between 1950 and 1968. In Huntsville, the shift in employment was away from agriculture and also away from textile manufacturing. The manufacturing industries that experienced employment growth in Brevard County and in Huntsville were: ordnance and guided missiles, instruments, fabricated metals, industrial controls, communications equipment, and electronic components. All of these industries manufacture components and hardware for defense and for aerospace programs. This kind of manufacturing employment is not likely to be maintained without sales to the federal government.

Between December 1965 and December 1968, employment in the ordnance and guided missiles industry in Huntsville dropped by 45 percent. At the peak of NASA's employment buildup in Huntsville, the ordnance and guided missiles industry accounted for almost two-thirds of total employment in the durable goods manufacturing industry in that area.

Since 1966, the greatest change in the structure of employment in Brevard County has been the decline in construction employment. In 1966, contract construction accounted for 11 percent of the county's labor force. In March 1969, 4.0 percent of the labor force was employed on construction projects. Tourism, especially high when the manned lunar launches occur, is the newest industry in Brevard County. The Brevard County Economic Development Council estimates that some 3 million visitors came to the Cape in 1970.

Employment Migration Patterns

Between 1960 and 1965, at least 45 percent of the labor force in Brevard County and 30 percent in Huntsville resulted from in-migration into the area. The largest number seeking employment in both areas came from out-of-state locations. Table 6–3 shows that only 21 percent of those employed in Brevard County in 1965 were also employed in the county in 1960. For Madison County the percentage was only slightly higher—25 percent. Although not shown in the data, a proportionately larger number of women have entered the labor force in each of these areas. Of those employed in each of the counties in 1965 but not in 1960, slightly more than 50 percent were women. A rate of 50 percent for new female entrants into the labor force is considerably above the national average of about one-third. The data do not distinguish between females who were primary workers and those who were secondary. However, because of the tight labor market that existed in these areas between 1960 and 1965, it is reasonable to assume that a large number of the women were secondary workers providing transitory income. Rising wage rates, which usually accompany increases in the demand for labor, can encourage females to substitute paid employment for household production or possible leisure.[9]

TABLE 6–3: EMPLOYMENT MIGRATION PATTERNS FOR BREVARD COUNTY, FLORIDA; MADISON COUNTY, ALABAMA; AND HANCOCK COUNTY, MISSISSIPPI, 1960 AND 1965 (*PERCENTAGE DISTRIBUTION*)

		PLACE OF EMPLOYMENT IN 1960			
		SAME COUNTY	SAME STATE	ELSEWHERE[a]	NOT EMPLOYED
Employed in Brevard County in 1965	100.0	20.8	18.2	28.4	32.6
Employed in Madison County in 1965	100.0	24.7	13.1	17.6	44.6
Employed in Hancock County in 1965	100.0	22.4	10.2	36.8	30.6

[a] Includes those in military service in 1960.

Source: U.S. Department of Commerce, Regional Economics Information System, Office of Business Economics.

Increases in Personal Income

The growth of total personal income in a community depends mainly upon increases in employment and increases in wages, salaries, and other kinds of remuneration. Changes in the level of personal income can reflect the economic growth and prosperity of a community. The rate of growth in total personal income between 1959 and 1966 in both Brevard County and Huntsville was more than double the national rate. In 1959, total personal income in Brevard County was $240 million and in Huntsville it was $280 million. By 1966, total personal income in Brevard County had soared to $694 million and to $590 million in Huntsville. Nationally, the level of total personal income rose from $384 billion to $587 billion, or just over 6 percent per annum. In Brevard County, the average rate of increase for the seven-year period was about 15 percent, whereas the rate in Huntsville was slightly less than 12 percent per annum. During the period, the relative share of national income earned in Brevard County rose from 0.06 percent to 0.12 percent. The relative share earned in Huntsville increased from 0.07 percent to 0.10 percent.

Because of the earlier slowdown of space spending in Huntsville, the upward spiral in the growth of total personal income stopped between 1966 and 1967. Total personal income rose by less than 1 percent during that year. If inflation were taken into account, total personal income in Huntsville actually was less in 1967 than it was in 1966. Total personal income in Brevard County, however, continued to rise at a rate in excess of 15 percent, throughout 1966, and by 1967 total personal income amounted to about $800 million. Between 1966 and 1967, the rate of growth in total personal income in Brevard County continued at a rate slightly more than twice the national rate of increase.[10]

Per capita personal income is a more useful index of economic growth and economic welfare because it reflects changes in the population. There can be little doubt that employment and income created by the federal government have helped to raise per capita personal income in these localities. For example, the average annual salary of civil service personnel employed at the Marshall Space Flight Center was $10,850 in December 1966. At the time, the average annual salary paid by NASA's prime contractors in the Huntsville area was about $10,000.

Despite the relatively high incomes earned in the area by some employees, per capita personal income in the Huntsville area remains below the national average. The difference, however, narrowed between 1950 and 1967. In addition, Huntsville's per capita personal income in 1967 was about 20 percent higher than the Alabama average. In 1950, per capita personal income in Huntsville was about $800, or almost 50 percent less than the national average of $1,500. As a result of defense spending in the 1950's, per capita disposable income in Huntsville amounted to about $1,600 in 1960 and was almost 75 percent of the national figure of $2,200. By 1967, per capita personal income in Huntsville was $2,600—about 80 percent of the national average of $3,200.[11]

In contrast, per capita personal income in Brevard County has been close to the national average since 1950. When it reached $1,400 in 1950, it was only $100 less than the national average. In 1967, when it had risen to $3,300, it was $100 more than the national average. Per capita personal income in Brevard County is considerably higher than the Florida state average. In 1967, per capita personal income in Florida amounted to $2,600, or $700 less than the Brevard County average.

Changes in the Distribution of Income

With the prosperity base established by the Department of Defense in the 1950's, NASA continued to provide relatively high incomes and opportunity for employment advancement for those working in Huntsville and Brevard County during the 1960's. Because of the secondary income and employment effects, these benefits were not limited to NASA employees or to employees of NASA contractors. Between 1960 and 1967, the space program contributed to a marked change in the distribution of household or family incom.: in these areas. The change in the distribution of income is given in Table 6–4. The data for Brevard County and for Huntsville are estimates, projections of U.S. Bureau of the Census benchmark figures, made by Sales Management, Incorporated. As a service to its clients, that organization annually publishes statistical information about the economic environment of all counties and major cities in the United States. As late as 1960, the income inequality in these areas was considerably greater than that in the United States in general. In 1960, more than 25 percent of the consuming units in both Brevard County and Huntsville received an annual income of less than $2,500. Although not directly comparable with the Sales Management data, Bu-

TABLE 6–4: CHANGES IN THE DISTRIBUTION OF INCOME IN BREVARD COUNTY, FLORIDA, AND MADISON COUNTY, ALABAMA, COMPARED WITH THE SOUTH AND THE UNITED STATES (*PERCENTAGE DISTRIBUTION*)

QUINTILE	1960	1965	1967
Less than $2,500			
Brevard	29	19	17[c]
Huntsville	27	16	16[c]
The South	31[a]	25[a]	21[a]
United States	22[b]	16[b]	13[b]
$2,500 to $3,999			
Brevard	26	20	12[c]
Huntsville	28	16	11[c]
The South	22[a]	19[a]	17[a]
United States	10[b]	8[b]	7[b]
$4,000 to $6,999			
Brevard	30	26	21[c]
Huntsville	31	26	19[c]
The South	20[a]	19[a]	19[a]
United States	34[b]	27[b]	25[b]
$7,000 to $9,999			
Brevard	10	14	15[c]
Huntsville	9	16	14[c]
The South	16[a]	20[a]	21[a]
United States	20[b]	24[b]	25[b]
Over $10,000			
Brevard	5	21	35[c]
Huntsville	5	26	40[c]
The South	11[a]	17[a]	22[a]
United States	14[b]	25[b]	30[b]

[a] Less than $3,000; $3,000–$4,999; $5,000–$6,999; $7,000 to $9,999; and over $10,000.
[b] Less than $3,000; $3,000–$3,999; $4,000–$6,999; $7,000 to $9,999; and over $10,000.
[c] Less than $3,000; $3,000 to $4,999; $5,000 to $7,999; $8,000 to $9,999; over $10,000.

Sources: Sales Management Incorporated, *Survey of Buying Power* (June 1961, June 1966, and June 1968); and U.S. Bureau of the Census, *Statistical Abstract of the United States: 1967, 1968, and 1969* (Washington, D.C.: U.S. Government Printing Office).

reau of the Census data show that the median family income in the United States in 1960 was $5,600. Between 1960 and 1965 the percentage of households in the lowest quintile dropped by 10 percentage points in Brevard County and by 11 percentage points in Huntsville. During the same period, the percentage of consuming units earning $10,000 or more rose by 16 percentage points in Brevard County and by 21 percentage points in Huntsville. As Table 6–4 shows, the result of that shift in the distribution of income made the communities surrounding Cape Kennedy and the Marshall Space Flight Center much more similar to the national distribution of income than to the distribution existing in the South. In 1967, the median income in the United States approached the $8,000 level. In Brevard County about 50 percent of the consuming units enjoyed an income of that amount or more. In Huntsville, 54 percent of all households received at least that amount. As the data show, by 1967 the distribution of income in Brevard County and in Huntsville was slightly more skewed in favor of the highest quintile than the national distribution of income.

It might be surprising to many that the percentage of consuming units earning less than $3,000 in Brevard County and in Huntsville was still a few percentage points higher than the national average in 1967. When compared with other southern cities of approximately the same size, however, the localities surrounding these two NASA centers have much less hardcore poverty. For example, between 25 percent and 27 percent of all households in Augusta, Georgia; Jackson, Mississippi; Lakeland, Florida; Montgomery, Alabama; and Savannah, Georgia, lived on a family income of less than $3,000 in 1967. In the Bristol-Kingsport–Johnson City, Tennessee area; in Macon, Georgia; and Pensacola, Florida, between 22 percent and 24 percent of all families earned less than $3,000 in 1967.[12]

Income Location Quotients

Income location quotients are one way to show how the structure of a locality differs from that of other localities and from the nation as a whole. Table 6–5 displays coefficients of specialization by industrial sector in Brevard County and in Huntsville, Alabama. A coefficient of 1.00 indicates that the relative importance of income generated from a given sector is identical with the national norm. The data confirm intuitive notions about the relative importance of the federal government to the economies of these areas. They also show the effect of recent changes (1965–1967) in the structure of federal procurement in these areas. Since 1950, each of the counties has been heavily dependent on federal civilian and military payrolls. Of the two areas, Huntsville is the more dependent. Despite the large increase in manufacturing in each of the areas, income created from

TABLE 6-5: LOCATION QUOTIENTS FOR TOTAL EARNINGS IN BREVARD COUNTY, FLORIDA, AND HUNTSVILLE, ALABAMA, SELECTED YEARS

INDUSTRIAL	1950		1962		1965		1967	
SECTOR	BREVARD	HUNTSVILLE	BREVARD	HUNTSVILLE	BREVARD	HUNTSVILLE	BREVARD	HUNTSVILLE
Total earnings	1.00	1.00	1.00	1.00	1.00	1.00	1.00	1.00
Farm	1.10	3.20	.28	1.73	.27	1.16	.18	1.15
Federal civilian	1.53	1.73	1.36	5.50	1.23	3.96	1.11	4.15
Military	2.88	1.50	2.52	2.47	2.24	1.83	1.25	2.58
Manufacturing	.40	.54	.80	.54	.70	.72	.96	.61
Contract construction	1.48	.68	1.16	.88	2.55	.81	1.10	.57
Transportation, communications, and public utilities	.35	.50	.26	.34	.29	.35	.26	.31
Wholesale and retail trade	1.19	1.00	.63	.68	.65	.64	.54	.67
Services	2.06	1.09	2.61	1.50	2.37	2.07	2.80	2.12

Source: U.S. Department of Commerce, Regional Economics Information System, Office of Business Economics.

that industrial sector remains below the national norm. Each of the areas has a larger than average service sector. The high specialization coefficients in the service sector result in large part from income generated by NASA's service support contractors who provide scientific and engineering services. The impact of NASA's large construction activity at Cape Kennedy is shown clearly in the data. The coefficient rose from slightly above the national norm in 1962 to 2.55 in 1965. In 1967, the coefficient once again approximated the national average. Because of the mission of the Marshall Space Flight Center, the Huntsville area did not witness such massive expenditures for the construction of facilities. Finally, the decline in the relative importance of agriculture in Brevard County can be gleaned from the data. Between 1950 and 1967, the coefficient dropped from about the national norm to a low 0.18 in Brevard County.

Other Economic and Social Changes Accompanying the Space Program

The opening of new employment opportunities on a large scale had far-reaching effects on the social and economic structures of Brevard County and Huntsville. The areas have enjoyed prestige and economic prosperity, although the communities have had to accommodate population growth. Table 6–6 presents selected indicators that reflect some of the economic and social changes that occurred in these areas.

TABLE 6–6: SELECTED INDICATORS OF ECONOMIC GROWTH AND SOCIAL CHANGE IN BREVARD COUNTY, FLORIDA, AND HUNTSVILLE, ALABAMA, 1950–1967

	1950	1960	1965	1967
Brevard County, Fla.				
Population	23,700	111,400	224,500	247,300
Education:				
Total public school enrollment	3,900	23,000	48,200	57,100
Percent federally related	10.0	44.7	52.4	51.8
Number of teachers	140	900	2,000	2,300
Number of classrooms	117	650	1,500	1,900
Building permits (millions of dollars)	N.A.	$37.2	$91.0	$70.2
Retail sales (millions of dollars)	$21.2	$125.4	$291.3	$321.7
Huntsville, Ala.				
Population	16,400	72,400	143,700	148,400
Education:				
Total public school enrollment	2,700	16,300	34,600	36,900
Number of teachers	100	560	1,200	1,230
Number of classrooms	96	570	1,000	1,200
Building permits (millions of dollars)	$5.1	$19.4	$71.5	$38.0
Retail sales (millions of dollars)	$34.2	$111.3	$207.8	$193.4

Sources: Institute for Social Research, Florida State University, *Urban Data Abstract: Brevard 1968* (Tallahassee, Fla.: Florida State University, 1968); Sales Management, Incorporated, *Survey of Consumer Buying Power,* various issues; Brevard County Planning Department; Brevard County Public School System; Madison County, Alabama, Chamber of Commerce; and Office of Huntsville City Schools.

Population

Between 1940 and the 1950 Census, the population of Brevard County grew at an average annual rate of 3.5 percent; that of Huntsville grew at a rate of 2.3 percent during the decade of the 1940's. The average annual rate of population growth in Brevard County was almost 20 percent between 1950 and 1960. The population of Huntsville also grew at the spectacular rate of 16 percent each year during the same period. With the advent of NASA and its Manned Space Flight Program, the population of these areas doubled during the decade, although the average annual rate of growth dropped by a few percentage points. In Brevard County the population grew at the annual rate of 5 percent between 1965 and 1968, reflecting the steady growth in employment in the county. The earlier retrenchment of NASA employment in Huntsville dampened population growth. Between 1965 and 1967, population growth in Huntsville dropped to less than 1.5 percent per annum. That growth rate, however, was still above the national average.

Education

An increase in the school-age population usually is an integral part of any large in-migration of people. Since 1950, the public school system of Brevard County has had to provide facilities and faculties for a school-age population that doubled about every five years. The greatest increase in county school enrollment took place shortly after the inception of the Manned Space Flight Program at Cape Kennedy; between the school years 1962–1963 and 1963–1964, almost 7,000 additional children enrolled in the county's public school system. Students whose parents had federally related employment comprised over half of that increase.

To serve the increasing school population, Brevard County greatly increased its school facilities. Total expenditures for primary and secondary education amounted to $10 million in the 1960–1961 school year and to about $38 million in the 1966–1967 school year. This compares with about $760,000 in 1950–1951. Because of the county's rapid building program, the number of students per classroom has averaged about 30 since the early 1960's.[13] Funds for the operation of Brevard County's school system came from federal, state, and local government sources. Although the federal government's contribution has averaged about 15 percent during the last 10 years, the county financed about 40 percent of the total by a property tax.

In Huntsville, school enrollment in 1965 was more than twice the 1960 level. To accommodate that increase, the Huntsville city school system also doubled its teaching staff and its classroom facilities. Although fluctuating yearly, the total capital outlay by the Huntsville city school system has averaged about $3 million since 1960. That outlay reached a high of about $6.4 million in 1966. The school construction program has been sufficiently large in Huntsville, as in Brevard County, to keep the average number of students per classroom at a fairly constant level of 30 since 1950. Similarly, the Huntsville school system has managed to keep the student-teacher rate at approximately 30.

Construction Activity

Another barometer of economic activity and growth is the number and value of building permits issued each year. Data on building permits in Brevard County include residential units, nonresidential and commercial construction, alterations and additions, and public buildings. No federal government construction data are reported. After NASA announced that it would establish a facility in Brevard County, building activity began to increase markedly. The value of all building permits reached a peak of $122 million in 1963. Between 1963 and 1967, about 15,000 new homes were built in Brevard County. During the same period, about 8,000 new apartments were constructed. Almost one-fourth of the total number of new housing units were built in 1964.

The housing market in Brevard County changed from a seller's market to a buyer's market in 1968. A survey conducted by the Federal Housing Administration indicated that 315 of the 1,712 houses constructed in 1968 remained unsold on January 1, 1969. Of the 315 unsold houses, 102 had been completed for three to six months and 109 had been completed six to twelve months earlier.

The majority of the 181 motels in Brevard County are less than ten years old. Although the total motel capacity of some 5,200 units is not sufficient to accommodate demand during the manned space launches, most of the motels operate at less than 50 percent capacity during normal periods.

A pronounced change in the urbanization of Huntsville, as well as of Brevard County, has been the rapid increase in the number of apartment units. In 1963, permits for multi-dwelling units in Huntsville comprised about 45 percent of the total number issued. This compares with about 6 percent in 1960. As the data in Table 6–6 show, the value of building permits issued in Huntsville in 1967 was about 45 percent below the 1965 level. The value of building permits in Huntsville reached a high of $81 million in 1963. The contraction of the space program was not the only cause of the decline in construction activity in Huntsville and in Brevard County. Part of the decline must be attributed to the national policy of tight money and high rates of interest after the escalation of the Vietnam War.

Retail Sales

By creating new employment and generating larger incomes, the Department of Defense in the 1950's and NASA in the 1960's had a catalytic effect on the service industries in these areas. In Brevard County, retail sales more than doubled between 1950 and 1955, increasing at an average annual rate of about 18 percent during the period. With a substantially larger base in 1960, retail sales again more than doubled between 1960 and 1965. During the later period, the average annual rate of increase continued at the high rate of about 18 percent. Although retail sales continued to rise between 1965 and 1967, the rate of increase dropped to about 5 percent per annum in Brevard County.

The increase in retail sales in Huntsville was slightly greater between 1960 and 1965 than between 1950 and 1955. The two periods approximate the time when the Department of Defense expanded its program in Huntsville (1950–1955) and when NASA's Manned Space Flight Program accelerated (1960–1965). Retail sales in Huntsville increased by about 90 percent between 1960 and 1965, compared with about 80 percent 10 years earlier. Merchants in Huntsville began to witness the decrease in space spending in 1966 when retail sales began to drop. By 1967, retail sales were $14 million below the level reached two years earlier.

Huntsville Under the Model Cities Program

In November 1967, Huntsville was selected as one of the first cities to receive a planning grant from the U.S. Department of Housing and Urban Development under the Model Cities Program. Authorized by Congress, with the Demonstration Cities and Metropolitan Development Act of 1966, the Model Cities Program provides technical and financial assistance to help communities develop and implement programs to improve the social, economic, and physical environment of blighted neighborhoods. In January 1969, Huntsville received approval of its five-year Model Cities Program and funds from four federal agencies—Department of Housing and Urban Development; Department of Health, Education and Welfare; the Office of Economic Opportunity; and the Department of Labor—to carry out the program for the first year. Huntsville's model neighborhood has a population of about 13,000, about 65 percent of whom are nonwhite. The blighted neighborhood is one of the oldest sections of Huntsville, as it is in most other cities. The target area contains about 36 percent of the city's school dropouts, but only 12 percent of the school population. Crime and delinquency rates in the area are substantially above those for the rest of the city. The goals established for the program include: revision and upgrading of curriculum, teaching methods, and facilities; vocational and on-the-job employment training;

employment information services; establishment of a light industry park; expanded public transportation facilities; a housing counseling service; three multi-purpose health centers; and neighborhood parks and recreation facilities.[14] Probably as a direct spin-off of the space program, the city of Huntsville plans to establish a computerized management support system to administer the program.

COMMUNITIES SURROUNDING THE MISSISSIPPI FACILITY

The geographic isolation of the Mississippi Test Facility inevitably resulted in a wide dispersion of the primary and secondary effects generated by the space program. Because of this, the effects are somewhat more difficult to measure. A substantial number of those employed at the Facility commute or commuted from New Orleans, Gulfport, Biloxi, and other communities on the Gulf Coast. The Mississippi Test Facility (MTF) is bounded by Picayune (Pearl River County) in the north, Bay St. Louis (Hancock County) in the southwest, and Pass Christian (Hancock County) in the southeast.

Historical Employment Growth Patterns

During the twenty years between 1940 and 1960, total employment in Hancock County grew by only 1,400. Almost two-thirds of that employment growth took place during the decade of the forties. During the entire period, the structure of employment remained stable. Historically, the lumber and wood manufacturing industries dominated the economic life of Hancock County. Before the early 1960's, there was also some employment in water-based industries, such as boat building and repair. The pre-NASA employment base of Pearl River County was similar to that of Hancock County. Between 1940 and 1960, employment in Pearl River County grew by only 1,300.[15]

In 1961, the rate of unemployment was 15 percent in Hancock County and 10 percent in Pearl River County. At that time, Hancock County had the highest rate of unemployment of any county in Mississippi. The amount of disguised unemployment or under-employment existing at that time is not known, but substantial under-employment usually exists in under-developed rural economies, such as this area was before the early 1960's.

The Space Program and Employment Growth

The direct increase in employment resulting from the establishment of the Mississippi Test Facility occurred only in Hancock County. Between 1960 and 1965, total employment in that county doubled, rising from about 4,000 to 8,100. In contrast, employment rose by 4 percent in Pearl River County—from about 6,900 in 1960 to 7,200 in 1965.

In the early 1960's, the temporary construction work force that came into the area to build the Mississippi Test Facility was more important to employment growth than the employees hired by NASA's test and service support contractors. The data displayed in Table 6–7 show that the increase in civil service and NASA contractor employment was greater than the increase in total employment in Hancock County between 1960 and 1965. The explanation for this is that expanded employment opportunities created by the space program pulled workers away from other employment. Although there was an influx of a large number of highly skilled people to the area, data provided by NASA staff located at the Mississippi Test Facility indicate that about one-half of those employed at the Facility in 1966 were recruited from the local population. It can be conjectured that employment commuting patterns changed markedly after the establishment of NASA's Mississippi Test Facility. Bureau of the Census data show that in 1960, almost one-third of the Hancock County's working population were employed at jobs located in other counties. Almost all of those who had employment outside of the county commuted to New Orleans and to Biloxi, Mississippi. About one-fourth of the working population of Pearl River County also commuted to out-of-county employment in 1960.

TABLE 6–7: CHANGE IN NASA EMPLOYMENT AT THE MISSISSIPPI TEST FACILITY, 1962–1969

FISCAL YEAR	NASA CIVIL SERVICE	TEST AND SERVICE CONTRACTOR	CONSTRUCTION WORK FORCE	TOTAL
1962	0	0	0	0
1963	10	150	202	362
1964	120	500	1,760	2,380
1965	255	3,500	2,145	5,900
1966	215	3,700	1,292	5,207
1967	170	3,100	203	3,473
1968	143	2,853	33	3,029
1969	143	2,650	–	2,793

Source: NASA Staff, Mississippi Test Facility.

In view of the history of economic stagnation in Hancock and Pearl River Counties, it is clear that all employment growth since the early 1960's resulted from the space program. As the data in Table 6–8 show, the space program generated sufficient employment to increase the combined average annual rate of employment growth in Hancock and Pearl River Counties from the dismal rate of 0.6 percent in the 1950's to 6.5 percent between 1960 and 1966.

There can be little doubt that if the Mississippi Test Facility were

TABLE 6-8: CONTRIBUTION OF NASA-GENERATED EMPLOYMENT TO EMPLOYMENT GROWTH IN HANCOCK AND PEARL RIVER COUNTIES, MISSISSIPPI, 1960–1968

Increase in Total Employment: 1960 to 1966		5,000
Increase in NASA Employment: 1960 to 1966		3,900[a]
Decrease in NASA Employment: 1966 to 1968		−1,000[a]
Actual Average Annual Rate of Change:	1940's	+2.1
	1950's	+0.6
	1960 to 1966	+6.5
	1966 to 1968	−6.4

[a] Includes employees of NASA and employees of NASA contractors. The number does not include the temporary construction work force, which amounted to 1,300 in June 1966 and to 600 in December 1966.

Sources: U.S. Department of Commerce, Office of Business Economics, *Growth Patterns in Employment by County: 1940–1950 and 1950–1960,* Vol. 5: *Southeast* (Washington, D.C.: U.S. Government Printing Office, 1966); Mississippi State Employment Commission, and National Aeronautics and Space Administration.

closed, employment, as well as rates of unemployment, would revert to the patterns existing prior to the establishment of the Facility. Between 1966 and 1968, NASA civil service and service support contractor employment dropped by about 1,000. If employment resulting from construction activities is included, the decrease amounts to 2,000. The drop in NASA employment was not offset locally. In each of the years between 1966 and 1968, combined employment in Hancock and Pearl River Counties dropped by more than 6 percent. The decline in space employment at the Mississippi Test Facility is a combination of two forces operating within the space program. One is the reduced space budget, and the other is the culmination of test procedures before launch activities. A number of NASA contractor employees at the Mississippi Test Facility, like those at Huntsville, shifted to projects located at Cape Kennedy.

Employment Security

In Hancock County, the rate of unemployment dropped from 15 percent to 2 percent between 1960 and 1966. In Pearl River County the drop was from 10 to 4 percent between 1960 and 1965. The trend toward lower rates of unemployment reversed in Pearl River County between 1966 and 1967. During that year the rate of unemployment approached 6 percent. A rate of unemployment in excess of 5.9 percent of the labor force is defined by the U.S. Department of Labor as substantial unemployment. The rate of unemployment remained at a low 2 percent in Hancock County in 1967, unlike Pearl River County, where NASA spending generated induced or secondary employment rather than direct employment. This results from the fact that NASA employees and employees of NASA contractors separating from space projects at the facility left the area. Indigenous manpower resources, providing services, do not usually shift geographic locations rapidly.

Another measure of employment security is the percentage of the labor force covered by the federal-state employment insurance system. Workers who are not covered by the system include mainly the self-employed in agriculture and small business, as well as unpaid family workers. In Hancock County, the percentage of the labor force covered by employment insurance rose from 20 percent in 1960 to 62 percent in 1966. Because the employment impact of the Mississippi Test Facility on Pearl River County has been indirect, i.e., by small businesses providing services, the percentage change in insurance coverage was lower than in Hancock County. The percentage of the labor force covered by employment insurance in Pearl River County rose from about 35 percent in 1960 to 40 percent in 1966. In Hancock County, the percentage of the labor force covered by employment insurance dropped by about 5 percentage points between 1966 and 1967. This trend can be expected to continue as the structure of employment reverts toward its historical pattern.

Changes in Personal Income

It is not surprising that total personal income in these areas rose sharply between 1962 and 1966. In Hancock County, total personal income increased fivefold—rising from $14 million in 1962 to $71 million in 1966. Although the increase was not as dramatic in Pearl River County, total personal income there rose from $26 million in 1962 to $37 million in 1966. This compares with an increase of about $4.5 million in Hancock County during the entire decade of the 1950's, and of about $7.0 million in Pearl River County during the same period.[16] In Hancock County, increased employment and income were accompanied by a sizable drop in welfare payments. In 1960, welfare payments made to persons living in Hancock County totaled $188,000; these payments dropped to $134,000 in 1966.

The average annual salary paid to civil service personnel at the Mississippi Test Facility was about $10,000 in 1966. The average annual salary paid by MTF construction contractors was about $9,800 at the same time. Data collected by the Social Security Administration show that the median earned wage in Hancock County rose from $1,744 in 1960 to $4,757 in 1965. The highest wages earned in 1965 were in those industries directly dependent on the space program. For example, the median wage earned was $7,600 in contract construction and $9,300 in manufacturing in 1965. This contrasts with a median wage of $4,000 in the wholesale and retail trade industry in the same year.

Change in the Structure of Income

Location or specialization quotients indicate how the space program changed the structure of income in Hancock and Pearl River Counties. The story told by the location quotients displayed in Table 6–9 is that NASA's impact in Hancock County was direct, whereas its impact operated indirectly through income and employment multiplier effects in Pearl River County. Before the establishment of the Mississippi Test Facility, income from farming and state and local government activities accounted for a disproportionately large part of total income earned in Hancock County. By 1965, earnings from contract construction and from services (NASA's service support contractors) were much higher in Hancock County than the national norm. Between 1962 and 1965, farm income in Hancock County became relatively unimportant. To provide for public safety, five agricultural towns were removed to make way for the Mississippi Test Facility. The towns, with a combined rural population of 850 families, were Gainsville, Logtown, Napoleon, Santa Rosa, and Westonia. Agricultural income in the late 1960's continued to be more important to Pearl River County than to Hancock County and other counties throughout the nation. The location quotients show that income generated in Pearl River County from service industries, wholesale and retail trade, and contract construction rose from slightly below to slightly above the national norm between 1962 and 1967. Income from the federal civil service and from the military plays a small role in the economic bases of these counties, compared with those in other parts of the country.

The Distribution of Income

Although these sparsely populated counties in southeast Mississippi have prospered markedly since the inception of NASA's Mississippi Test Facility, per capita disposable income remains substantially below the national average. The percentage of all consuming units earning less than $3,000 is also much higher than the national average, as well as that in the South. To be sure, the percentage change in the lowest income quintile was larger in Hancock and Pearl River Counties than the national change. The result of this was to narrow the gap between these counties and the national and southern averages. In 1951, 55 percent of all consuming units in Hancock County and 52 percent of those in Pearl River County earned less than $3,000 a year. By 1967, 35 percent of all families in Hancock County and 32 percent in Pearl River County earned less than $3,000. In each of the areas, this amounted to a decrease of about 20 percentage points in six years. During the same period, the national change was 9 percentage points. The decrease for the South was 10 percentage points. Nevertheless, the fact that about one-third of the families

TABLE 6–9: LOCATION QUOTIENTS FOR TOTAL EARNINGS IN HANCOCK AND PEARL RIVER COUNTIES, MISSISSIPPI, FOR SELECTED YEARS

	1950		1962		1965		1967	
INDUSTRIAL SECTOR	HANCOCK	PEARL RIVER	HANCOCK	PEARL RIVER	HANCOCK	PEARL RIVER	HANCOCK	PEARL RIVER
Total earnings	1.00	1.00	1.00	1.00	1.00	1.00	1.00	1.00
Farm	1.81	2.09	1.84	2.42	.06	1.42	.11	2.38
Federal civilian	.65	.90	.79	.48	.70	.61	.70	.45
Military	.52	.48	.81	.45	.17	.46	.22	.38
Manufacturing	[a]	1.20	[a]	1.10	[a]	.89	[a]	.80
Contract construction	1.28	.50	1.09	.52	4.35	1.10	1.89	1.31
Transportation, communications, and public utilities	.91	.50	1.51	.49	.34	.55	.75	.46
Wholesale and retail trade	.94	.82	1.00	.88	.25	1.14	.30	1.01
Services	1.19	.83	1.44	.83	2.63	1.02	2.55	1.32

[a] Not available in order to avoid disclosure of data for individual reporting units.

Source: U.S. Department of Commerce, Regional Economics Information System, Office of Business Economics.

in these counties earned less than $3,000 in 1967 puts the counties well below the national average of about 13 percent for the below-$3,000 family income level.[17] See Table 6–4. For the highest quintile, an income in excess of $10,000, the relative change was less for Hancock and Pearl River Counties than that for the nation. In 1961, only 6 percent of all consuming units enjoyed an income in excess of $10,000 in both Hancock and Pearl River Counties. Six years later, 13 percent of all households in Hancock County and 12 percent in Pearl River County earned more than $10,000. Between 1960 and 1967, the percentage of all United States families in the highest quintile rose from 14 to 30 percent.

Other Indicators of Economic Growth and Social Change

Widening the employment base had an immediate impact on the drowsy towns in these counties. Descriptions painted by the first wave of NASA settlers into the area liken the atmosphere then existing in those counties to that of most nineteenth-century boom towns. Construction workers, scientists, engineers, and managers poured into a half dozen or so small Mississippi communities from all parts of the country. The demand for the most simple conveniences could not be met with the given facilities of the rural towns. Streets were torn up to modernize and enlarge outdated sewer and water systems. Understandably, many community leaders became bitter and antagonistic toward the space program after the sharp drop in NASA employment at the Mississippi Test Facility in 1966.

Population

During the twenty years between 1930 and 1950, the population of Hancock County grew by only 500. Population growth in Pearl River County was slightly higher (1,200) during the twenty-year period. The combined population of the two counties grew by less than 4,000 between 1950 and 1960. Then suddenly within a four-year period the population of Hancock County expanded by about 200 percent, rising from 6,800 in 1962 to 20,600 in 1966. The population explosion in Pearl River County was only slightly less dramatic. During the same period, the population of Pearl River County rose from 16,700 to 31,700. Then, as employment opportunities contracted, the population of Hancock County dropped by 2,400 between 1966 and 1967; the decrease in Pearl River County amounted to 2,700 during the same year.

School Enrollment

Although the school-age population of Hancock and Pearl River Counties increased after NASA employees and their families moved into the counties, the relative growth was not as large as that which took place after NASA established a facility in Brevard County, Florida. The reason for this, of course, is that NASA and NASA contractors hired a large number of local workers. Their children were already attending school in the two Mississippi counties. Nonetheless, data from the Mississippi State Department of Education show that combined enrollment in public and private schools in the two counties rose by about 2,200 between the 1962–1963 and the 1965–1966 academic years. To maintain a minimum of 35 students per classroom and per teacher, about 63 new classrooms would have had to have been constructed and the same number of new teachers would have had to have been hired by the public and private school systems of the two counties. Predictably, between the 1965–1966 and 1966–1967 academic years, combined public and private school enrollment in the two counties dropped by 270 students.

Residential Construction

With a history of population stability in the county, there can be little doubt that there were few residential facilities available for rent or for sale when the influx of MTF employees began. NASA personnel estimate that between 1962 and mid-1968, about 750 new single-family dwelling units and eight new apartment complexes were built in the several communities near the Mississippi Test Facility. The apartments average between 50 and 100 units each. Personnel at MTF estimate that the occupancy rate in the newly constructed apartment units was no more than 50 percent in 1968 and 1969. In part, the low occupancy rate is explained by the decrease in employment at the Mississippi Test Facility. It is also explained by undue optimism and speculation on the part of builders. After NASA announced that it would put its launch vehicle test site in Hancock County, local newspaper reports speculated that as many as 10,000 NASA and NASA contractor employees would move into the area. At its peak, NASA employment was about half that amount.

Retail Sales

Rising retail sales generally coincide with increases in income. After a decade of little growth, retail sales more than doubled in Hancock and Pearl River Counties in the early 1960's. During the entire decade of the 1950's, combined retail sales in Hancock and Pearl River Counties rose by only $2.7 million. This compares with an increase of about $25.8 million during the three years between 1962 and 1965. The growth rate dampened in 1966. The increase during that year amounted to 7 percent in Hancock County, compared with 36 percent a year earlier. In Pearl River County, the rate of growth in retail sales dropped from 24 percent to 6 percent during comparable periods. Combined retail sales in the two counties totaled $49.3 million in 1965 and $52.3 million in 1966. The growth trend reversed in 1967, when the value of retail sales in the counties declined to $46.2 million.[18]

Sales Tax Revenue

The increase in economic activity in the service industries was mirrored by the change in sales tax revenues. States, especially the southern states, are dependent on revenue from the sales tax to help finance the public services that necessarily increase when population grows. The nation-wide recession was reflected in a decrease in revenue from sales taxes in Hancock and Pearl River Counties between 1960 and 1961. After 1962, sales tax revenues increased substantially, particularly in Hancock County. Between 1962 and 1965, sales tax revenue collected in Hancock County rose from $380,000 to $1.9 million. During the same period the increase was from $600,000 to $1.3 million in Pearl River County. Between 1965 and 1967, the value of sales tax collections dropped by 45 percent in Hancock County and by 15 percent in Pearl River County. In each of the counties, sales tax revenue had dropped to about $1.1 million.[19]

Given the lack of an economic base without NASA, there can be no doubt that economic activities in these counties will continue to contract in response to decreased employment demand at the Mississippi Test Facility.

HOUSTON, TEXAS, AND NEW ORLEANS, LOUISIANA

Because of significant differences in the economic environments of Houston and New Orleans, the commonality of NASA operations in the two cities ends with the fact that the Space Agency introduced a new kind of basic industry into large urban economies. Although NASA employment comprised a small fraction of total employment in each of the cities, a new source of employment has a greater impact on a stationary economy (New Orleans) than on a rapidly expanding one (Houston).

Historical Growth Patterns

During the past quarter of a century, Houston grew from a city with a population of 385,000 to a major metropolitan area, with a population approaching two million. As a sizable metropolitan area, New Orleans is a much older city. By the end of World War I, the population of New Orleans was as large as that of Houston in 1940. Yet, it was not until the mid-1960's that the population of New Orleans reached one million. Despite a number of common factors, such as geographic orientation to the Gulf Coast, major port cities, and mineral wealth, it is difficult to imagine more dissimilar economies than those of Houston and New Orleans in the early 1960's.

Houston

Houston's rapid growth is a point of community pride. By 1960, it had become the seventh largest city in the nation, having grown during the fifties at nearly twice the average rate of the ten largest cities. Table 6–10 gives the average annual rate of employment growth in Houston between 1940 and 1968. As the data show, the growth of the economy of Houston has been broadly based. In the 1950's much of the impetus for economic growth came from the expansion of the petroleum, gas, and petrochemical industries. During that period, employment grew more rapidly in the mining sector (petroleum and natural gas exploration) than in any other sector of the economy. In the 1960's, however, that sector became less important as a source of employment growth in Houston. Expansion of petroleum-oriented ancillary activities in manufacturing (oil field equipment) and transportation (gas transmission companies) contributed to the diversification of the economic base of Houston. In support of high rates of growth in other sectors, the service industries in Houston also grew at least as fast as the rate of total employment growth.

TABLE 6–10: AVERAGE ANNUAL RATE OF EMPLOYMENT GROWTH IN HOUSTON, TEXAS, 1949–1968

INDUSTRIAL SECTOR	1940–1950	1950–1960	1961–1966	1966–1968
Mining	2.1	5.6	0.7	3.1
Contract construction	7.9	0.5	3.3	9.4
Manufacturing	5.4	2.8	4.0	3.5
Transportation, communications, and public utilities	4.6	1.8	1.2	4.2
Wholesale and retail trade	4.9	3.1	4.3	7.3
Finance, insurance, and real estate	4.8	5.0	4.2	5.9
Services	2.4	4.5	5.0	8.9
Government	0.4	4.3	8.7	7.6
Total	4.6	3.7	4.2	6.5

Sources: 1940 to 1960: U.S. Department of Commerce, Office of Business Economics, *Growth Patterns in Employment by County, 1940–1950 and 1950–1960,* Vol. 5: *Southeast* (Washington, D.C.: U.S. Government Printing Office, 1966); 1961 to 1968, U.S. Department of Labor, Bureau of Labor Statistics.

New Orleans

Table 6–11 summarizes employment growth patterns in New Orleans since the beginning of World War II. During most of the period, employment in New Orleans increased at a rate substantially lower than that of total nonagricultural employment in the United States. The average growth rate in New Orleans during the 1940's and the 1950's was about 90 percent of the national growth rates in those decades. The 1961 to 1966 growth rate of 4.2 percent marked a radical departure from the long-term employment growth trend. During that short period, employment in New Orleans grew at a rate 27 percent higher than the national growth rate. Unlike Houston, New Orleans is and has been dependent on its mining sector (petroleum, gas, and sulfur extraction).

TABLE 6-11: AVERAGE ANNUAL RATE OF EMPLOYMENT GROWTH IN NEW ORLEANS, LOUISIANA, 1940-1968

INDUSTRIAL SECTOR	1940–1950	1950–1960	1961–1966	1966–1968
Mining	19.6	10.4	9.0	4.1
Contract construction	4.4	1.0	11.6	−6.5
Manufacturing	1.3	1.8	6.7	−2.0
Transportation, communications, and public utilities	3.6	−0.2	1.9	1.9
Wholesale and retail trade	3.6	0.3	3.6	0.4
Finance, insurance, and real estate	2.8	3.7	2.6	2.6
Services	1.3	1.9	5.4	3.8
Government	5.8	1.4	3.8	5.7
Total	2.8	1.8	4.2	1.2

Sources: 1940 to 1960: U.S. Department of Commerce, Office of Business Economics, Growth Patterns in Employment by County, 1940–1950 and 1950–1960, Vol. 5: *Southeast* (Washington, D.C.: U.S. Government Printing Office, 1966); 1961 to 1966, U.S. Department of Labor, Bureau of Labor Statistics.

The Employment Buildup for the Space Program in Houston

The 1,620-acre site selected for the Manned Spacecraft Center borders on Clear Lake, about 20 miles south of downtown Houston. The facility, about a $300 million capital investment in the late sixties, rises on a plain which had been used for cattle grazing as late as 1962. The data in Table 6–12 show the timing of the employment buildup at the Manned Spacecraft Center. As the construction work force completed its tasks, and as civil service employment began to stabilize, NASA's technical and service support contractors became the principal source of employment

TABLE 6-12: NASA EMPLOYMENT AT THE MANNED SPACECRAFT CENTER IN HOUSTON AND THE MICHOUD ASSEMBLY FACILITY IN NEW ORLEANS, 1962-1968

	1962	1964	1966	1968
Manned Spacecraft Center				
Civil service	2,043	3,862	4,323	4,400
Technical and service contractor	0	2,300	6,300	8,400
Construction contractor	800	1,500	750	0
Total	2,843	7,662	11,373	12,800
Michoud Assembly Facility				
Civil service	127	276	250	158
Technical and service contractor	4,233	11,135	9,000	6,135
Construction contractor	1,500	2,000	0	0
Total	5,860	13,411	9,250	6,293

Sources: 1962–1966: Ronald M. Konkel, *Space Employment and Economic Growth in Houston and New Orleans, 1961–1966,* unpublished thesis, Tulane University, 1968, pp. 47 and 52; 1968, National Aeronautics and Space Administration.

growth in the space program. It was during the years 1964 to 1966 that the center became fully operational. The data show that reductions in space spending between 1966 and 1968 had not affected the size of the NASA labor force at the Manned Spacecraft Center in 1968. This results from the fact that the main mission of the center in Houston is manned space flight control operations.

In 1966, the average salary of all civil service personnel at the Manned Spacecraft Center was $10,000. For scientists and engineers, the average salary was $12,300. Of the total civil service work force in 1966, about half was classified by NASA as scientists and engineers. The same general wage pattern existed for employees of NASA contractors. There is no doubt that the salaries of employees at the Manned Spacecraft Center are higher than the average paid in the Houston area. In 1966, the average wage paid to nonsupervisory workers in manufacturing establishments in Houston was $6,700.[20]

The Employment Buildup for the Space Program in New Orleans

Vacant since the end of the Korean War, the Michoud Assembly Facility was re-opened in 1962. One of the most significant aspects of the NASA employment buildup at the Michoud Facility is the short time that elapsed between the opening of the facility and its employment peak. From the date of its opening to the end of 1964, employment increased by more than 13,000. During that period, the average monthly increase in employment was 360. By comparison, total employment in the entire New Orleans metropolitan area increased at an average monthly rate of only 425 during the decade of the 1950's. By the end of 1964, NASA employment at Michoud began to decline. The 1966 employment level was 30 percent below the peak, and by the end of 1968, employment was 50 percent below the 1964 level.

The mission of the Michoud Facility is the production of the rocket launch vehicles. Manufacture, of course, takes place before test and launch activities. By the end of fiscal 1968, all Saturn I-B vehicles had been manufactured. Three of the Saturn I-B vehicles had been launched, and nine had been put into storage. Of the fifteen S-IC stage units used on the Saturn V, twelve were delivered before the end of fiscal 1970. Only two were in manufacture in fiscal year 1970. The Nixon budget for fiscal 1970 called for three additional Saturn V vehicles using the S-IC stage, which will continue production at the Michoud Facility at a low rate into the 1970's.

NASA employment in New Orleans, as in Houston, was an important source of wage and salary income. The average salary of Michoud employees in 1966 was about $8,000, whereas the average for all manufacturing employment in New Orleans was about $6,800 in 1966.

The Michoud Assembly Facility was the largest industrial employer in the state of Louisiana in 1966. At that time, the combined civil service and contractor payroll was about $80 million. A payroll of that size accounted for about 8 percent of total earnings in manufacturing in Louisiana.[21]

Contribution of NASA to Employment Growth

Table 6–13 shows how much the employment growth rate in Houston was affected by the opening of the Manned Spacecraft Center. Although the space program had a discernible influence on the growth of total employment in Houston, it was by no means a dominant force. Even with the secondary employment effects considered, the rate of growth would not have been substantially different if the Center had not opened during that period. Without NASA, the rate of employment growth in Houston would have been 3.4 percent instead of 4.2 percent. This estimate is based on an employment multiplier of 2.45 computed for the Houston metropolitan area.[22]

TABLE 6–13: CONTRIBUTION OF NASA-GENERATED EMPLOYMENT TO EMPLOYMENT GROWTH IN HOUSTON, TEXAS, AND NEW ORLEANS, LOUISIANA, 1960–1966

CHANGE IN EMPLOYMENT	HOUSTON	NEW ORLEANS
Increase in total employment: 1960 to 1966[a]	135,100	59,800[b]
Increase in NASA employment: 1960 to 1966	11,400	10,800[c]
Actual average annual rate of growth: 1960 to 1966	4.2	3.1[e]
Estimated growth rate without NASA direct: 1960 to 1966[d]	3.9	2.6
Estimated growth rate without NASA direct and induced: 1960 to 1966[d]	3.4	1.9

[a] Nonagricultural wage and salary employment.

[b] Excludes increase from revision of New Orleans Standard Metropolitan Statistical Area in 1964.

[c] NASA employment in mid-1966.

[d] Estimate based on an employment multiplier of 2.45 for Houston and 2.27 for New Orleans.

[e] The average annual rate of growth is higher between 1961 and 1966 than between 1960 and 1966 because of a drop in employment between 1960 and 1961.

Source: Mary A. Holman and Ronald M. Konkel, "Manned Space Flight and Employment in the South," *Monthly Labor Review* (March 1968), p. 33.

If the employment impact of the space program is excluded, Houston's rate of employment growth (3.4 percent) would still have been about 50 percent higher than the national rate between 1960 and 1966. That adjusted rate is only slightly lower than the average rate experienced during the 1950's. As an approximation, NASA's contribution to total employment in Houston between 1960 and 1966 was equal to about one year's growth from all other basic sectors in the Houston economy.

Although small reductions in NASA employment will take place, the prominent role played by the Center in the Apollo Program assures a high level of employment at the Manned Spacecraft Center through the completion of the Apollo Program. A change in employment after completion of the Apollo lunar exploration series will depend on the role of the Center in future space programs.

When the Michoud Assembly Facility opened in 1962, New Orleans was at the bottom of an employment slump that began in 1957. Although it was precipated by the national recession that began in that year, the slump in New Orleans employment was more protracted. Employment in New Orleans remained low, unlike employment in the rest of the nation, which recovered from the recession by 1959. Total employment in New Orleans did not begin to approach its 1957 level until 1963, partly in response to the opening of new employment opportunities at Michoud. The depressed nature of the New Orleans economy can be seen by unemployment statistics. From 1957 to 1961, total employment rose from 9,000 to 21,000, and the unemployment rate rose from 2.7 to 6.2 percent.

The expansion of economic activity in New Orleans between 1961 and 1966 was not broadly based. During that period New Orleans had a short-lived construction boom. After a decade of virtually no employment growth in the contract construction industry, the rate of employment growth rose to 12 percent per annum. Most of the construction that took place in New Orleans during the early 1960's was for office buildings and hotels, as well as for the Michoud Assembly Facility. Even if the contract construction industry is included, it is not an exaggeration to state that virtually all of the increase in the rate of employment growth in New Orleans can be attributed to the space program. As Table 6–13 shows, the employment growth rate without the employment buildup at Michoud would have been 1.9 during the mid-1960's. To be sure, there were sources of employment growth other than employment at Michoud and in the construction industry during this period. However, these increases reflected the recovery of the New Orleans economy rather than its expansion.

The increased rate of employment growth in New Orleans was not maintained after 1966. Employment in the construction industry decreased markedly between 1966 and 1968. About 25 percent of that decrease in employment can be attributed to the completion of construction at the Michoud Assembly Facility. The decrease in manufacturing employment at Michoud was slightly more than the decrease in total manufacturing employment in New Orleans. By the end of 1968, the rate of employment growth in New Orleans had dropped below the low rate existing during the 1950's. The rate of unemployment in New Orleans, as in other areas affected by the space program, did not rise sharply after the reduction in NASA employment. In March 1969, the rate of unemployment in New Orleans was 3.9 percent. The explanation for this is that many of those separating from the space program moved out of the New Orleans area.

Because of the different roles of the Manned Space Flight Center and the Michoud Assembly Facility, the long-term employment impact of the installations probably will differ. Being a production facility, Michoud is more specialized than the Manned Spacecraft Center. The specialized nature of the facility makes New Orleans more vulnerable than Houston to any sporadic shifts in the level and project-mix of the space program.

THE ROLE OF THE SPACE PROGRAM IN LOS ANGELES[23]

Federal procurement has been an important factor contributing to the economic growth of California, particularly southern California. Yet because a large proportion of federal contracts go to firms in the aerospace industry, that industry has been more vulnerable than many others to shifts in federal procurement. The assumption of declining defense and aerospace employment has frequently been built into economic forecasting models for the Los Angeles area, as well as for the entire nation.[24]

It is estimated that between 1951 and 1962, an average of 16 percent of total defense purchases were supplied by firms and other organizations located in California.[25] Between fiscal 1962 and 1966, the Department of Defense let about 22 percent of its prime contract awards to firms in California. During that period, these awards averaged about $5.6 billion annually. California has received a larger percentage of NASA prime and subcontract awards than any other state. Statistics on the geographic distribution of NASA procurement are given in Chapter 8.

Population Growth

To understand the role of the space program in Los Angeles, certain general trends and patterns of economic growth must be analyzed. For the last two decades, population growth in Los Angeles County has accounted for about one-third of the total population increase in the state. Because Los Angeles outgrew its earlier boundaries, a five-county area, which includes Los Angeles, Orange, San Bernardino, Riverside, and Ventura Counties, might provide a better identity for that amorphous urbanization called Los Angeles. During the mid-1960's, the five-county area accounted for over 50 percent of the population growth in California. In 1966, the population of the five-county area totaled almost 10 million.

Personal Income

Per capita personal income in Los Angeles, as well as California in general, has been well above the national average for several decades. In 1950, per capita personal income was $1,900; by 1965, it had risen to $3,900. In each year, per capita income in Los Angeles was almost 30 percent higher than the national average. The high level of per capita income in Los Angeles has induced much of the long-term population growth in the area. Since 1950, in-migration into Los Angeles accounted for about two-thirds of the increase in population.

Long-Term Employment Growth

Although long-run trends show dynamic and spectacular growth, short-run trends show some signs of stagnation. This is true particularly of the aerospace industry. Since 1940, slightly more than 40 percent of employment growth in California has occurred within Los Angeles County. Table 6–14 shows the rapid rise in employment between 1940 and 1965. Although during that 25-year period about 72,000 new jobs were created annually, the rate of employment growth has been continually falling. In the decade of the 1940's, the average annual rate of employment growth in Los Angeles County was 5.2 percent. This compares with 3.9 percent in the decade of the 1950's, and 2.6 percent in the first half of the 1960's. The slowdown in the employment growth rate in Los Angeles during the first half of the 1960's was offset somewhat by faster employment growth in the counties surrounding Los Angeles, i.e., Orange, Riverside, San Bernardino, and Ventura Counties. Along with the nation-wide increase in economic activity, the rate of employment growth in Los Angeles rose to 3.5 percent between 1965 and 1968.

Within the non-agricultural sector, the structure of employment in Los

TABLE 6–14: LONG-TERM GROWTH IN NON-AGRICULTURAL EMPLOYMENT IN LOS ANGELES COUNTY, 1940–1965 (*THOUSANDS*)

	1940	1950	1960	1965
Contract construction	65	126	136	142
Manufacturing	205	435	760	778
Wholesale and retail trade	249	421	546	622
Other services	413	510	742	882
Government	41	156	237	330
Other	26	13	11	11
Total non-agricultural[a] employment	999	1,661	2,432	2,765

[a] Excludes armed forces.

Source: Mary A. Holman and Ronald M. Konkel, "Space Employment in Los Angeles: A Declining Role in the Aerospace Industry?", *California Management Review* (Fall 1968), p. 24.

Angeles has changed slightly, following the same pattern as California in general since 1940. Small increases in the relative importance of employment in government and in manufacturing have been offset by small decreases in the relative importance of employment in some of the service industries.

Employment Growth in the Manufacturing Sector

Firms located in California have made a significant contribution to the increase in employment in the manufacturing sector of the national economy. Table 6–15 gives data on the change in manufacturing employment in California as a percentage of the total national change between 1950 and 1965. The employment change in Los Angeles is given as a percentage of the employment growth in California. Almost one-fourth of the total national increase in manufacturing employment since 1950 occurred in California; about 12 percent of that change in employment was in Los Angeles. The table clearly shows that the relative importance of employment growth in Los Angeles and in California has been greatest in the manufacture of ordnance, primary metals, aircraft and parts, and instruments. Firms in these industries, of course, are more dependent on sales to the federal government than firms in other industries.

TABLE 6–15: RELATIVE GROWTH OF EMPLOYMENT IN MANUFACTURING IN LOS ANGELES AND IN CALIFORNIA, 1950–1965

	EMPLOYMENT CHANGE IN CALIFORNIA AS A PERCENTAGE OF NATIONAL CHANGE	EMPLOYMENT CHANGE IN LOS ANGELES AS A PERCENTAGE OF NATIONAL CHANGE	EMPLOYMENT CHANGE IN LOS ANGELES AS A PERCENTAGE OF CHANGE IN CALIFORNIA
Total manufacturing	23.6	12.2	51.7
Durable goods	22.7	12.0	52.9
Ordnance	44.2	22.3	51.0
Primary metals	31.1	11.1	35.7
Fabricated metals[a]	6.3	2.4	37.5
Non-electrical machinery[a]	6.4	2.6	40.0
Electrical machinery[a]	10.2	[b]	[b]
Aircraft and parts	23.9	14.9	62.5
Instruments	12.9	6.2	50.0

[a] Change between 1960 and 1965.

[b] Decrease in employment from 117,000 in 1960 to 97,000 in 1965.

Source: Mary A. Holman and Ronald M. Konkel, "Space Employment in Los Angeles: A Declining Role in the Aerospace Industry?", *California Management Review* (Fall 1968), p. 25.

Short-Term Employment Trends in Manufacturing

Employment in the manufacturing sector of California rose steadily after recovering from the 1957–1958 recession, except in 1964. In contrast, employment in the manufacturing sector of the Los Angeles-Long Beach metropolitan area did not regain the peak reached in 1957 until 1966. Recovery lagged in the durable goods rather than the nondurable goods industries. After rising rapidly in 1966, employment remained stable in the durable goods manufacturing sector between 1967 and 1968. An index of manufacturing employment is given in Table 6–16. The base is the average level of employment in 1959 and 1960. The 1959–1960 average is used because of the cyclical change in employment in those years. That cyclical change was more pronounced in the Los Angeles area than in California generally.

TABLE 6–16: SHORT-TERM CHANGES IN EMPLOYMENT IN MANUFACTURING IN LOS ANGELES–LONG BEACH, 1957–1968 (*1959–1960 AVERAGE=100*)

INDUSTRIAL SECTOR	1957	1960	1961	1962	1963	1964	1965	1966	1967	1968
Durable goods, manufacturing	109	99	96	102	102	99	100	113	119	119
Ordnance	83	105	107	133	154	158	156	156	140	146
Primary metals	106	98	101	104	102	99	101	114	117	117
Fabricated metals	[a]	99	99	103	103	103	105	114	116	122
Machinery	[a]	99	100	107	108	108	113	127	135	136
Electrical machinery	[a]	106	105	110	100	88	88	101	110	108
Aircraft and parts	144	92	83	84	84	83	87	104	116	111
Instruments	[a]	101	104	111	112	108	113	120	120	129
Nondurable goods, manufacturing	100	101	100	102	102	105	106	110	113	117
Total manufacturing	107	100	98	102	102	101	102	112	116	118

[a] Not available.

Source: U.S. Department of Labor, Bureau of Labor Statistics.

Despite the increase in the military demand for aircraft and parts for the Vietnam War, employment in the aircraft industry through the 1960–1968 period has been well below the peak reached in 1957. Employment in the electrical machinery industry reached a peak in 1962 and then declined steadily until 1966. The decline in employment in the electrical machinery industry in the early 1960's resulted from decreased procurement by the Department of Defense for missile and space systems in California. In fiscal year 1962, prime procurement awards to firms in California for space and missile systems totaled \$3.2 billion. By fiscal 1966, immediately before the large increase in defense spending for Vietnam, Department of Defense prime procurement awards for space and missile systems had declined to \$1.9 million.[26] During that period the decrease in defense purchases for space and missile systems was, in part, offset by increased defense procurement of electronics and communications equipment.

Employment in manufacturing in the Anaheim-Santa Ana-Garden

Grove area (Orange County), in contrast with the Los Angeles-Long Beach area, expanded rapidly during the 1960's. Using the 1959–1960 average as 100 percent, employment in all manufacturing industries in that area was 217 percent of the base in 1965. By 1968, employment in the manufacturing sector was 290 percent of the base. The size of the manufacturing sector, however, was only slightly more than one-tenth of that in Los Angeles County in 1968. Employment in the durable goods manufacturing sector of the Anaheim-Santa Ana-Garden Grove area, in contrast with the Los Angeles-Long Beach area, continued to rise sharply between 1967 and 1968.

Rates of Unemployment

Los Angeles and other parts of California have enjoyed significantly higher secular population and employment growth rates than the nation as a whole. Employment opportunities generally have expanded at a sufficiently rapid rate to absorb the large number of workers annually entering the labor market. One way to look at employment absorption capacity is through rates of unemployment. During the 1950's the pattern of change in the rate of unemployment was the same in Los Angeles as nationally. Decreasing rates of unemployment in Los Angeles were coincident with high national levels of employment.

Since the early 1960's, however, the rate of unemployment in Los Angeles has been substantially above the national average. In fact, the rate of unemployment in the Los Angeles-Long Beach metropolitan area was essentially the same in 1967 (5.6 percent) as it was in 1960 (5.7 percent), and about 50 percent higher than the national average.[27] By 1968, the rate of unemployment in Los Angeles-Long Beach dropped to 4.7 percent. This compares with a national average of 3.4 percent in 1968. The high rate of unemployment in Los Angeles reflects, in part, the short-term downtrend in employment in the durable manufacturing sector of the Los Angeles economy during the first half of the 1960's. It also reflects the high rate of in-migration into the area.

Employment in the Aerospace Industry

Aerospace employment in California is heavily concentrated in the southern part of the state, mainly in the Los Angeles-Long Beach-Anaheim area and in San Diego. In 1968, those areas accounted for about 75 percent of total aerospace employment in California.

The definition of aerospace employment used in this study is the same as that employed by the California Department of Industrial Relations, Division of Labor Statistics and Research. That agency uses four broad standard industrial classifications (SIC) to measure trends in aerospace employment. The industry groups are ordnance and missiles (SIC 19);

electrical machinery, equipment, and supplies (SIC 36); aircraft and parts (SIC 372); and instruments and related parts (SIC 38).[28]

A large proportion of non-agricultural wage and salary employment in California is directly related to defense and space production. In 1968, employment in these industries comprised about 35 percent of the total wage and salary employment in 1968. In the Anaheim-Santa Ana-Garden Grove area, these industries comprised about 57 percent of total manufacturing employment and 19 percent of total wage and salary employment in 1968.

Although the secular trend in aerospace employment in California has been upward, year-to-year fluctuations have been numerous and frequently wide in magnitude. As Table 6–17 shows, the greatest employment changes occurred between 1957 and 1959, between 1961 and 1962, and between 1965 and 1966.

The Los Angeles-Long Beach area has been gradually losing its relative importance as the center of aerospace employment in California. Not until the 1965–1966 military buildup did aerospace employment in Los Angeles regain its 1957 level. The decline in the share of California aerospace employment in the Los Angeles-Long Beach area reflects the general shift in aerospace employment away from the aircraft and missile industry and into the electronics industry. Existing electronics firms have been expanding, and new ones have been moving into both the Anaheim-Santa Ana-Garden Grove and the San Jose areas. The decrease in DOD procurement between fiscal year 1963 and fiscal year 1965 did not affect these areas much.

Aerospace employment in California rose by about 66,000 between 1965 and 1966. The increase in the Los Angeles-Long Beach area amounted to about 41,000. That employment gain, largely the result of defense requirements, comprised slightly more than half of the increase in manufacturing employment in the Los Angeles-Long Beach area. The impact of the increase in aerospace employment in the durable goods manufacturing sector of the economy in the Los Angeles-Long Beach area was much stronger. The gain in aerospace employment amounted to almost two-thirds of the total increase in durable goods employment between 1965 and 1966.

In mid-1968, employment in California's aerospace industry began to decline. In the Los Angeles-Long Beach area, the decrease in aerospace employment between 1968 and mid-1969 was the largest since the late 1950's. The decline resulted from the sharp decrease in employment on NASA projects, as well as a downward realignment in Department of Defense requirements after the rapid escalation of the war in Vietnam in late 1965. In addition, employment in the commercial sector of the industry dropped as a result of the completion of several commercial contracts for jet airliners.

TABLE 6–17: EMPLOYMENT IN AEROSPACE INDUSTRIES IN CALIFORNIA AND SELECTED METROPOLITAN AREAS, 1957–1969 (*ANNUAL AVERAGE, THOUSANDS*)

	1957	1958	1959	1960	1961	1962	1963	1964	1965	1966	1967	1968	1969[a]
California	448	418	463	461	469	509	510	483	482	548	597	595	568
Percentage change		−6.7	+10.8	−0.4	+1.7	+8.5	+0.2	−5.3	−0.2	+13.7	+8.9	−0.3	−4.5
Los Angeles-Long Beach													
Number	332	283	303	296	284	301	295	282	287	328	348	343	322
Percent of California	74.1	67.7	65.4	64.2	60.5	59.1	57.8	58.4	59.5	59.9	58.3	57.6	56.7
Percentage change		−14.8	+7.1	−2.3	−4.1	+6.0	−2.0	−4.4	+1.8	+14.3	+6.1	−1.4	−6.1
Anaheim, Santa Ana, and Garden Grove													
Number	10	13	18	21	32	46	54	53	54	61	73	73	69
Percent of California	2.2	3.1	3.9	4.6	6.8	9.0	10.6	11.0	11.2	11.1	12.2	12.3	12.1
Percentage change		+30.0	+38.5	+16.7	+52.4	+43.8	+17.4	−1.9	+1.9	+13.0	+19.7	0	−5.5
Total above as percent of California	76.3	70.8	69.3	68.8	67.4	68.2	68.4	69.4	70.7	71.0	70.5	69.9	68.8

[a] Data are for June, 1969

Source: State of California, Department of Industrial Relations, Division of Labor Statistics and Research.

NASA Employment in Los Angeles

It is estimated that total NASA employment in California amounted to about 166,000, or approximately 42 percent of the total NASA work force in June 1966.[29] Employment by NASA in California is concentrated in the Los Angeles-Long Beach-Anaheim metropolitan area. NASA employment in that area probably comprised between 85 and 95 percent of total NASA employment in California, or between 140,000 and 158,000 people in June 1966. The exact number cannot be readily determined because the geographic distribution of hundreds of lower-tier subcontractors, vendors, and materials suppliers is uncertain. An estimated 80,000 to 90,000 people worked in the Los Angeles area on NASA's Manned Space Flight Program alone in June 1966.

Table 6–18 presents an estimate of what the average annual rate of employment growth in Los Angeles might have been between 1960 and 1965 without NASA employment and also without NASA's large Manned Space Flight Program. The estimate, of course, assumes that in the Los Angeles economy all factors except the space program remained unchanged between 1960 and 1965. During that period, employment in the Los Angeles area actually grew at an average of 3.2 percent, or by about 83,000 annually. With an estimated increase in total NASA employment of about 150,000 between 1960 and 1965, the annual contribution of NASA to employment growth in Los Angeles was about 30,000. Similarly, the estimated total increase in employment on the Manned Space Flight Program was 85,000; the annual increase was about 17,000. These estimates do not include any induced employment in the service or nonbasic industries that must have been generated via

TABLE 6–18: ESTIMATED CONTRIBUTION OF NASA EMPLOYMENT TO EMPLOYMENT GROWTH IN THE LOS ANGELES METROPOLITAN AREA[a]

EMPLOYMENT GROWTH	WITH TOTAL NASA EMPLOYMENT	WITHOUT TOTAL NASA EMPLOYMENT	WITHOUT MANNED SPACE FLIGHT EMPLOYMENT
Average annual rate of employment growth[b]	3.2	2.1[c]	2.6[d]
Average annual increase in employment	83,000	53,000	66,000

[a] Includes the Los Angeles-Long Beach and Anaheim-Santa Ana-Garden Grove area.

[b] Includes only wage and salary workers.

[c] Based on an estimated increase in NASA employment of 150,000 between 1960 and 1965.

[d] Based on an estimated increase in employment on the Manned Space Flight Program of 85,000 between 1960 and 1965.

Source: Computed from data collected by the U.S. Department of Labor, Bureau of Labor Statistics, and the National Aeronautics and Space Administration.

the multiplier from the increase in NASA employment. Without NASA employment, the annual rate of employment growth in Los Angeles would have been 2.1 percent rather than the 3.2 percent actually experienced. This means that although NASA employment comprised a relatively small percentage of total employment in Los Angeles, it contributed about one-third of the growth in employment between 1960 and 1965. NASA's Manned Space Flight Program alone contributed about one-fifth of the growth in employment in the area. Although operating at a low level, the space program was in existence in 1960. It was the large buildup for NASA's Manned Space Flight Program that gave impetus to employment growth in California. Because of this, the contribution of the space program to employment growth in Los Angeles probably is closer to one-fifth than to one-third.

Impact of Space Employment in the Aerospace Industry

To be sure, NASA employment has been more important to the aerospace industry in Los Angeles than to employment in general. The main impact of NASA employment during the first half of the 1960's was as a stabilizing influence on aerospace employment in the Los Angeles-Long Beach-Anaheim area. The employment buildup for NASA's Manned Space Flight Program was a major factor, if not the most important, in preventing a large contraction in aerospace employment between 1960 and 1965.

By piecing together available but incomplete data, it is possible to discern the impact of NASA employment, and more important, employment on the Manned Space Flight Program, on the Los Angeles aerospace industry between 1960 and 1965. Data collected by the joint NASA-DOD Economic Information System show that about 88 percent of total NASA employment in the mid-1960's was located at plants grouped in one of the four industry divisions used here to define the aerospace industry. During the same period, about 91 percent of those employed on projects for the Manned Space Flight Program were located at plants in one of the four industry divisions. The percentage of those working in the aerospace industry is higher for manned space flight employment than for NASA employment in general because of the large work force at plants producing launch vehicles. Those plants are in the ordnance and missiles component of the aerospace industry.

The data displayed in Table 6–19 provide an estimate of the stabilization role played by NASA in the aerospace industry in Los Angeles. Actual employment in the aerospace industry rose by about 23,600 between 1960 and 1965. Without the large increase in employment at plants engaged on projects for the Manned Space Flight Program, aerospace employment might have dropped by about 63,300 during the five-year period. It would be unrealistic to assume that there would have been no offset employment for the total space effort in Los Angeles. Nevertheless, the estimate of a possible decrease of 108,400 in aerospace employment without NASA employment is provided as a high range for an estimate of the stabilization role played by the Space Agency in the aerospace industry in Los Angeles.

TABLE 6-19: STABILIZATION ROLE OF NASA EMPLOYMENT IN THE AEROSPACE INDUSTRY IN METROPOLITAN LOS ANGELES[a] 1960-1965

Actual change in aerospace employment	+ 23,600[b]
Without Manned Space Flight employment	− 63,300[c]
Without total NASA employment	−108,400[d]

[a] Includes wage and salary employment in aerospace industry in the Los Angeles-Long Beach and Anaheim-Santa Ana-Garden Grove area.

[b] Average annual rate.

[c] Approximately 91 percent of estimated Manned Space Flight employment of 85,000 in Los Angeles.

[d] Approximately 88 percent of estimated NASA employment of 150,000 in Los Angeles.

Sources: Data on aerospace employment provided by the State of California, Department of Industrial Relations, Division of Labor Statistics and Research. Industry distribution of total NASA and Manned Space Flight employment estimated from data in the joint NASA-DOD Economic Information System.

Employment on the Space Program, 1966–1969

Employment on the space program ceased to be an offset to the decline in other aerospace employment, or perhaps more correctly, competitive with other aerospace employment after the escalation of military requirements for Vietnam, no later than early 1966. Between June 1966 and June 1968, an estimated 52,000 people in the Los Angeles area separated from NASA projects. By June 1968, NASA employment in Los Angeles amounted to about 98,000. At that time, it was approximately 60 percent of the June 1966 NASA work force. Some of the contractors witnessing a substantial decrease in NASA work at plants located in the Los Angeles area include: Hughes Aircraft, Marquardt, McDonnell Douglas, North American Rockwell, and TRW. There can be no doubt that transition to alternative employment within the aerospace industry was made with ease between June 1966 and June 1968. The increase in other aerospace employment was more than enough to absorb those separating from the space program. Yet by mid-1968, reduced defense requirements and weakened commercial demand provided little or no alternative employment opportunities for those separating from NASA projects. Between June 1968 and June 1969, aerospace employment in the Los Angeles area dropped by about 25,000. By mid-1969, the aerospace industry was the weakest component in the durable goods manufacturing sector of the Los Angeles economy.

SUMMARY: THE ABILITY OF LOCALITIES TO ABSORB CHANGING LEVELS OF NASA EMPLOYMENT

The nation can decide to allocate any one of a number of different resource levels for the space effort during the 1970's and beyond.[30] The matter of future NASA funding levels will be discussed in the last chapter. It suffices here to say that NASA funding levels since fiscal 1966 indicate that community leaders in those localities most committed to the space program should take stock of their abilities to adjust to rapidly decreasing levels of space activity.

Costs of Economic Re-adjustment

The two main economic costs, or problems of economic re-adjustment, that would result from a continued contraction of the space program are those associated with community dislocation and those associated with re-employment of displaced workers. Both kinds of costs can be small if the communities surrounding the space installations can absorb displaced workers in jobs commensurate with existing levels of skill. The ability of the several local economies to absorb displaced employees varies by geographic area and by the skill characteristic of workers who would separate from the space program. Also, it must be emphasized that the economic displacement of communities and of individuals would change if offset programs were initiated by federal, state, or local governments. On the basis of the economic environments existing in the late 1960's, Table 6–20 presents an appraisal of the ability of localities to

TABLE 6–20: ESTIMATE OF EMPLOYMENT ABSORPTION CAPACITY OF LOCALITIES SURROUNDING MSF CENTERS

LOCALITY	NASA RESEARCH AND DEVELOPMENT PERSONNEL	NASA SUPPORT PERSONNEL	INDIRECT EMPLOYMENT
Brevard County, Fla.	Poor	Poor	Poor
Hancock and Pearl River Counties, Miss.	Poor	Poor	Poor
Huntsville, Ala.	Poor	Poor	Fair
Houston, Tex.	Poor	Excellent	Excellent
New Orleans, La.	Poor	Fair	Fair

absorb those who would be directly and indirectly affected were the space program cut back sharply. Because the economic base of Los Angeles is much broader than that of the five southern communities, that area is discussed separately. The three main factors that affect the ability of these areas to absorb displaced work forces are: the size of employment displacement in relation to total employment; the structure of employment; and the growth potential of employment.

Huntsville, Brevard, and Hancock

The outlook for re-employment in the areas surrounding the Marshall Space Flight Center, the Kennedy Space Center, and the Mississippi Test Facility was not favorable in the late 1960's.[31] In each of these areas, the expansion of NASA had been the predominant growth influence during the preceding five or six years. In each area, NASA employment had accounted for a large proportion of total employment. In each of the areas, nondefense and nonspace manufacturing was small and had been declining in relative importance. In none of the areas could employment growth from other sectors of the economy be expected to cushion much of the impact of declining space employment.

Houston and New Orleans

NASA employment in both of these areas is small in comparison with total employment. Local re-employment prospects for research and development personnel, however, are poor in these areas. Local science-based firms are mainly in the petroleum and chemical industries. Separating workers with general clerical and technical skills could be easily absorbed into Houston's diversified and dynamic economy. New Orleans' rates of unemployment could rise unless a high rate of employment growth from nonspace sources materialized. By the late 1960's, however, New Orleans had already witnessed a 50 percent decrease in NASA employment. The decrease in NASA employment, coupled with the drop in employment in the construction industry, made the employment base of New Orleans revert to the stagnant pattern of the 1950's.

NASA Research and Development Personnel

The answer to job displacement for research and development personnel, professional employees, and skilled industrial workers is geographic mobility. The relocation of these employees, however, would be of major importance to several local economies (mainly Huntsville, Brevard County, and Hancock County). If many should stay, it would enhance the potential for future industrial development of the given area; but it would also add to the problem of short-term employment re-adjustment. If most should leave, it would mitigate the problem of short-term re-adjustment by holding down the number of unemployed persons; but it would also strip the local economy of a valuable asset—skilled manpower.

NASA Support Personnel

This group of NASA employees is composed of clerical, administrative, and technical support personnel—usually recruited from the local area. For these, mobility is more difficult. Few would be able to relocate at employers' expense. Many would be tied to the area because of the employment of spouse and home ownership. This group is generally dependent on the local economy for alternative employment. If communities experienced continued cutbacks in NASA employment, released support personnel could be expected to swell the ranks of the temporarily unemployed. If new industrial employment did not materialize, these workers would be forced to relocate.

Indirect Employment

The third group affected by changes in NASA employment is less homogeneous. It includes small businessmen; professionals, such as doctors and lawyers; store clerks and other employees of diverse service establishments; and state and local government employees. Employment within this group is indirectly tied to the space program as part of the expansion of the local economy that resulted from the NASA employment buildup. The severity of the adverse impact on this group would be related to the size of direct NASA employment compared with total employment in the given area. Many workers in this group would also have to relocate if the local industrial base were not restored.

Los Angeles Metropolitan Area

There can be no doubt that transition to alternative employment for those first separating from the space program was made with ease. The increase in other aerospace employment between June 1966 and June 1968 was more than enough to absorb those separating from the space program. However, if DOD aerospace procurement were to revert to the pattern of the early 1960's, and if there were a substantial decrease in NASA employment, commercial employment demand would have to increase substantially to stabilize employment in the aerospace industry and to prevent unemployment rates from rising to pre-Vietnam levels. By 1968, employment in the aerospace industry in Los Angeles already had begun to revert to the pattern of the early 1960's.

NOTES

1. The United States Arms Control and Disarmament Agency has supported such research for a number of years. See, for example: *Community Adjustment to Reduced Defense Spending,* Report No. 33, 1965; *Adjustments to Reduced National Defense Expenditures in New Mexico,* 1965; *Post Lay-Off Experiences: Republic Aviation Workers,* Report No. 35, 1966; *The Dyna-Soar Contract Cancellation,* Report No. 29, 1965; and *Reemployment Experience of Martin Company Workers Released at Denver, Colorado,* Report No. 36, 1966.
2. Any discussion of socio-economic impact necessarily reflects the perspective taken by the analyst. The principal frame of reference given here is on the economic changes witnessed by these communities. For a description of the social environment of the Cape

Kennedy area that is somewhat analogous to the biblical description of the last days of Sodom and Gomorrah, see "Behavior, Life in the Space Age," *Time*, July 4, 1969, pp. 38–43.

3. U.S. Department of Commerce, *Growth Patterns in Employment by County: 1940–1950 and 1950–1960*, 8 vols. (Washington, D.C.: U.S. Government Printing Office, 1965 and 1966).

4. *Ibid.*, Vol. 5, *Southeast*.

5. Broadly conceived, the economist has a single analytical tool for measuring the economic effects of a change in spending. The technique, multiplier analysis, is applicable to changes in consumer spending, investment spending, government spending, exports, as well as changes in tax rates. The objective of multiplier analysis is to measure the indirect or secondary impacts to determine the total effect. Multiplier analysis is applicable to problems at the local, state, regional, and national levels. The empirical content of the model can be based on income data or employment data, but preferably both. The income multiplier shows that an increase in spending causes income to rise by more than the initial injection. This results from secondary increases in income and consumption. The multiplier is merely the coefficient that shows the relation between the change in investment and the change in income. The size of the coefficient depends on the magnitude of leakages (principally savings) from the circular flow of total income. The multiplier is expressed in terms of what economists call the "marginal propensity to consume," or the incremental increase in consumption that results from an incremental increase in income. The conventional expression for the multiplier is:

$$k = \frac{1}{1 - \frac{\Delta C}{\Delta Y}};$$

where k is the coefficient for the multiplier; ΔC is the change in consumption; and ΔY is the change in income. A high marginal propensity to consume $\left(\frac{\Delta C}{\Delta Y}\right)$, results in a large coefficient for the multiplier. The multiplier effect is then:

$$\Delta I \cdot k = \Delta Y;$$

where ΔI is the initial autonomous change in spending and ΔY is the ultimate change in national income.

The development of the employment multiplier owes its origin to the work of city planners and geographers rather than to economists. The economic base of a locality can be divided into two kinds of employment: basic and service. The first is employment associated with the production of goods and services for people and for firms outside the locality. This is the basic or export sector of the employment base. Employment in the basic sector includes those activities that bring new money into a community. Basic employment can include employment in agriculture, manufacturing, and by the federal government. The second kind of employment in a locality is nonbasic or service employment. This includes employment in such industries as retail trade, insurance, real estate, public utilities, and local transportation. Workers in these industries produce goods and services for the local market. A simple form of the employment multiplier then is:

$$k = \frac{E_s + E_b}{E_b};$$

where k is the coefficient for the multiplier; E_s is employment in the nonbasic or service industries; and E_b is employment in the basic industries.

A community with a large and diversified economic structure tends to have larger indirect impacts than a small locality that imports most of its goods and services from other cities. An employment, or an income, multiplier coefficient of 1 implies that an autonomous change in spending has no indirect impact in a community.

6. Mary A. Holman and Ronald M. Konkel, "Manned Space Flight Employment in the South," *Monthly Labor Review* (March 1968), pp. 30–36.

7. The employment multipliers were calculated at intervals for the period between 1940 and 1966. The resulting coefficients were quite stable for Brevard County. However, the ratio between total employment and basic employment in Huntsville rose slightly, but consistently, during the period. This follows the national trend and means that employment in the service industries became increasingly important as the economy of Huntsville grew.

8. Alabama Department of Industrial Relations and Florida State Employment Commission.

9. For a discussion of the economic motives that induce women to enter the labor force, see Glen G. Cain, *Married Women in the Labor Force: An Economic Analysis* (Chicago: University of Chicago Press, 1966).

10. Data come from the *1969 Economic Report of the President,* and also from the U.S. Department of Commerce, Office of Business Economics, Regional Economics Information System.

11. Per capita disposable personal income, which takes into account personal income taxes, in the United States was $2,700 in 1967.

12. Sales Management, Incorporated, *Survey of Consumer Buying Power* (June 1968).

13. Data from the Brevard County Public School System.

14. U.S. Department of Housing and Urban Development, *HUD News,* January 8, 1969.

15. U.S. Department of Commerce, Office of Business Economics, *Growth Patterns in Employment by County: 1940–1950 and 1950–1960,* Vol. 5: *Southeast* (Washington, D.C.: U.S. Government Printing Office, 1966).

16. Regional Economics Information System, Office of Business Economics, U.S. Department of Commerce.

17. Sales Management, Incorporated, *Survey of Consumer Buying Power* (June 1962 and June 1968).

18. *Ibid.,* June issues, 1960–1968.

19. Bureau of Business and Economic Research, Mississippi State University, State College, Mississippi.

20. Ronald M. Konkel, "Space Employment and Economic Growth in Houston and New Orleans, 1961–1966," unpublished thesis, Department of Economics, Graduate School, Tulane University, 1968, pp. 45–50.

21. *Ibid.,* pp. 50–55.

22. Holman and Konkel, *Monthly Labor Review,* p. 33.

23. Much of the analysis in this section was published in the Fall 1968 issue of the *California Management Review.* See Mary A. Holman and Ronald M. Konkel, "Space Employment in Los Angeles: A Declining Role in the Aerospace Industry?", pp. 23–30.

24. See, for example, the comments of participants at a symposium entitled "California's Search for New Sources of Economic Growth," in *California's Future Economic Growth,* ed. by Warner Z. Hirsh and Richard N. Baisden (San Francisco, Calif.: Diablo Press, 1965); also Emile Benoit and Kenneth E. Boulding, eds., *Disarmament and the Economy* (New York: Harper and Row, 1963).

25. Roger E. Bolton, *Defense Purchases and Regional Growth* (Washington, D.C.: Brookings Institution, 1966), pp. 152–53.

26. Office of the Secretary of Defense, *Military Prime Contract Awards by Region and State: Fiscal Years 1962–1966,* January, 1967. (mimeographed)

27. Paul O. Flaim, "Jobless Trends in 20 Large Metropolitan Areas," *Monthly Labor Review* (May 1968), p. 23.

28. This definition of aerospace employment necessarily includes some employment

within the four industry groups that is not aerospace employment. For example, establishments producing certain kinds of household appliances (e.g., cooking equipment, laundry equipment, and sewing machines) are grouped in the electrical machinery industry. Also the definition excludes some aerospace employment in other industry groups that cannot be readily identified. For example, grouped in the industry miscellaneous services (SIC 89) are engineering services and scientific research. On balance, the errors are probably compensating.

29. Estimated by the Management Information Systems Division, National Aeronautics and Space Administration, Washington, D.C.

30. The range of funding levels projected for the post-Apollo period usually incorporates a wide margin. For example, three possibilities were selected by a panel of experts, the President's Science Advisory Committee, to illustrate the possible scope of future spending for the exploration of space in 1972. The range went from a low of $3.5 billion to a high of $7.0 billion. See *The Space Program in the Post-Apollo Period* (Washington, D.C.: U.S. Government Printing Office, 1967), pp. 45–53. Leonard A. Lecht, with the National Planning Association, also projects a wide range for expenditures between $3.5 billion and $6.0 billion in 1975. See *Goals, Priorities and Dollars, the Next Decade* (New York: The Free Press, 1966), pp. 270–85.

31. In the late 1960's, community leaders were hopeful that a major expansion of the Pascagoula, Mississippi facility of the Ingalls Shipbuilding Division of Litton Industries would help ease the impact of declining employment at NASA's Mississippi Test Facility. Similarly, community leaders in Huntsville were optimistic that five large national firms (Automatic Electric Company, Barber Coleman, PPG Industries, Dunlop Tire and Rubber, and U.S. Corrugated Fiber Box Company) would locate new plants in Huntsville and create a large number of new jobs in Huntsville during the 1970's.

7 Performance: Economic Stability

The Employment Act of 1946 established economic stability as a national economic goal, and with that legislation, Congress declared that it would be the continuing responsibility of the federal government to minimize fluctuations in the United States economy. Although in the post-World War II period the United States witnessed four periods of contracting economic activity, most economists would agree that compared with the prewar cyclical record, the recessions since 1946 have been shorter, milder, and less frequent. Statistical determination of the nature and causes of cyclical instability has been the subject of many books on business cycle theory.[1] If a single major factor is to be blamed, it was overexpansion of inventories, encouraged by expectations of rising prices, that precipitated the downturn that occurred in 1948. The sharp drop in defense expenditures probably was the principal cause of the 1953 contraction. The decline that took place in 1957–1958 can be attributed to an unjustified investment boom that finally resulted in an abrupt decline in spending for plant and equipment. Restrictive monetary and fiscal policy, which dampened consumer and business spending, was the most probable cause of the cycle that occurred during the 1959–1960 election year.[2]

EMPLOYMENT INSTABILITY IN THE AEROSPACE INDUSTRY

As measured by magnitude of change, swings in federal spending during the postwar period have been considerably wider than changes in consumer or business purchases. Fluctuations in the federal sector, however, have been less frequent. As indicated in Chapter 1, instability in the federal sector has mirrored autonomous factors, primarily changes in military requirements. Yet the aerospace industry, which is heavily dependent on government spending, has witnessed less severe fluctuations than most other durable manufacturing goods industries. One feature of a capitalistic industrial economy is that fluctuations in production in the producers-durable goods industries tend to be wider than changes in the production of consumer goods or the sale of services. Table 7–1 displays data showing percentage changes in the level of employment in the aerospace, motor vehicle, non-electrical machinery, and primary metals industries. The data also show changes in total employment in all durable goods manufacturing industries. In addition, Table 7–1 provides information on the number of cyclical fluctuations that occurred in these industries during the period.

TABLE 7-1: CHANGE IN EMPLOYMENT IN THE AEROSPACE INDUSTRY AND IN OTHER DURABLE GOODS MANUFACTURING INDUSTRIES, 1950–1967

INDUSTRY	NUMBER OF CYCLES: EXPANSION PERIODS	NUMBER OF CYCLES: CONTRACTION PERIODS	PERCENTAGE CHANGE IN EMPLOYMENT: TROUGH TO PEAK[a]	PERCENTAGE CHANGE IN EMPLOYMENT: PEAK TO TROUGH[a]
All durable goods manufacturing industries	4	3	16.4	−9.7
Aerospace[b]	4	3	23.1	−5.0
Motor vehicles[c]	4	4	18.7	−10.5
Non-electrical machinery[d]	4	3	20.6	−7.8
Primary metals[e]	4	4	9.5	−9.6

[a] Median change in average annual employment levels.
[b] SIC 1925 and 372.
[c] SIC 371.
[d] SIC 35.
[e] SIC 33.
Source: U.S. Department of Labor, Bureau of Labor Statistics.

A glance at the data clearly shows that the growth of employment in the aerospace industry during the four expansion periods was about 50 percent greater than that in the total durable goods manufacturing sector. This is another way of saying that employment demand was much stronger in the aerospace industry during the period than in the durable manufacturing industries in general. The employment buildup was especially strong in the aerospace industry in the early 1950's. In response to requirements caused by the Korean War, employment in the aircraft and parts component of the aerospace industry almost doubled. In contrast, total employment in the durable goods manufacturing sector rose only about one-fourth during the Korean War.

Using the median values for the three periods of contraction between 1950 and 1967, employment decline in the aerospace industry was only half as severe as the contractions experienced by the durable goods manufacturing industries in general. The same is true of aerospace employment compared with the other heavy industries listed in Table 7–1. The most severe contraction in aerospace employment occurred between 1957 and 1960. During that period, employment dropped by 140,000, or approximately 16 percent.

Hidden in the average figures displayed in the table is the fact that swings in employment in the aerospace industry have not necessarily coincided with the business cycles that took place in the economic system during the seventeen-year period. Employment recovery in the aerospace sector lagged behind that of employment in all durable manufacturing industries by about one year in the immediate post-Korean War period. Again, employment contraction in the aerospace industry lagged behind that in durable manufacturing by about one year during the recession in the mid-1950's. Between 1956 and 1957, employment in all durable manufacturing industries dropped by about 120,000. Between those years, however, aerospace employment rose by 60,000. Unlike total employment demand in the durable manufacturing industries, which increased between 1958 and 1959, employment demand in the aerospace industry continued to decline until June 1960. The aerospace industry also suffered a decline in employment between mid-1963 and mid-1965 that was not experienced by other sectors of the economy.

At that time, defense spending had stabilized and had thus become a declining proportion of the gross national product. Before the military buildup for Vietnam, the President and others were concerned about the possible adverse effects that disarmament might have on the economy. In 1963, the President established an inter-agency committee to implement adequate federal policies to help individuals, industries, and communities adjust to economic change and dislocation that might result from changes in defense requirements. Gardner Ackley, then Chairman of the Council of Economic Advisers, headed the Interagency Committee on the Economic Impact of Defense and Disarmament.[3]

THE STABILIZATION ROLE PLAYED BY NASA EMPLOYMENT IN THE MID-1960's

One economic aspect of the space program has been its stabilizing effect on employment demand in the aerospace industry in the mid-1960's. In this context, such basic economic questions as whether employment provided by NASA in the aerospace industry was warranted, in terms of an optimum allocation of resources, must be disregarded. The same is true of any possible conflicts between the goals of stability and an equitable distribution of income. As indicated in Chapters 1 and 3, it is difficult for the policy-maker to weigh possible economic and social trade-off costs between national goals that are as diverse as stability, efficiency, and equity. Allowing for this omission in the analysis, the data displayed in Table 7–2 show the magnitude of the contribution of the space program to employment stability in several sectors of the aerospace industry.

TABLE 7–2: STABILIZATION ROLE PLAYED BY THE SPACE PROGRAM IN THE AEROSPACE INDUSTRY, SELECTED PERIODS

INDUSTRY SECTOR	ACTUAL CHANGE IN EMPLOYMENT	WITHOUT NASA EMPLOYMENT
	Change: 1961 to 1965[a]	
Aircraft and parts	+15,500	−76,600
Guided missiles and space vehicles	+ 3,300	−64,600
Communication equipment	+12,100	−42,600
Electronic components	+61,900	+54,600
Instruments[b]	+ 200	−11,800
	Change: Employment Peak to 1965[a]	
Aircraft and parts (1963–1965)	−14,000	−106,100
Guided missiles and space vehicles (1963–1965)	−17,000	− 84,900
Communication equipment (1962–1965)	−28,100	− 82,700
Electronic components	[c]	[c]
Instruments (1960–1965)[b]	− 600	− 12,600

[a] Average annual employment levels.
[b] Includes SIC 281 and 282.
[c] Except for minor dip between 1962–1963 steady growth.

Source: Industry data, U.S. Department of Labor, Bureau of Labor Statistics, *Employment and Earnings Statistics for the United States, 1909–1966,* Bulletin No. 1312–4 (Washington, D.C.: U.S. Government Printing Office, 1966). NASA employment estimated from data collected by the NASA-DOD Economic Information System and from other NASA reports and records.

A conservative estimate of the stabilizing effect of space employment is given by the data in the top part of the table. The actual change in employment in these sectors of the aerospace industry between 1961 and 1965 hides the impact of declining defense expenditures during the years between 1961 and 1965. Also much of the employment growth that took place in these segments of the aerospace industry materialized during the last half of 1965 as a result of the employment buildup for the commitment in Vietnam. Nevertheless, the data show that without employment provided by NASA, and assuming no alternative offset employment, four of the five segments of the aerospace industry would have witnessed a sizable drop in employment. In terms of aggregate employment, the space program did not play the same role of providing offset (for the Department of Defense) employment in the electronic components segment of the aerospace industry. Without the space program, the rate of employment growth in that segment would have been imperceptibly lower.

The change in employment during the years displayed in the data in the lower part of Table 7–2 provides a measure of the effect of declining Department of Defense procurement during the 1960's. For example, between 1963 and 1965, total employment in the guided missiles industry actually dropped by 17,000 or 10 percent. Without NASA business, the drop in employment in that industry might have been as large as 84,900, which would have been a 50 percent decline. In the aircraft and parts industry the actual decline in employment between 1963 and 1965 was only 14,000, or 2 percent; without the space program it could have amounted to as much as 106,100, or 17 percent. It is, of course, impossible to know what the level of employment would have been in the aircraft industry without the space program during the early 1960's. It seems unlikely, however, that commercial demand alone could have generated an equal amount of offset employment. It was in the early 1960's that the commercial airlines converted to turbine and jet-powered planes on a large scale.

SPACE EMPLOYMENT AS A RESOURCE RESERVOIR

Again, disregarding any trade-off analysis between the goals of stability and efficiency and between stability and an equitable distribution of income, the space program once more proved to have a stabilizing effect on the aerospace industry during the late 1960's. There can be little doubt that the existence of the space program increased the ease with which the United States mobilized for the Vietnam conflict. The program had brought together and had maintained a large reservoir of skilled, technical, and scientific manpower in the aerospace industry during the early 1960's. With the phase-down in the space program in early 1966, those employees separating from NASA projects increased the supply of technical personnel available to satisfy defense requirements. A sizable proportion of those who had worked on the space program, however, separated from NASA plants, as well as from NASA projects. Table 7–3 shows the shift in employment that took place in 136 different plants of NASA contractors between June 1966 and June 1968. In the aggregate, about 42 percent transferred to defense employment. Because the increase in defense requirements was not high enough to maintain employment at the plants of some NASA contractors, about 9 percent of those separating from NASA projects shifted to non-NASA–non-DOD government work or to commercial projects. It is estimated that between

June 1966 and June 1968 about 49 percent of those separating from the space program transferred to other plants. It is most probable that the majority transferred to plants in the aircraft industry, because total employment in the guided missiles component of the aerospace industry witnessed a decline of 4 percent during the two-year period.

TABLE 7–3: SHIFT IN EMPLOYMENT AT THE PLANTS OF NASA CONTRACTORS, JUNE 1966–JUNE 1968

INDUSTRY SECTOR	CHANGE IN NASA EMPLOYMENT	ABSORPTION CAPACITY AT NASA PLANTS	CHANGE IN INDUSTRY EMPLOYMENT
Guided missiles and space vehicles[a]	−31 percent	DOD absorbed 79 percent; other government and commercial absorbed 15 percent; 6 percent left NASA plants.	−4 percent
Aircraft and parts[b]	−39 percent	Commercial and other government employment also down; DOD employment absorbed 40 percent of NASA, other government, and commercial decrease; 60 percent left NASA plants.	+15 percent
Electrical equipment[c]	−54 percent	DOD absorbed all of NASA decrease; other government and commercial up substantially; 12 percent rise in total employment at NASA plants.	+3 percent
Business and engineering services[d]	−22 percent	DOD employment absorbed 30 percent; other government and commercial absorbed 48 percent; 22 percent left NASA plants.	+9 percent[e]
Total employment at NASA plants	−43 percent	DOD employment absorbed 42 percent of decrease; other government and commercial absorbed 9 percent; 49 percent left NASA plants.	Total employment up 3 percent

[a] SIC 1925
[b] SIC 372
[c] SIC 36
[d] SIC 891 and 737
[e] SIC Only 891

Sources: Joint NASA/DOD Economic Information System, and U.S. Department of Commerce, Bureau of Labor Statistics.

GEOGRAPHIC EMPLOYMENT INSTABILITY

In addition to the fact that federal procurement cycles frequently do not correspond in amplitude or in timing to economy-wide business cycles, procurement cycles can induce severe fluctuations in the economic activity of localities. To be sure, communities with a single or a few large government contractor plants are most likely to experience employment instability and periodic rates of excessively high unemployment.

During the 1960's many localities in which aerospace employment was concentrated sustained wide fluctuations in aerospace employment. This reflects the fact that localities that are dependent on federal procurement witness fluctuations in the level of employment with the changing fortunes of specific companies in obtaining follow-on contracts, as defense and space projects are phased-down. San Diego enjoyed a rapid buildup in employment in the aircraft industry as a result of aircraft production contracts won by Convair during the Korean War. After the war, employment in the aircraft industry dropped. The next major buildup, generated by General Dynamics Corporation's participation in the Atlas Missile Program, came in the mid-1950's. The long decline in aerospace employment in San Diego during the 1960's mirrored the phase-down of the Atlas Program, as well as the failure of the Convair Division of General Dynamics to obtain much in the way of commercial orders for air transports. Between 1957 and 1965, employment in the aerospace industry in San Diego dropped from 56,000 to 26,000. Total non-agricultural employment also dropped during the early 1960's, a time when economic activity was increasing in the rest of the nation.

The case of Seattle offers another example. The Boeing Company plays an important role in the industrial sector of Seattle. After cancellation of the Dyna-Soar Program, Boeing laid off 5,000 employees within a three-month period in 1964. The fortune of Boeing reversed a year or so later, as a result of an increase in the demand for commercial transports produced by the company. Employment at the Seattle-Everett plants of Boeing doubled in the succeeding three-year period.

Two major aerospace companies are located in Fort Worth—the Convair Division of General Dynamics and the Bell Helicopter Division of the Textron Corporation. Following the phase-down in the production of the B-58 bomber at General Dynamics, employment in the aircraft industry in Fort Worth dropped sharply in the early 1960's. As a result of the award of the contract for the F-111 (the TFX) to General Dynamics, and following an increase in the demand for helicopters produced by Bell, aerospace employment began to rise markedly during the late 1960's. On Long Island, Republic Aviation was forced to lay off almost 15,000 employees after the completion of the F-105 fighter program between 1963 and 1965. That decline was partially offset by the successful entry of the Grumman Aircraft Engineering Corporation, also located on Long Island, into the space program.

There are a number of other examples of local employment instability resulting from government project cycles. Both Baltimore and Denver are affected by the fortune and the misfortune of the Martin Marietta

Company. Wichita, Kansas experiences instability brought on by fluctuation in business at the Boeing Wichita plant. And, of course, Sacramento is the home of Aerojet-General, a subsidiary of the General Tire and Rubber Corporation. Employment at Aerojet-General grew rapidly during the 1950's because of its capabilities in the field of nuclear rocket propulsion. Changing technological requirements for space and missiles programs, however, suddenly reduced the demand for nuclear rocket engines. As a consequence, corporate employment at Aerojet-General dropped by almost 50 percent between 1964 and 1967.

The fact that employment in the aerospace industry is unstable does not mean that the only communities subject to cyclical swings are those dependent on the federal government. Throughout the history of our country localities have had to adjust to changes in the level of their economic activity. Swings in the level of local economic activity have resulted from such diverse phenomena as depletion of natural resources, changes in distribution centers, displacement by imports, and from shifts in the demand for products.

THE DEPENDENCE OF STATES ON GOVERNMENT EMPLOYMENT

Economic dependence of localities and of states on federal employment depends more on the relative size of space and defense employment than on the absolute size of such employment. Data provided by the joint NASA-DOD Economic Information System show that the smaller states are most dependent on federal procurement. The data also indicate that those states tended to become even more dependent on federal procurement as a result of the defense buildup for Vietnam.

Table 7–4 provides information about the sizes of total labor markets and of federal employment in 19 states and the District of Columbia. Ten of the states had the highest federal dependency ratios in 1967, and ten of the states had the largest total labor forces. As used here, the dependency ratio is combined NASA and Department of Defense civil service and contractor employment as a percentage of the total labor force of the area. Only California is common to both lists. The amount of federal employment shown in the second column includes civil service personnel and employees of government contractors. If included, military personnel would double the dependency ratios for a number of the states listed in the table. Combined military and civilian employment would raise the 1967 dependency ratio of Alaska to 43, of Hawaii to 21, and of Georgia to 10. As the data show, the size of combined NASA and DOD employment in those states with the highest dependency ratios is about 50 percent smaller than total NASA and DOD employment in those states with the largest labor forces.

Excluding the District of Columbia, Utah was the most dependent of

TABLE 7-4: REGIONAL DEPENDENCE ON DEFENSE AND SPACE EMPLOYMENT, JUNE 1965 AND JUNE 1967

STATE	SIZE OF LABOR FORCE IN MAY 1967 (THOUSANDS)	AMOUNT OF SPACE AND DEFENSE EMPLOYMENT JUNE 1967 (THOUSANDS)	GOVERNMENT DEPENDENCE RATIO[a] JUNE 1967	JUNE 1965
Highest dependency ratios				
District of Columbia	410	42	10.3	8.6
Utah	404	40	9.9	7.6
Alaska	96	9	9.4	9.8
Hawaii	287	25	8.7	7.9
Virginia	1,697	143	8.4	7.1
Connecticut	1,280	96	7.5	5.7
Maryland	1,373	94	6.8	5.6
California	7,711	499	6.5	4.9
New Hampshire	284	18	6.3	4.5
Rhode Island	381	20	5.2	3.7
Georgia	1,723	83	4.8	3.6
Largest labor forces				
New York	8,115	170	2.1	1.7
California	7,711	499	6.5	4.9
Illinois	4,908	88	1.8	1.0
Pennsylvania	4,827	156	3.2	2.3
Ohio	4,295	107	2.5	2.0
Texas	4,244	183	4.3	3.0
Michigan	3,405	47	1.4	1.0
New Jersey	2,842	95	3.3	2.4
Massachusetts	2,444	106	4.3	3.1
Florida	2,372	82	3.5	3.1

[a] Excludes military personnel.

Source: Adapted from Roger F. Riffler and Paul B. Downing, "Regional Effect of Defense Effort on Employment," *Monthly Labor Review* (July 1968), pp. 1–8.

all states on federal procurement in 1967. Missiles and electronics manufacturing are the newest and also the largest industries in that state. Some of the largest manufacturing companies include Atlantic Research Company, Hercules, Litton Industries, the Marquardt Corporation, Sperry Utah, and Thiokol Chemical. Although each of the Utah plants of these companies was performing some work for NASA in the mid-1960's, combined NASA employment at the plants played a negligible role in contributing to Utah's relatively high dependence on federal employment. NASA contractor employment did not amount to more than 150 people at any of the plants in early 1967.

With the exception of California, and to a much smaller extent Connecticut, the same is true of the other states with the highest federal procurement dependency ratios. In Connecticut, NASA employment was about 5 percent of total direct employment provided by the federal government. On the average, the largest labor forces at NASA contractor plants and at NASA installations are in states with relatively low dependency ratios—the five Gulf Coast States, New York, New Jersey, Massachusetts, Missouri, and Washington.

Because federal employment is not distributed evenly within the states, dependency ratios for specific localities are higher than the state averages. In Connecticut, direct employment provided by the federal government is concentrated in Hartford, New London-Groton-Norwich, and Bridgeport. Similarly, the high concentration areas in Georgia are: Macon, Albany, Columbus, Atlanta, and Augusta. In Virginia direct federal employment is most heavily concentrated in the Newport-News—Hampton and the Norfolk-Portsmouth areas.

DEMAND SHIFT INFLATION

In addition to creating instability in local employment, shifts in the structure of federal procurement can give rise to inflationary pressures in the economy. Such inflationary pressures can exist in the absence of a fully employed economy. Contemporary economists usually distinguish between three kinds of inflation: demand-pull, cost-push, and structural inflation. Demand-pull inflation occurs when consumers, businesses, and federal, state, and local governments attempt to buy more than the economy can produce, given the availability of resources and the state of technology. Demand-pull inflation can be described loosely as too much money chasing too few goods.

Cost-push inflation originates in the productive sector rather than in the spending sector of the economy. Simply conceived, cost-push inflation results from an attempt by some or all of the factors of production to command a larger share of national income than would be warranted by increases in productivity. Increased income in excess of increased productivity can result from both wage-push or profit-push. In the mid-1960's, President Johnson, on the advice of his Council of Economic Advisers, issued guideposts for non-inflationary price and wage behavior in an attempt to curb cost-push inflation. The wage guidepost was, in effect, a plea to unions not to increase wages by more than the nation-wide productivity trend. The way in which productivity is measured is discussed in Chapter 9. For their pricing policies, the guideposts urged business leaders to pass high productivity gains along to consumers in the form of lower prices. For industries with low rates of productivity gains, price increases were to be just commensurate with increases in unit costs.

With structural inflation, elements of both demand-pull and cost-push inflation occur, but in specific sectors of the economy rather than across the board. Structural inflation is commonly called demand-shift inflation by economists, because it results from stresses and strains in particular sectors of the economy. Such pressures on prices and wages then permeate the system.[4] Two phenomena exist in most capitalistic industrialized economies that give rise to this kind of inflationary pressure. First, wages and prices tend to be flexible in an upward direction, but rigid downward. Second, resources cannot shift instantaneously in response to changes in demand because of differences in labor skill and materials requirements, and differences in the geographic locations of industries and plants.

Structural inflation, caused in part by changes in federal procurement, occurred after the shift in government demand necessitated by the Vietnam War. Beginning in early 1966, the government shifted away from the production of space systems and from R & D intensive weapons systems to the production of more conventional weapons. This resulted in an increase in demand in one sector of the government market and a decrease in demand in another sector. If resources could shift rapidly and if wages and prices were flexible, this shift in federal procurement might have been accomplished without causing a major problem. However, government contractors tend to segregate business by source of funding. In addition, resources required for space projects are not perfectly substitutable for resources required for more conventional defense projects. Furthermore, plants operated by the same contractor are frequently in different geographic locations. California, of course, is one notable exception.

To attract workers, firms offer premium wage rates, and in some instances firms pay transportation and moving expenses for workers to change geographic locations. The institutional structure of the American economy is such that producers faced with decreasing government business are usually forced to pay wages similar to those paid by plants subject to increased demand. The matter of rising labor and materials costs, no doubt, was compounded in the period immediately following the large defense buildup for Vietnam. This resulted from the probable expectations of managers of firms seeking follow-on or new space and other R & D contracts. At least when spending for the space program first began to contract, and spending for Vietnam began to rise, the expectations of many companies engaged in work for NASA must have been that the Vietnam buildup would be short-lived and that federal funds would soon be released for the space program. To be in a position to undertake anticipated space and other R & D projects, some contractors undoubtedly paid premiums to key personnel to insure that those employees did not change employment in response to the shift in federal procurement.

Although the share of the 3.1 percent per annum increase in the price level occurring between 1965 and 1968 that is attributable to demand-pull, cost-push, and structural inflation is not known, there can be little doubt that the shift in demand in the federal sector contributed significantly to the general increase in the price level that took place during the period.

THE DEPENDENCE OF BUSINESS FIRMS ON GOVERNMENT EMPLOYMENT

In Chapter 8, the structure of NASA and defense procurement will be analyzed in terms of its potential harm to the competitive environment of the American economy. Another way of looking at the defense-space industry is as a captive industry, an industry in which the federal government can affect the stability and growth of corporations by the selection of contractors.

Using prime contract awards data for the 100 largest defense contractors, Weidenbaum finds that contrary to many comments about the "military-industrial complex," the corporate giants do not dominate the defense market.[5] Using data on corporate assets in 1965, Weidenbaum finds that it is the medium-sized companies that receive the largest share of the dollar value of prime contract awards. Of the 69 percent of prime contract awards going to the top 100 contractors, companies with assets of between $250 million and $999 million received 39 percent of those awards. This compares with 17 percent of awards going to companies with assets of $1 billion or more, and with 13 percent to companies with assets of less than $250 million. Weidenbaum's research also shows that it is the medium-sized companies rather than the corporate giants that are most dependent on sales to the federal government (Department of Defense).[6]

Employment data provided by the joint NASA/DOD Economic Information System confirm and strengthen Weidenbaum's findings. As previously discussed, employment data are superior to data on prime contract awards because they are not biased by subcontracting activity.[7]

The Corporate Organization of NASA Contractors

Some important NASA suppliers are divisions, affiliates, or wholly owned subsidiaries of large corporations whose major product line tends to be specialized and outside of the defense-space market. The automotive Big Three stand out most clearly as NASA contractors whose major corporate commitments are in non-aerospace industries. AC Electronics and Allison are both divisions of General Motors; Philco Corporation is a subsidiary of Ford; and, of course, Chrysler has its Space Division. Other contractors are conglomerates, such as Textron, International Telephone and Telegraph, and AVCO, which have unrelated technologies, products, services, markets, and plants under one corporate identity. Still another large group of NASA contractors is composed of the airframe companies that entered into the production of guided missiles and space vehicles.

Dependence of NASA Contractors on Federal Employment[8]

Table 7–5 displays information about the relative importance of NASA and defense employment to firms engaged in production for NASA in June 1966. Company employment is used in the widest sense—that of the parent company and its subsidiaries.[9] The data, no doubt, overstate the degree to which management coordinates the cross-flow of employment between subsidiary companies and even between large divisions. The contractor groups displayed in Table 7–5 are not exclusive; it is possible for a given firm to be included in as many as four of the six categories. Grumman Aircraft Engineering, for example, is among the group of contractors with the largest NASA work forces (group 1); among those most dependent upon NASA (group 2); among those most dependent on combined NASA and DOD employment (group 3); and is among the middle-sized NASA contractors in terms of the size of total corporate employment (group 5).

The matrix given in Table 7–6 provides additional information about the groups of contractors: (1) the relative importance of the groups to all employment on the space program; (2) the number of contractors that have at least 50 percent of reported NASA employment under prime contract; and (3) the industry classification of each firm.

The information displayed in the two tables suggests several conclusions about the commitment of NASA contractors to the space program and to federal employment in general.

First, contractors with the largest NASA work forces are not, as a group, among the corporate giants in American industry. On the average, total corporate employment of contractors with the largest NASA work forces (group 1) is between 35 and 50 percent smaller than that of the group of contractors with the largest total corporate work forces (group 4).

Second, contractors that are most dependent on NASA employment (group 2) tend to be about one-tenth the size of the group of NASA contractors with the largest total corporate work forces (group 4). Only six companies (Bendix, Boeing, Douglas, General Tire and Rubber, North American Aviation, and TRW) are common to both groups. Third, contractors with the smallest total corporate work forces (group 6) are more dependent on NASA and on total government employment than contractors with large (group 4) and middle-sized corporate work forces (group 5).

Weidenbaum found that the majority of the smaller firms, as measured

TABLE 7–5: DEPENDENCE OF SELECTED GROUPS OF NASA CONTRACTORS ON NASA AND TOTAL GOVERNMENT EMPLOYMENT, JUNE 1966

NASA CONTRACTOR GROUP	NASA EMPLOYMENT AS A PERCENTAGE OF TOTAL CORPORATE EMPLOYMENT		NASA AND TOTAL DOD EMPLOYMENT AS A PERCENTAGE OF TOTAL CORPORATE EMPLOYMENT		CORPORATE WORK FORCE	
	MEAN	MEDIAN	MEAN	MEDIAN	MEAN	MEDIAN
(1) 30 contractors with largest NASA work forces	12.3	6.5	44.0	42.6	98,900	49,100
(2) 30 contractors most dependent on NASA employment	18.7	12.9	51.1	49.8	21,600	7,800
(3) 30 contractors most dependent on total government employment	16.9	6.9	70.1	66.9	22,600	7,400
(4) 30 contractors with largest corporate work forces	5.4	2.4	25.4	14.4	155,700	93,400
(5) 31 contractors with middle corporate work forces	5.4	2.7	36.1	30.8	17,300	14,100
(6) 31 contractors with smallest corporate work forces	11.2	5.4	45.5	49.0	3,300	3,000

Sources: Computed from data supplied to the Joint NASA-DOD Economic Information System; *Fortune Directory of 500 Largest Industrials;* and various other corporate directories. Mary A. Holman, "The Dependence of Business Firms on Government Employment," *Quarterly Review of Economics and Business,* 8,4 (Winter, 1968), pp. 73–79.

TABLE 7–6: INFORMATION ABOUT SELECTED GROUPS OF NASA CONTRACTORS, JUNE 1966

NASA CONTRACTOR GROUP	PERCENT OF TOTAL NASA EMPLOYMENT	NO. UNDER PRIME CONTRACT	NO. OF AIRCRAFT COMPANIES	NO. OF ELECTRONICS COMPANIES[a]	NO. OF ENGINEERING AND BUSINESS SERVICE COMPANIES	NO. OF MOTOR VEHICLE COMPANIES	NO. OF INSTRUMENT COMPANIES	NO. OF COMPANIES IN OTHER INDUSTRIES
(1) 30 contractors with largest NASA work forces	50.5	20	16	7	1	3	1	2
(2) 30 contractors most dependent on NASA employment	38.6	16	12	8	3	0	2	5
(3) 30 contractors most dependent on total government employment	31.2	18	14	9	4	1	1	1
(4) 30 contractors with largest corporate work forces	42.4	19	10	11	0	4	1	4
(5) 31 contractors with middle corporate work forces	9.3	14	10	14	0	0	0	6
(6) 31 contractors with smallest corporate work forces	3.3	15	5	13	4	1	4	3

[a] Includes 35 companies with primary product lines in: radio and TV equipment; communications equipment and electronic components; and electrical transmission and industrial equipment.

Source: Computed from data supplied to the Joint NASA-DOD Economic Information System. Mary A. Holman, "The Dependence of Business Firms on Government Employment," *Quarterly Review of Economics and Business,* 8, 4 (Winter, 1968), pp. 73–79.

by assets, received defense contracts exceeding 50 percent of their total sales in 1965. This compares with less than 25 percent for the largest firms in his sample. Weidenbaum contends that the high ratio for the smaller firms is not typical of the thousands of small businesses participating in the defense market. He attributes the finding to the nature of his sample, the 100 DOD contractors receiving the largest absolute amounts of prime contract awards.[10] The data used here are not subject to that limitation. The cut-off point used is low—plants with at least 10 employees engaged on work for the Space Agency. It seems reasonable to believe that Weidenbaum's data do, in fact, portray the relative dependence of smaller companies, compared with larger ones, on the defense market. Of course, this does not mean that each and every small business that occasionally supplies goods and services to the government is heavily dependent upon sales to the Department of Defense or to NASA.

TABLE 7-7: NASA AND DEFENSE EMPLOYMENT AS A PERCENTAGE OF TOTAL CORPORATE EMPLOYMENT,[a] JUNE 1966

30 CONTRACTORS WITH LARGE CORPORATE WORK FORCES	31 CONTRACTORS WITH MIDDLE CORPORATE WORK FORCES	31 CONTRACTORS WITH SMALLEST CORPORATE WORK FORCES
	75 to 100 Percent Federal Employment	
McDonnell Aircraft[b,c]	Curtiss-Wright	Hayes International[b,c]
North American Aviation[b]	Grumman Aircraft Engineering[b,c]	Marquardt Corp.[c,d]
		Moog, Inc.[c,d]
		Raymond Eng. Labs.[c,d]
		Sanders Assoc.[c]
		Vitro Corp. of America[c]
	50 to 74 Percent Federal Employment	
General Dynamics[b]	Avco Corp.[c]	Atlantic Research
Lockheed Aircraft[b]	Collins Radio[b,c,d]	Babcock Electronics[d]
	Martin Marietta	Bogue Electric
	Northrop Corp.[b,c,d]	Brown Engineering[b,c]
	Raytheon Company[b,c,d]	EDO Corp.[d]
	Rohr Corp.[d]	Electronic Communications[d]
	Thiokol Chemical[b,c]	Mason-Rust[c]
		Murdock Mach. & Eng.[d]
		Spaco, Inc.[c]
	25 to 49 Percent Federal Employment	
Bendix Corp.[b,c]	Beech Aircraft[c,d]	Aeronca, Inc.[d]
Boeing Company[b,c]	Control Data[c]	Cook Electric[d]
Douglas Aircraft[b,c]	Excello Corp.[d]	Dynamics Corp. of Amer.
General Tire and Rubber[b,c]	Fairchild Hiller	Menasco Mfg.[d]
TRW, Inc.[b,c,d]	General Precision[b]	Perkin Elmer
Textron, Inc.[b,c]	Hercules, Inc.	Scientific Data Systems[c,d]
United Aircraft[b]	LTV, Inc.[b]	United Industrial
	Lear Siegler[d]	
	Ryan Aeronautical[b,c,d]	
	Standard Kollsman Inst.[c,d]	
	Westinghouse Airbrake[d]	

	Less Than 25 Percent Federal Employment	
AT&T (Bellcomm)	American Mach. & Foundry	American Bosch Arma[c,d]
Borg Warner[d]	Bechtel Corp.[c]	Ball Brothers[c]
Chrysler Corp.[b]	Cutler Hammer	Conrac Corp.[c,d]
Ford Motor Co.[b]	Emerson Electric[d]	General Signal[d]
General Electric[b]	Fairchild Camera & Inst.[d]	ITEK Corp.
General Motors[b]	Federal Mogul Bower[d]	Packard-Bell Elec.
General Tel. and Elec.	General Instrument[d]	Pneumodynamics, Inc.[d]
Goodyear Tire & Rubber	Odgen Corp.	Sangamo Electric
Honeywell, Inc.[b,c]	Schlumberger, Ltd.[d]	Sun Shipbuilding & Drydock[d]
Int'l Bus. Machines[b]	Signal Gas & Oil[b,c,d]	
Int'l Harvester	Varian Associates	
IT&T[d]		
Litton Industries		
Motorola, Inc.[b,c]		
Radio Corp. of Amer.[b,c]		
Sperry Rand[d]		
Texas Instruments[d]		
Western Electric		
Westinghouse Electric[d]		

[a] Excludes Hughes Aircraft; before mergers of McDonnell Aircraft and Douglas Aircraft and of North American Aviation and Rockwell Standard.

[b] Among 30 contractors with largest NASA work forces.

[c] Among 30 contractors most dependent on NASA.

[d] Less than 50% of NASA employment under prime contract.

Sources: Joint NASA-DOD Economic Information System: *Fortune 500;* and various corporate directories. Mary A. Holman, "The Dependence of Business Firms on Government Employment," *Quarterly Review of Economics and Business,* 8, 4 (Winter, 1968), pp. 73–79.

Fourth, the data show that the importance of government employment resulting from subcontracts is not limited to the smallest contractors. Among the group of contractors with the largest NASA labor forces (group 1), one-third of the firms are principally subcontractors, i.e., less than 50 percent of their total NASA employment originated from prime contracts. The importance of employment resulting from subcontracts has, no doubt, increased in recent years because of the systems approach to federal procurement. Many of NASA's subcontractors provide sophisticated subsystems, such as communications and data systems for the Apollo spacecraft (Collins Radio); stabilization and control systems (Honeywell, Inc.); and radar altimeter and velocity sensors for surveyor (Ryan Aeronautical). As Table 7–6 shows, almost one-half of the companies with the highest NASA dependency ratios (group 2) are primarily subcontractors.

Table 7–7, which disaggregates the data displayed in the preceding tables, lists 92 NASA contractors[11] according to the size of their total corporate employment and to their dependence on combined NASA and DOD employment. The table also identifies companies that: are among those most dependent upon NASA employment; are among those contractors with the largest NASA work forces; and those that are primarily NASA subcontractors. The resulting distribution confirms in greater detail Weidenbaum's finding that the medium-sized and small firms are more dependent on federal procurement than the giants in American industry. About one-third of the medium-sized companies have at least 50 percent of their corporate work forces committed to NASA and DOD projects. About one-half of the smallest companies have at least a 50 percent dependency ratio. This compares with about 10 percent for the largest companies.

Combined NASA and DOD dependency ratios also show that contractors with the largest NASA work forces tend to be less dependent on government-generated employment than contractors with smaller NASA work forces. As the table indicates, many of the smaller contractors provide goods and services to the Space Agency under subcontract.

THE AEROSPACE FIRMS AND DIVERSIFICATION

Diversification can improve the abilities of companies to absorb shocks resulting from changes in government procurement. Company diversification can be accomplished in different ways, such as acquisition, merger, or internal development. Diversification can also take different forms—concentric versus conglomerate, for example. Concentric diversification is based on the expansion of corporate activity around established resources. Concentric diversification has been compared with a spiral moving continually outward from a central core of products, services, and technologies.[12] Conglomerate diversification comes about when unrelated technologies, products, services, markets, and plants come under one corporate identity. In recent years, conglomerate diversification has become more common, and, in response, the Department of Justice has become concerned that conglomerate diversification can become a restraint to competition.

One of the best examples of concentric diversification is that of the rapid entry and deep involvement of aircraft companies in the production of guided missiles and space vehicles. No doubt, this kind of diversification lessened the dependence of airframe companies on the Department of Defense, but not on federal procurement in general. Another example of concentric diversification by some aircraft companies takes the form of accepting government contracts to apply systems analysis to urban problems. The State of California and several federal agencies have let a number of contracts to firms in the aerospace industry to solve problems in such fields as transportation, crime prevention, education, and pollution control. Again, this is a move away from dependence on the Department of Defense, but not from reliance on several civilian federal agencies and also on state and local governments.

Almost 70 percent of employment on the space program at its peak, resulted from contracts let to companies with primary product lines in the aircraft and parts industry. For the most part, firms in the aircraft industry have had a discouraging history of failure after failure in their attempts at commercial diversification.

A few companies producing for the aerospace industry, however, exemplify success with commercial diversification. Much of the commercial business resulted from conglomerate diversification. The names of some such companies include: AVCO, International Telephone and Telegraph, Ling-Temco-Vought, and Textron. NASA's largest contractor, North American Rockwell, is a new company. The merger between North American Aviation and Rockwell Standard, one of the nation's largest automotive components producers, was the first large space-age merger.[13] In part, the merger was undertaken to lessen the dependence of each of the companies on the cyclical nature of its business.[14] The success of North American Rockwell in converting highly specialized aerospace hardware and technologies into commercial items will be known only after a substantial time lag.[15]

The economic and social policy issue is whether or not firms heavily dependent on federal procurement should be helped by the federal government to prepare themselves to absorb swings in federal procurement. This question would become particularly important if the nation could successfully reduce defense requirements and procurement. Diversification, of course, should not be fostered or encouraged if it reduces economic efficiency by restraining competition.

THE SPACE PROGRAM AND DISCRETIONARY FISCAL POLICY

It is clear, ex post, that the space program was a stabilizing factor in the American economy during the 1960's. This is true in terms of both absorbing those separating from defense programs in the early 1960's and providing a reservoir of skilled manpower to be used by the Department of Defense for the Vietnam War. Ex ante, however, it is doubtful whether the space program could be labeled as countercyclical. During the postwar period, economic instability has never been the only factor, and it has seldom been the most important factor, in determining the inception and level of new government programs or in formulating federal budgetary policy. The federal fiscal process is such that the Executive Branch and Congress justify most countercyclical actions on grounds other than economic stabilization. Most programs commonly labeled as "counter-recessionary" would probably have been initiated anyway.[16]

Discretionary Fiscal Action Versus Automatic Stabilizers

Government tools for the maintenance of economic stability include monetary policy, debt management policy, direct controls on prices and wages, and fiscal policy. Fiscal policy is the concern here because it deals with changes in the level of government spending, as well as, of course, changes in taxation. The fiscal system of the United States has built-in stabilizers that automatically counteract changes in the economic climate. The most obvious of these stabilizers is unemployment compensation insurance and other forms of welfare transfers. The progressive nature of the personal and corporate tax structure is another such force. As personal and corporate income receipts rise or fall, without any change in tax rates, government tax revenues rise or fall. The various agricultural subsidy programs are another cushioning element in the economy.

Another kind of corrective fiscal action is discretionary, i.e., one that requires Presidential and Congressional decision-making and approval. Discretionary fiscal policy includes changing tax rates, varying transfer payments, and varying federal expenditure programs. A number of economists argue that during the post-World War II period, discretionary fiscal actions came too late and were too modest in their proportions to play much of a role in changing the severity of cyclical economic behavior. In most of the recessions, as some argue, discretionary anti-cyclical actions were not applied until after the lowest point in the economic contraction had been reached. The major impact of discretionary actions has been to speed change rather than to initiate reversal.[17]

Lags in the Process of Discretionary Action

Unlike the built-in stabilizers that go into effect automatically, discretionary fiscal action can be a time-consuming process. To be sure, there is a lag between the time that economic activity reverses and the time that reversal is recognized by fiscal authorities. The advent of the computer, along with more comprehensive and sophisticated statistical reporting systems, has done much to reduce this lag. Another lag occurs because the budget process itself is long. That is the lag between recognition time and the time fiscal action materializes. Still another lag occurs between the time that fiscal action takes place and the time the desired effects on the economy materialize. The length of the third lag, of course, varies with the kind of discretionary fiscal action that Congress initiates.

Lead-time in the Procurement of R & D and Space Hardware

The space program, like defense R & D projects, is not suitable as a tool of discretionary fiscal policy.[18] This results from the fact that most space hardware can be obtained only with a long lead time between commitment and delivery. Table 7–8 displays an estimate of the production lead time requirement for the major hardware components going into the completed Apollo spacecraft. The lead time estimates are based on NASA spending rates during late 1966. The requirements are for production and delivery to the launch site. The estimates do not take into account time lags resulting from the several echelons of decision-making that deemed the hardware essential for the Apollo spacecraft; nor do the estimates include the time required for research and development before the components became operational. As the data show, the shortest lead time is eighteen months, and the hardware component is the instrument unit produced by International Business Machines. The longest procurement lead time is for the first stage (S-IC) of the giant Saturn V booster. The lead time for most of the other major hardware items is about three years.

TABLE 7–8: PROCUREMENT LEAD TIME FOR SELECTED APOLLO HARDWARE ITEMS[a], 1966

HARDWARE ITEM	CONTRACTOR	PRODUCTION LEAD TIME (MONTHS)
Command and Service Module	North American Rockwell	30
Lunar Excursion Module	Grumman Aircraft Engineering	22
Components of the Saturn V launch vehicle:		
S-IC stage	Boeing Aircraft	39
F-1 engine	Rocketdyne[b]	37
S-II stage	North American Rockwell	30
J-2 engine	Rocketdyne[b]	33
S-IVB stage	McDonnell Douglas	34
J-2 engine	Rocketdyne[b]	30
Instrument unit	International Business Machines	18

[a] Approximate time in months from ordering item to delivery at launch site. Excludes time lapse for contract competitions, preproduction research and development, and prelaunch checkout.

[b] A division of North American Rockwell.

Source: Office of Manned Space Flight, National Aeronautics and Space Administration.

As Novick points out, it is easy to lose sight of the time dimension in the R & D process, as well as the fact that money cannot be fully substituted for time. Because of the long-lead procurement, continuity of effort in large-scale R & D projects buys more than spasmodically large bursts of expenditures.[19] Fluctuating with changes in international affairs, federal spending for national defense has been the least stable component of total spending in the United States during the postwar period. At times defense spending has initiated adverse economic tendencies, and at times it has mitigated them. Because the space program is so new, it is not yet clear whether outlays for space will also enlarge and contract in response to changes in international circumstances. If space spending does fluctuate as defense spending does, and one set of auguries indicates that it will, the space program will become another component of federal spending that can give rise to inflationary and deflationary shocks. Again it is too soon to know, but another set of auguries indicates that the space program can play an offset role to changes in defense employment. If spending for space performs such a function, there will be large disruptions in the space program. In addition, such a policy will reduce the coherence and consistency of achieving space goals.

SUMMARY

With the Employment Act of 1946, economic stability was established as a national goal. Although changes in levels of federal spending have been larger than changes in consumer and business investment spending during the postwar period, the aerospace industry witnessed milder fluctuations than most other durable goods manufacturing industries. One reason for this is the stabilizing role played by the space program during the decade of the sixties.

As a result of a decrease in defense requirements, actual employment in the guided missiles and space vehicles component of the aerospace industry dropped by 17,000 between 1963 and 1965. Without the space program, the decrease in employment at plants throughout the country producing guided missiles and space vehicles might have amounted to about 85,000. In the aircraft and parts segment of the aerospace industry, employment dropped by 14,000 during the same period; but without the space program, the decrease could have been as large as 106,000. Between 1962 and 1965, employment at plants producing communications equipment dropped by 28,000; without NASA offset employment, the decline could have amounted to 83,000.

After the phase-down of the space program in early 1966, NASA again played a stabilizing role in the aerospace industry. Employees of NASA contractors separating from space projects increased the supply of technical and scientific personnel available to meet rising defense requirements for the Vietnam War. During the mid-1960's, the space program had brought together and maintained a large reservoir of skilled, technical, and scientific manpower in the aerospace industry. Because of the change in the procurement-mix required for Vietnam, however, many of those transferring from space to defense projects left NASA plants in one geographic location to go to work on defense projects located in another part of the country. Because wages and prices tend to be flexible upward but rigid downward, and because resources cannot instantaneously shift in response to changes in demand, the marked shift in the structure of federal procurement, after the rapid buildup for the Vietnam War, must have contributed to a significant part of the inflationary spiral witnessed by the economy in the late 1960's.

Federal procurement cycles do not affect communities throughout the nation evenly. Changes in employment in a number of localities mirror the fortune or the misfortune of specific companies in obtaining contracts for defense and space projects. The localities that have experienced the greatest fluctuations in aerospace employment during the postwar period include: Boulder, Denver, Fort Worth, Los Angeles, Wichita, Sacramento, San Diego, and Seattle.

Federal procurement also has differential effects on the stability and growth of corporations. The more captive or dependent a firm, the more vulnerable it is to shifts in the level of federal procurement. Contrary to many reports about the "military-industrial complex," the corporate giants do not dominate the military market. In addition, contractors with the smallest total corporate work forces are more dependent on NASA and on total government employment than contractors with large and middle-sized corporate work forces. The dependence of many small contractors on federal procurement results from their sales of goods and services to the government under subcontracts. The abilities of companies to absorb fluctuations in federal procurement improve by diversification into commercial product lines. Most firms in the aerospace industry, however, have failed in their postwar attempts to diversify.

Although the space program was a stabilizing force in the American economy in the 1960's, it is not a suitable tool of discretionary fiscal policy. The major components of space hardware are obtained only after long-lead times. If used as a countercyclical tool, spending for space could reinforce and aggravate economic instability rather than counteract instability. In addition, if spending for space is used to offset changes in defense or private spending, there would be large disruptions in the space program which would increase the cost and reduce the cohesiveness of achieving space goals.

NOTES

1. For example, see Bert G. Hickman, *Growth and Stability of the Postwar Economy* (Washington, D.C.: Brookings Institution, 1960); and Maurice W. Lee, *Macroeconomics: Fluctuations, Growth and Stability* (4th ed.; Homewood, Ill.: Irwin, 1967).

2. U.S. Council of Economic Advisers, "The Employment Act: Twenty Years of Policy Experience," *1966 Annual Report of Council of Economic Advisers*, pp. 170–86.

3. *Report of the Committee on the Economic Impact of Defense and Disarmament*, Gardner Ackley, Chairman (Washington, D.C.: U.S. Government Printing Office, 1965).

4. The name usually associated with the concept of structural inflation is Charles L. Schultze, former Director of the Bureau of the Budget. See Charles L. Schultze, "Recent Inflation in the United States," Joint Economic Committee, Study Paper No. 1, *Study of Employment, Growth and Price Levels*, 86th Cong., 1st Sess., 1960.

5. Murray L. Weidenbaum, "Concentration and Competition in the Military Market," *Quarterly Review of Economics and Business*, 8, 1 (Spring, 1968), pp. 7–17. Also, Murray L. Weidenbaum, *An Exploratory Analysis of Defense/Space Companies*, Working Paper No. 6513 (St. Louis, Mo.: Washington University, 1965).

6. Weidenbaum, "Concentration and Competition in the Military Market," pp. 7–17.

7. The data used here consist of a sample of 93 contractors that had at least 10 employees engaged on projects for NASA in June 1966. Because defense employment is so much larger than space employment, many of these contractors had plants engaged on defense projects but not on NASA projects. To show dependence on defense and space employment, the data also include employment at the plants of the 93 contractors that produced solely for the Department of Defense. Although the sample used here includes 275 different plants, only 198 had at least 10 employees engaged on projects for NASA.

8. Part of the following discussion of contractor dependence was published in the *Quarterly Review of Economics and Business:* Mary A. Holman, "The Dependence of Business Firms on Government Employment," 8, 4 (Winter, 1968), pp. 73–79.

9. Although Western Electric is a subsidiary of American Telephone and Telegraph, it is treated as a separate company. Western Electric is the research and manufacturing arm of AT & T. The *Fortune Directory of 500* (1966) lists Western Electric as a separate corporate entity.

10. Weidenbaum, "Concentration and Competition in the Military Market," p. 12.

11. Hughes Aircraft is excluded because of the company's well-known reluctance to disclose information about its corporate activities.

12. John S. Gilmore and Dean C. Coddington, *Defense Industry Diversification,* U.S. Arms Control and Disarmament Agency Publication No. 30 (Washington, D.C.: U.S. Government Printing Office, 1966).

13. The merger of two other large NASA contractors, McDonnell Aircraft and Douglas Aircraft, in 1967 was quite different. Although at the time of merger Douglas had a larger commercial component, both aircraft companies have similar technologies and capabilities.

14. John S. Gilmore and Dean C. Coddington, "Diversification Guides for Defense Firms," *Harvard Business Review* (May–June, 1966), pp. 144–59.

15. "North American Rockwell: The First Space-Age Merger," *Forbes* (July, 1968), pp. 67–70.

16. Wilfred Lewis, Jr., *Federal Fiscal Policy in the Postwar Recessions* (Washington, D.C.: Brookings Institution, 1962), pp. 16–17.

17. See for example, Lewis, *ibid.,* pp. 20–21 and 270; also, Arthur F. Burns, "Progress Toward Economic Stability," *American Economic Review* (March, 1960), pp. 1–19.

18. Writing in the late 1950's, Keezer forecast that the research and development industry would operate as a stabilizing force in the American economy. Dexter M. Keezer, *The New Forces in American Business: An Analysis of the Economic Outlook for the Sixties* (New York: McGraw-Hill, 1959), pp. 64–67.

19. U.S. Congress, "Lead Time in Modern Weapons," by David Novick, *Federal Expenditure Policy for Economic Growth and Stability, Hearings,* before the Subcommittee on Fiscal Policy of the Joint Economic Committee, 85th Cong., 1st Sess., November, 1957, pp. 374–77.

8 Performance: Concentration, Monopoly, and Economic Control

It is well known that of the tens of thousands of business firms in the United States, a relatively small number provide the vast majority of goods and services procured by the federal government. This fact has given rise to much public debate and controversy for decades. During the period immediately preceding and following World War II, the debate centered on the role of the government in contributing to monopoly and economic control in the manufacturing sector of the economy. The strongest indictment against the government was probably made by two economists, Walter Adams and Horace M. Gray, in the mid-1950's.[1] On the other side of the coin, it was argued that government purchases served to eliminate monopoly during World War II. An example of this was the role played by the government in the aluminum industry. By means of federal purchases, the government induced the entry of two major aluminum producers as part of its effort to destroy the Alcoa monopoly.[2]

Sharp increases in federal funds for complex space and weapons hardware, requiring the development and use of sophisticated technology, heightened the controversy and widened the number of controverted issues. One has only to peruse the voluminous testimony of public officials, industry representatives, and academicians before a half-dozen different Congressional committees to see the gamut of public concern.[3] Although the issues are multifarious, they can be divided into two principal groups. One group of issues, which are discussed in the first part of this chapter, pertains to equity and efficiency in the regional and geographic distribution of federal procurement. The other set of issues relates to the concentration of economic power in American industry. The second part of this chapter presents an analysis of the concentration of NASA's research and development effort among large corporations, the extent to which NASA and the Department of Defense rely on the same companies for the provision of goods and services, and also the effect of NASA patent policies on monopolistic control of American industry.

THE GEOGRAPHIC DISTRIBUTION OF FEDERAL PROCUREMENT

As is true of most subjects discussed in this book, the economics and the politics of the geographic distribution of federal procurement, especially R & D procurement, are inseparably inter-related, and there are no generally accepted criteria by which actual performance can be measured. From the taxpayer's point of view, however, the solution is clear. The federal government's policy should be that of getting the most and the best for the least. This means that any regional distribution of federal procurement should be based solely on the criterion of economic efficiency. Yet, the taxpayer has other roles: he is a constituent, or possibly a local or state government official, or a member of Congress. From his other perspective, there is some notion that federal procurement should be allocated on a basis of a "fair share."

Sense-of-the-Congress

The view that R & D procurement should reflect an "equitable" geographic distribution is spelled out in the recommendations made by several Congressional committees during the mid-1960's.[4] One of the Congressional inquiries was undertaken by the Subcommittee on Science, Research, and Development of the House Committee on Science and Astronautics.[5] After extensive hearings in the spring of 1964, the Subcommittee, chaired by Emilio Q. Daddario, recommended that ways be explored to achieve a more even geographic distribution of federal R & D funds. Further, the Subcommittee recommended that the White House sponsor a government-industry conference to: (1) identify the unused technical capability of "research-poor" sections of the country; and (2) implement a broader R & D subcontracting pattern.[6] Research undertaken by the Subcommittee yielded a number of findings, some unexpected, some ridiculous. In 1963, for example, Nevada ranked 1st in the per capita distribution of federal R & D funds, but 20th, when ranked by prime contract awards. On a per capita basis, both New York and Pennsylvania were below the national average, whereas on the basis of amount of total awards they ranked 2nd and 5th, respectively. Another finding was that Nebraska, on a per scientist basis, ranked first, with more than seven times the national average, but only 47th when ranked by the amount of prime contract awards. Still another fact discovered by the Subcommittee was that New Mexico received more extra-mural research money per $1,000 contribution to federal taxes than did any other state in 1963. In that year, New Mexico enjoyed a 97 percent R & D award return on every federal tax dollar paid. California, which annually has received over one-fourth of all federal R & D procurement awards during the last decade, did not place first when ranked by any of these measures of concentration.[7]

The Congressional hearings led to some Congressional and Presidential action. Senate Resolution 110, 90th Congress, 1st Session, requested the National Science Foundation to investigate legal actions that might be required to provide for a more equitable distribution of federal R & D funds to all qualified institutions of higher learning.[8] In NASA's Authorization Acts for fiscal years 1966 and 1967, Congress specified that it is in the national interest that the Space Agency consider the geographic distribution of its research funds, whenever feasible.[9] In the fall of 1965, President Johnson directed the heads of executive agencies to administer all mission-oriented research programs so as to strengthen and increase the number of academic institutions capable of performing research of high quality.[10]

The reasons why communities and regions want to develop a local R & D complex are clear. In the first place, any federal procurement is exogenous income, which, through the multiplier effect, can increase employment, increase income, and stimulate economic growth. In addition, research and development creates industries that are "clean," compared with manufacturing activities and some other forms of industrial development. The income and educational characteristics of those engaged in research and development tend to be higher and in some ways "superior" to those of the more conventional business endeavors. Localities have also viewed the research and development industry as having a competitive advantage over other industries. With a strong R & D base, the industrial structure of a community or of a region is less likely to become obsolete and thus lead to an exodus of the youthful population. As the preceding chapter indicated, these benefits do not come without cost. A strong research and development industry necessarily implies local or regional dependence on the federal government. Also, it cannot be assumed that a large inflow of federal R & D dollars will develop into a permanent local research and development complex. Seattle, with Boeing, and St. Louis, with McDonnell Douglas, are classic examples of localities that have received billions of dollars of federal R & D funds; and yet these communities failed to develop research and development complexes. As discussed in Chapter 6, the same is true of the communities surrounding the NASA manned space flight field centers.

Reasons for the R & D Complexes

In recent decades, five major research and development complexes have sprung up in the United States. They are the Boston-Cambridge area, the New York-Northern New Jersey area, Washington, D.C., the San Francisco-Palo Alto complex, and the Southern California complement in Los Angeles-Long Beach. With the exception of the Washington metropolitan area, each has at least one university that has been a center of academic excellence for a number of years.

Traditional economic location theory offers little to explain the geographic concentration in research and development activity. Location theory emphasizes the relationship of industries to their sources of materials, to their labor supplies, and to their markets.[11] In comparison with most manufacturing industries, research laboratories have small materials requirements. Manpower is the most important input in the R & D industry. Scientific, engineering, and technical manpower is highly mobile. According to one economist, Charles T. Stewart, the R & D industry is essentially footloose; yet it exists in a limited number of areas.[12]

Stewart believes that over time, the geographic agglomeration of research and development results in substantial economies and advantages over those areas which do not have concentrated research facilities. One kind of economy that arises from geographic concentration is "internal" to the research and development industry. Concentrated research results in the development of a diversified and self-contained labor pool from which the numerous establishments can draw. The labor pool, which develops auxiliary services as well as technical and engineering services, increases productivity in the conduct of research. The second economy is what economists call an "external economy." The economy is external to the research and development industry and it tends to be environmental. The large research complex, according to Stewart, provides an intellectual and cultural atmosphere that attracts and holds qualified R & D personnel. Because of these economies, research tends to breed research. This simply means that R & D personnel have strong residential preferences for the areas where research is concentrated. The geographic characteristics, of course, can be quite diverse, i.e., the Cambridge-Boston area does not have much in common with that of Southern California.[13]

Shapero, another researcher in the dynamics of the R & D industry, believes that no one factor can be singled out as the cause for the growth of an R & D complex. According to Shapero, however, the availability of local financial support for companies engaged in research and development is an essential ingredient in research agglomeration. The availability of a local pool of financial capital creates economies similar to those that result from a large R & D labor pool. An R & D-oriented financial community provides the impetus to support new, small, and growing companies.[14]

National Goal Trade-off Analysis

Although not important in the past, R & D procurement can be used as a tool of economic and social policy today. The net benefits expected from federal spending for research and development depend on the relationship between the positive and negative effects that such spending has on the attainment of national goals, as well as on the priorities set for the national goals. A given geographic distribution of R & D funds implies that the funds are, in fact, being used to accomplish a given set of pertinent national goals. The opportunity cost of that existing geographic distribution of funds is simply the foregone return on other national objectives that cannot be pursued because the funds are used in another way. A re-allocation of a given R & D budget among localities or regions would be advantageous if the funds were being used inefficiently or if there were a change in the priorities established for national goals. It has been argued that the existing distribution of federal R & D funds, with its great concentration, is inefficient because it grossly underestimates the scientific capabilities of some areas and grossly overestimates the scientific capabilities of other areas.[15] This argument is weakened by the fact that private R & D endeavors, which are subject to the market and price mechanism, have also tended to flow into the agglomerated R & D complexes.

The use of federal R & D procurement to stimulate economic growth in depressed and low income sections of the country would represent a new set of national goals. It would also represent a change in the relative priorities established for federal procurement policy. Some prominent economists, including Harry G. Johnson, believe that it should be a deliberate federal policy to locate scientific research in the backward areas of the country to encourage their industrial development. Johnson contends that in the long run such a policy would provide a socially and economically more attractive attack on poverty than many of the area development policies used during the 1960's. Johnson agrees, however, that there is a natural tendency for scientific activity to concentrate around established centers of scientific accomplishment. And, according to Johnson, that agglomerative tendency is probably the most efficient way of conducting scientific research, from the point of view of science itself.[16]

GEOGRAPHIC DISTRIBUTION OF NASA'S R & D EFFORT

NASA made a deliberate effort to expand the geographic participation of states in the space program. As indicated in Chapter 2, the 1961 Silverstein Memorandum to NASA's site selection committee explicitly stated that preference for the location of the Manned Spacecraft Center would be given to a mid-continent state. In mid-1961, NASA's research and development effort was heavily concentrated on the East and West Coasts. This avowed policy was coupled with the logistics requirement that the research, production, test, and launch facilities for NASA's enormous boosters be in close proximity to one another and also be located on year-around navigable water. With the then existing NASA research facility at Huntsville, on the Tennessee River, and the established launch facility at Cape Canaveral (Kennedy), it was inevitable that NASA would also put its manufacturing and test facilities in the South. The Manned Space Flight Program, as well as NASA in general, has been responsible for expanding and for developing new aerospace research complexes in the South. The significance of the shift in aerospace employment into this area can be seen from the data displayed in Table 8-1. In June 1967, the five Gulf Coast States accounted for about 26 percent of total NASA employment, compared with slightly more than 11 percent of the Department of Defense's industrial and civil service labor force. The numbers in the table do not include military personnel. Although those working on NASA projects comprised about 10 percent of the national total, NASA personnel and employees of NASA contractors accounted for about 20 percent of the combined NASA and DOD work force in the South.

TABLE 8-1: GEOGRAPHIC DISTRIBUTION OF SPACE AND DEFENSE EMPLOYMENT[a], JUNE 1967 *(PERCENTAGE DISTRIBUTION)*

GEOGRAPHIC LOCATION	DEFENSE	NASA[b]	TOTAL	NASA AS A PERCENTAGE OF TOTAL
Gulf Coast[c]	11.5	25.7	13.0	19.9
New England and Middle Atlantic[d]	35.3	21.0	33.8	6.2
Pacific Coast[e]	18.5	44.2	21.0	21.0
Central and other[f]	34.7	9.1	32.2	2.8
Total	100.0	100.0	100.0	10.0

[a] Excludes military personnel.

[b] The distribution displayed above includes civil service employment, which is concentrated in the Gulf Coast States and the New England and Middle Atlantic regions.

[c] Alabama, Florida, Louisiana, Mississippi, and Texas.

[d] Connecticut, Maine, Massachusetts, New Hampshire, Rhode Island, Vermont, Delaware, District of Columbia, Maryland, New Jersey, New York, Pennsylvania, Virginia, and West Virginia.

[e] California, Oregon, Washington, Alaska, and Hawaii.

[f] All other states.

Source: Computed from data in Roger F. Riefler and Paul B. Downing, "Regional Effect of Defense Effort on Employment," *Monthly Labor Review* (July 1968), pp. 1–8.

The data in the table also show that even though the space program resulted in a wider national R & D base, the program also increased the share of federal procurement going to the Pacific Coast, especially to California. The Midwest and the Mountain States have benefited the least from employment created by the space program.

SUBCONTRACTING AND THE GEOGRAPHIC DISTRIBUTION OF PROCUREMENT

It has been correctly argued that the geographic distribution of prime contract awards can be a misleading indicator of the concentration of procurement because of substantial subcontracting activities. As more information becomes available, however, the evidence mounts that the geographic distribution of federal procurement is not significantly changed by subcontract awards. Although the information is not conclusive, one study conducted by the Department of Defense suggests that the pattern of defense subcontracting might be geographically more concentrated than that of prime awards. The study indicated that ten states accounted for 64 percent of prime contract awards, whereas the same ten states subsequently received 76 percent of subcontract awards.[17]

Data collected by NASA on the impact of subcontract flows are more complete than similar information for defense activity. NASA has information about the net contract positions of states for the seven-year period between July 1, 1961 and June 30, 1968. That information is displayed in Table 8–2. Data on prime contract awards going to the top 22 states are also presented. A glance at the table clearly shows that the redistribution effect of subcontract activity among NASA contractors is slight. The rank order of each state does not change much as a result of subcontract awards. This is true despite the fact that nine of the locations listed in the table (California, New York, Louisiana, Alabama, Texas, Missouri, Wisconsin, District of Columbia, and Washington) are "net export" states. For those locations, subcontract awards to other states exceeded subcontract awards received from other states during the period. The reason for the relatively small impact of subcontract flows on the geographic distribution of NASA procurement is straightforward. States and regions with agglomerative productive capacity in any given industry can compete effectively for subcontract awards, as well as for prime contract awards.

In his research on NASA's subcontracting activities, Burgess finds that specialized industrial capabilities of states participating in the space program vary considerably by geographic area.[18] According to Burgess, California has a broad and general capability in almost all of the relevant industries—airframes, electronics, communications, instruments, and research laboratories. That state has a competitive advantage for the

TABLE 8-2: GEOGRAPHIC DISTRIBUTION OF NASA PROCUREMENT AWARDS, JULY 1, 1961–JUNE 30, 1968

	PRIME CONTRACT AWARDS				NET CONTRACT AWARDS[a]			
STATE	AMOUNT (MILLIONS OF DOLLARS)	PERCENTAGE	CUMULATIVE PERCENTAGE	RANK	AMOUNT (MILLIONS OF DOLLARS)	PERCENTAGE	CUMULATIVE PERCENTAGE	RANK
Total awards	22,663	100.0	100.0	[c]	22,283	100.0	100.0	[c]
California	9,916	43.7	43.7	1	9,148[b]	41.0	41.0	1
New York	2,256	10.0	53.7	2	1,360[b]	6.1	47.1	3
Louisiana	1,669	7.5	61.2	3	1,507[b]	6.7	53.8	2
Alabama	1,357	6.0	67.2	4	1,265[b]	5.7	59.5	5
Florida	1,293	5.7	72.9	5	1,352	6.0	65.5	4
Texas	1,061	4.7	77.6	6	1,052[b]	4.7	70.2	6
Missouri	831	3.7	81.3	7	823[b]	3.7	73.9	7
Maryland	611	2.7	84.0	8	652	2.9	76.8	9
New Jersey	494	2.3	86.3	9	636	2.9	79.7	11
Massachusetts	436	1.9	88.2	10	744	3.3	83.0	8
Wisconsin	390	1.7	89.9	11	257[b]	1.2	84.2	16
Pennsylvania	358	1.6	91.5	12	638	2.9	87.1	10
District of Columbia	280	1.2	92.7	13	267[b]	1.2	88.3	15
Ohio	260	1.1	93.8	14	350	1.6	89.9	13
Virginia	235	1.0	94.8	15	241	1.1	91.0	17
Washington	195	0.9	95.7	16	151[b]	0.7	91.7	20
Minnesota	150	0.7	96.4	17	353	1.6	93.3	12
Connecticut	138	0.6	97.0	18	320	1.4	94.7	14
Michigan	120	0.5	97.5	19	152	0.7	95.4	19
Colorado	103	0.5	98.0	20	144	0.6	96.0	21
Illinois	97	0.4	98.4	21	156	0.7	96.7	18

[a] Prime contract awards plus the difference between subcontract awards from other states and those awarded to other states.
[b] Subcontract awards to other states exceeded awards from other states, i.e., there was a net outflow of subcontract awards.
[c] Not applicable.

Source: National Aeronautics and Space Administration, *Annual Procurement Report: Fiscal Year 1968*, pp. 79–80.

production of such aircraft-related items as thrust chambers, landing and recovery systems, propellant tanks, fuel feed and exhaust assemblies, motors, and engines. Connecticut possesses a similar, but less developed, capability for aircraft-related components, subsystems, and systems used for the space program. Massachusetts has a competitive advantage for the production of many electronics components. The Middle Atlantic States tend to concentrate on the production of measuring instruments, communications devices, computers, and electronics components. Finally, fabricated metal products and industrial machinery come mostly from firms located in Pennsylvania and the East North Central region of the country.[19]

Rightly, William Miernyk argues that the economic effects of federal research and development are much more widely diffused than one would suppose by looking at only the areas where awards are let and sublet. Because of the lack of adequate data, it is not readily possible to identify the indirect production induced by federal R & D spending. As an example, however, Miernyk estimated the indirect effects of government R & D spending on West Virginia. Among the "haves" and the "have nots," in terms of federal prime contract awards, West Virginia has been a "have not" state. The estimate, using the national input-output table constructed by the United States Department of Commerce, is for 1967. West Virginia received four-hundredths of one percent of total direct procurement awards. Indirect production, however, amounted to an estimated four-fifths of one percent of total federal R & D spending in 1967. This means that indirect production generated in West Virginia from federal R & D procurement is almost twenty times as large as that of direct procurement awards. Unfortunately, Miernyk has not made a similar estimate for each of the fifty states.[20]

MONOPOLIZATION AND ECONOMIC CONCENTRATION IN INDUSTRY

According to contemporary economic theory, the most efficient distribution of productive resources in an economy will occur when the number of firms supplying a product is sufficiently large to preclude any firm from influencing the market price.

Competition Versus Monopoly

The next six paragraphs provide those working outside the field of economics with a brief description of market structures and the benefits of competition. The sections can be scanned or skipped without loss of continuity. The market structure for optimum economic efficiency is called perfect competition. If a product were produced under conditions of perfect competition, consumers would get the product at the lowest possible price. The notion of perfect competition, however, is a theoretical ideal. A number of market conditions are required if perfect competition is to exist. The characteristics include: (1) a large number of buyers and sellers, (2) production of a standardized or homogeneous product, (3) complete knowledge about prevailing prices and future events, (4) perfect and instantaneous mobility of resources producing the product, and (5) freedom of firms to enter and to exit from the industry. Prior to the 1930's and the intervention of the federal government in the market, many agricultural products were characterized as produced under conditions of perfect competition. Today, no product can be described as produced under conditions of perfect competition.[21]

Perfect competition is only one competitive situation in economic theory. Other kinds of competition, or more aptly, adjectives used by economists to describe a market environment, include: atomistic, free, pure, imperfect, monopolistic, oligopolistic, effective, potential, workable, cut-throat, predatory, and unfair competition.[22] With the exception of pure competition, each of the adjectives describe a market structure that can, and usually does, have monopolistic aspects. As a matter of fact, some of the monopolistic features that characterize a free enterprise system are those that make it diversified and dynamic. The best known of these monopolistic elements include advertising, discounts, warranties, quality and style changes, as well as other sales promotion techniques.

It is as difficult to identify a perfect monopoly as it is to identify a perfectly competitive industry. At one time in economic theory, a monopolist was defined as the sole producer of a product. With such a definition, the firm and the industry were identical. For years the Aluminum Company of America was cited as an example of a monopoly. To a lesser extent, the same was true of Dupont. Another definition of a monopolist in the economic literature is any firm that has any control over its own price. Using that definition, nearly every firm is a monopolist, each competing with other monopolists.

The legal definition of monopoly is no better than the several definitions used in economics. Under Section 2 of the Sherman Act, the offense of monopoly has two aspects: (1) the possession of monopoly power in the relevant market, and (2) the willful acquisition or maintenance of that power, as distinguished from growth or development as a consequence of a superior product, business acumen, or historic accident.[23] A shorthand legal definition of monopoly is, the power to raise prices or to exclude competition when one desires to do so.[24] Each antitrust case, of course, is different. The only common ground is that the law requires the examination and analysis of the evidence about market structure and about market behavior. On the basis of the legal evidence, i.e., "facts," market behavior and action may or may not be considered harmful to competition.[25]

Although passed into law in 1890, Section 2 of the Sherman Act highlights the fact that the two economic and social policies of the maintenance of competition and economic growth need not coincide. From a legal point of view the problem is to distinguish a superior competitive position that is the result of growth and development from one that is the result of the willful acquisition and maintenance of monopoly power. At present, the antitrust and regulatory agencies provide few criteria by which to clarify the legality or the illegality of business conduct that affects both competition and technological change.[26]

The Benefits of Competition

Although it is impossible to give a universally agreed-upon definition of competition, it is not impossible to outline some of the essential ingredients of a competitive environment. They include: independence of rivals, responsiveness of price to changes in demand and cost conditions, availability of market information, countervailing power, existence of potential competitors, innovation and progressive behavior, turnover among firms, and consumer choice. At least three broad sets of economic and social benefits accrue to the competitively structured economic system. First, because prices reflect the wants of consumers and the costs of producers, resources flow toward their most productive uses. The first benefit gives rise to the second. Under competition, income created in the production process tends toward an equitable diffusion among consumers. Finally, competition creates a persuasive incentive for improved managerial techniques that results in cost reductions and in product and process improvements and innovations.[27]

The Competitive Structure of the Defense-Space Market

Broadly conceived, the structure of the defense-space industry is what economists call a "monopsonistic" market, where a number of producers sell to a single buyer, the federal government. Given the existence of several federal agencies engaged in procurement activities and the possibility of inter-agency and inter-service rivalry, the defense-space industry could, perhaps, be called an "oligopsonistic" market. This simply means that there are a number of producers selling to a small number of buyers, i.e., the different federal agencies and departments.

The competitive environment of the defense-space industry has been comprehensively analyzed and described by Peck and Scherer, by Stekler, and by Weidenbaum.[28] Scherer describes the defense procurement process as a non-market, quasi-administrative buyer-seller relationship, in which the government must deliberately structure its relations with contractors so as to assure the successful execution of weapons programs. Scherer believes that there are no simple rules that permit the government decision-maker to know when and how much competition is appropriate. Each procurement presents a different case. Competition among those in the defense industry has, according to Scherer, provided benefits, such as: the encouragement of innovation, efficiency, and increased probability of project success. But Scherer believes that some kinds of competitive behavior can also stifle rather than stimulate effective procurement. The negative aspects that can result from too much competition include: destruction of the morale of technical personnel whose projects are threatened; reduction in the flow and exchange of technical information between firms that must cooperate and at the same time compete with one another; and the unwarranted resource expenditure of supporting more than one approach to a given technical objective.[29]

On the selling side of the market, Stekler observes that although there are a substantial number of firms in the aerospace industry, sales of most products tend to be concentrated among a relatively small number of companies. He believes that there is an absence of active rivalry among firms after the initial design competition for a given weapons or space system. That lack of rivalry is attributed, in part, to the complexity and cost of modern weapons and space systems. According to Stekler, the principal barrier against new firms entering into the industry is the cost of the facilities. In addition, firms in the industry are required to spend immense amounts of money on research to compete for and to obtain new contracts. Such expenditures, of course, must also be incurred by new entrants. On the buying side of the market, Stekler observes that the existence of several federal agencies and some inter-service rivalry have probably increased competition in the aerospace industry.[30]

Weidenbaum believes that the high technology government market is a crossroad between the public and the private sectors of the American economy. Some of the characteristics of this unique market include: its monopsonistic structure, the fact that price is not the controlling determinant of sales, that technical capability is a principal requirement in the competitive process, that firms must cope with rapid obsolescence resulting from rapid technological change, that firms produce after the sale (i.e., contract) and not for inventory, and finally, that the majority of work in the high technology market is performed by corporations oriented to public requirements rather than to market demands.[31]

Concentration in American Industry

It is an undeniable fact that American industry is concentrated. An indirect way to measure the degree of competition in industry is by means of concentration ratios. Conventionally, concentration ratios show the percent of assets, say, of the 4 largest, the 8 largest, and the 20 largest firms in a given industry or in the entire manufacturing sector of the economy. The use of concentration ratios is not limited to assets. Data on employment, value added, sales, and income can also be used to measure concentration. Concentration ratios provide an indirect indicator of the competitive environment because they do not reveal the intensity of competition between, say, the four largest firms in an industry. Such information would require a case study of the competitive relationship among firms in a given industry. At least one economist in the field of industrial organization, however, believes that, in general, the smaller the concentration ratio, the greater the odds that a competitive environment exists.[32]

If total assets are used as a measure of economic concentration, it can be seen that the 5 largest companies in 1962 owned 12 percent of the total assets of all manufacturing companies in the United States. The first 10 companies owned 18 percent, the first 50 owned 36 percent, and the first 100 owned 46 percent of total assets in 1962.[33] Several researchers have found that concentration in American industry has not increased since the turn of the century; if anything, there has been a slight decline during the period.[34] One explanation for the lesser degree of economic concentration is that large companies have tended to diversify across industry lines.

Concentration of Government Procurement and the Government R & D Effort

The concentration of government procurement among firms, as well as the concentration of R & D effort, has been measured from several sets of data. The roughest, and, no doubt, the most frequently used measure, is derived from data on prime contract awards. These data show that the first 100 contractors receive about 65 percent of total DOD prime contract awards and about 90 percent of total NASA awards. Prime contract awards data give no view of the net contract position (in- and out-flow of subcontracts) of firms at a point in time or over time. Data published by the National Science Foundation show that research and development performed by American industry is also concentrated. In 1966, the 4 largest manufacturing companies engaged in research performed 22 percent of the total. The first 100 companies perform 82 percent of total R & D and 95 percent of federally financed R & D in industry.[35] It is not possible to discern from NSF data the source of federal funds or the concentration of R & D employment among firms. From data obtained in surveys made by the National Research Council, James S. Worley estimates that the 100 largest companies employed about 40 percent of R & D personnel in industry in 1955.[36] The National Research Council data do not readily permit a distinction between those employed on federal projects and those working on company projects, nor do the data distinguish between employment provided by the various federal agencies.

The Concentration of NASA Employment Among Contractors

As indicated, prime contract awards data show that the 100 largest NASA contractors receive about 90 percent of total NASA awards to industry. The picture of the distribution of NASA's procurement effort, however, is quite different when measured by employment rather than by prime contract awards. The impact of multi-tier subcontracting is reflected in the concentration ratios displayed in Table 8–3. Although more recent data are available, the distribution of NASA employment is for June 1966, the approximate time when NASA contractor employment reached its maximum size. Subcontract employment lags behind prime contract employment in the initial phases of a program and then it drops off more rapidly than prime contract employment as a program approaches completion.

When measured by employment, the share of NASA's procurement going to the first 4 contractors, and also to the first 8 is reduced by almost 50 percent. The concentration ratio for the first 100 contractors drops by slightly more than one-third. Cost information provided to the joint NASA-DOD Economic Information System by contractors shows that 40 percent of the dollar volume of NASA work performed by industry in 1966 was spent on subcontracts, materials, and services.[37] The data in the table also show that concentration in the distribution of NASA's

TABLE 8–3: CONCENTRATION RATIOS FOR NASA PRIME CONTRACT AWARDS. TOTAL EMPLOYMENT, AND SCIENTIFIC EMPLOYMENT[a], 1966

NASA CONTRACTORS RANK BY SIZES OF PRIME CONTRACT AWARDS AND BY NUMBER OF EMPLOYEES	PRIME CONTRACT AWARDS, FISCAL 1966	TOTAL NASA EMPLOYMENT, JUNE 1966	NASA SCIENTISTS AND ENGINEERS, JUNE 1966
		Percent of Total	
1st 4 contractors	51.0	27.4	32.8
1st 8 contractors	65.0	34.5	45.3
1st 20 contractors	79.8	47.9	60.2
1st 40 contractors	88.2	55.5	68.5
1st 100 contractors	93.0	59.5	70.8[b]
Total	$4,087,700	$338,200	$71,800

[a] Includes only NASA contracts with business firms.

[b] Includes the first 80 contractors.

Sources: National Aeronautics and Space Administration, Procurement Office, *Status of Contracts and Grants,* Fiscal 1966, Management Information System, *NASA Manpower Information Digest,* January 1968; and Joint NASA-DOD Economic Information System.

scientific talent in industry is much less than that for prime contract awards.[38]

The use of concentration ratios to measure the participation of a certain number of firms in the space effort, or more generally, the country's R & D effort, can be deceptive. For example, NASA's employment in industry is more concentrated than that of the Department of Defense or that of R & D employment in general. Data from the joint NASA-DOD Economic Information System show that the 4 largest defense contractors account for about 7 percent of total defense employment in industry. This compares with 27 percent for NASA. Yet, defense employment of the 4 largest DOD contractors was more than twice the size of the 4 largest NASA contractors in June 1966. The concentration ratios reflect differences in relative size of DOD and NASA employment. Procurement by the Department of Defense resulted in 2.8 million jobs in industry in June 1966, whereas NASA employment was 360,000 at the same time.[39] The concentration ratio for DOD employment corresponds more closely with National Science Foundation data (9 percent for 1965) and with National Research Council data (6 percent for 1955).[40]

Pareto Slope Coefficients

Another way of looking at the effect of subcontracting on the distribution of NASA employment is by means of what economists call the "Pareto slope coefficient of inequality, alpha." The main advantage of the Pareto slope coefficient, when compared with concentration ratios, is that it provides a single number that expresses the degree of inequality; the principal disadvantage is that it gives a good fit only for the upper ranges of a distribution. Alpha measures the degree of inequality within a given distribution; the higher the coefficient, the more nearly equal the distribution.[41]

Table 8–4 displays Pareto alphas computed for NASA employment and for NASA prime contract awards. The data in the table tell two additional things about the concentration of NASA employment. The additional dimensions are: the concentration of NASA employment under prime contract and the concentration of NASA employment under subcontract.[42] The several alphas tell approximately the same story as the conventional concentration ratios: total employment is less concentrated than prime contract awards; and employment of scientific personnel is less concentrated than prime contract awards but more concentrated than total employment. In addition, the alphas show that employment under prime contract is about as concentrated as prime contract awards (for the first 50 contractors), and that NASA employment under subcontract is less concentrated than either prime contract employment or total employment.

TABLE 8–4: PARETO SLOPE COEFFICIENTS FOR NASA CONTRACT AWARDS AND NASA EMPLOYMENT, 1966

FIRST 50 CONTRACTORS[a]	PARETO ALPHA	R^2	STANDARD ERROR
Prime contract awards	.632	.986	.0152
Total employment	.805	.992	.0336
Prime contract employment	.632	.974	.0640
Subcontract employment	.848	.953	.0929
Scientists and engineers	.730	.945	.0984

[a] Includes contracts with business firms only.

Sources: National Aeronautics and Space Administration, *Status of Contracts and Grants,* Fiscal 1966; and Joint NASA-DOD Economic Information System.

Space Employment Under Prime Contract and Subcontract

The subcontracting activity of NASA has had the largest effect on the communication equipment and the electronic components industries. Table 8–5 shows that impact for NASA's Manned Space Flight Program. The simple matrix given in the table outlines the relationship between the economic role of NASA suppliers, the kind of procurement action, and the industry distribution of employment. The names and the economic functions of the contractors were given in Chapter 2.

TABLE 8–5: DISTRIBUTION OF MANNED SPACE FLIGHT EMPLOYMENT BY ECONOMIC FUNCTION OF MAJOR PRODUCERS[a]

		INDUSTRY DISTRIBUTION OF EMPLOYMENT		
ECONOMIC ROLE OF SUPPLIERS	PERCENT OF EMPLOYMENT UNDER PRIME CONTRACT	PERCENT IN AIRCRAFT AND MISSILES	PERCENT IN COMMUNICATIONS EQUIPMENT AND ELECTRONIC COMPONENTS	PERCENT IN OTHER INDUSTRIES
Space system producers	90.4	86.3	0	13.7
Subsystems producers	23.2	41.4	43.6	15.0
Overflow producers	18.7	24.6	42.9	32.5
Components and parts producers	33.1	6.8	79.5	13.7
Service firms	86.8	20.4	51.2	28.4

[a] Includes business firms with plants having 10 or more NASA employees.

[b] Based on primary product line of parent company rather than individual establishments.

Source: Compiled from data in the Joint NASA-DOD Economic Information System.

As Table 8–5 shows, almost all space employment of the systems firms is under prime contract and is in the aircraft and missiles industries.[43] In contrast, most of the space employment of the overflow and parts producers is under subcontract and is outside the aircraft industry. To a lesser extent the same is true of the subsystems firms. As indicated, the systems firms contract to deliver complete systems. Although the subsystems firms are usually subcontractors, a few are prime contractors who provide subsystems directly to NASA. The systems approach to procurement, frequently with prime and subcontract team proposals, has changed subcontracting patterns during the last decade or so.[44]

NASA Contractors and the Commonality of NASA Employment

Several economists have hypothesized that NASA and DOD contractors constitute a closed network of establishments, linked together through specialized technological capability. Contract awards data show that NASA contractors usually perform both prime and subcontract work, although any given business firm tends to function mainly as a prime contractor or as a subcontractor. For example, in fiscal 1963 and 1964, NASA's 12 largest contractors let 68 percent of their first-tier subcontract awards to firms listed among the 100 largest prime contractors.[45] Employment data, however, show that the link between NASA prime and subcontractors, as well as between NASA and DOD contractors, is not strong.

The eight coefficients of rank correlation displayed in Table 8–6 give measures of the commonality between NASA and other employment at plants engaged in work for the space program. Item 1, the coefficient for the rank order of NASA employment under prime contract and NASA prime contract awards, is given to show how closely the data from the joint NASA-DOD Economic Information System fit the relevant contract awards data. With the exception of the first item, the coefficients show some commonality because they are positive; but that association is not strong.

Employment under NASA Prime and Subcontracts

Items 2 and 3 in Table 8–6 show that the rank correlation between the size of NASA work forces under prime contract and under subcontract is low. The correlation is low because the systems firms, with large NASA work forces under prime contract, have primary product lines in the aircraft and missiles industries; whereas the subsystems, components, and parts producers, with large work forces under subcontract, are in the communication equipment, electronic components, and instruments industries.[46]

The low coefficients for NASA prime and subcontract employment show that transition into systems procurement for many of the airframe companies was accomplished by the establishment of electronics divisions, with electronic capabilities limited to systems integration.[47] Such a structure helps to maintain the pattern of inter-industry subcontract flows because many of the airframe companies did not acquire electronics production capabilities.

TABLE 8-6: MEASURES OF COMMONALITY OF NASA EMPLOYMENT, JUNE 1966

ITEM	UNIVERSE	NUMBER OF OBSERVATIONS	COEFFICIENT OF RANK CORRELATION
1	NASA employment under prime contract NASA prime contract awards	93	+.9893
2	NASA employment under prime contract NASA employment under subcontract	93	+.3977
3	NASA employment under prime contract NASA employment under subcontract	67[a]	+.5415
4	Total NASA employment Total corporate employment	92[b]	+.5287
5	Total NASA employment Total commercial employment[c]	93	+.3572
6	NASA scientists and engineers Commercial scientists and engineers[c]	91	+.4063
7	Total NASA employment Total DOD employment[d]	93	+.5202
8	NASA scientists and engineers DOD scientists and engineers[c]	93	+.6114

[a] The smaller number of observations excludes several large aircraft companies and service support contractors that do little work for NASA under subcontract. It also excludes those contractors with little or no employment reported under prime contract.

[b] Excludes Hughes Aircraft Company.

[c] Includes some non-NASA and non-DOD government employment.

[d] DOD employment at EIS plants with 10 or more NASA employees.

Sources: Joint NASA-DOD Economic Information System (plants with 10 or more NASA employees); and NASA Procurement Report for Fiscal Year 1966.

Dissecting the information given in the aggregate data, Table 8–7 displays coefficients of rank correlation by industry. Predictably, the coefficients of rank correlation by industry are lower than those calculated without the industry distinction. This is simply another way to show that NASA's subcontracting flows tend to be of an inter-industry rather than intra-industry character. The somewhat higher association between NASA's prime contract employment and subcontract employment in the communication equipment and electronic components industries shows that more electronics firms provide subsystems directly to NASA, under prime contract, than firms in other industries.

NASA, Total Corporate, and Commercial Employment

Although NASA contractors rank among the largest industrial corporations in the United States, NASA employment is not concentrated among the very largest corporations (Item 4 in Table 8–6). Of the 92 companies in the universe used here, 56 were included on the *Fortune* list of 500 for

TABLE 8-7: COEFFICIENTS OF RANK CORRELATION IN SELECTED INDUSTRIES, JUNE 1966

UNIVERSE	AIRCRAFT AND PARTS		COMMUNICATION EQUIPMENT AND ELECTRONIC COMPONENTS		OTHER INDUSTRIES	
	NO. OF OBSERVATIONS	RHO	NO. OF OBSERVATIONS	RHO	NO. OF OBSERVATIONS	RHO
NASA employment under prime contract NASA employment under subcontract	26	+.2981	35	+.5353	32	+.3103
Total NASA employment Total commercial employment[a]	26	+.4680	35	+.3896	32	−.1967
Total NASA employment Total DOD employment[b]	26	+.6974	35	+.6018	32	+.0252

[a] Includes some non-NASA and non-DOD government employment at EIS plants with 10 or more NASA employees.

[b] Includes DOD employment at EIS plants with 10 or more NASA employees.

Source: Compiled from data in the Joint NASA-DOD Economic Information System.

1966. One-fourth of the 92 companies had fewer than 5,000 employees and total corporate employment was less than 10,000 for 40 percent.

The relationships are also weak between NASA and commercial employment at the plants of these contractors. This reflects the historical tendency of government contractors to segregate work by source of funding. Company diversification, especially when in the direction of autonomous divisions, leads to the segregation of government and commercial effort. For example, International Business Machines has its Federal Systems Division; Westinghouse Electric Company has a Defense and Space Center; the Boeing Company has its Aero-Space Division; and AC Electronics and Allison are both divisions of General Motors.

NASA and DOD Employment

There can be little doubt that the Space Agency has used part of the network of contractor capability created by the Department of Defense; but NASA and DOD employment in industry have not been concentrated at the same plants. For that matter, NASA and DOD prime contract awards have not been concentrated among the same firms. Of the 100 largest recipients of DOD prime contract awards in fiscal 1966, fewer than half (48) were among the first 100 companies receiving NASA awards. For those companies common to both lists, the coefficient of rank correlation is not strong—+.4243.

Most conspicuously absent from the NASA list are the large steel, metals, petroleum, and rubber companies. Firms in these industries do not do as much research for the Department of Defense as firms in other industries. NASA, of course, is engaged in relatively more research and development work than the Department of Defense. Yet, when NASA prime contract awards are compared with DOD awards for research, development, test, and evaluation activities, there is still not much commonality among the 100 largest. Forty-nine companies are on both lists (subsidiaries included with parent companies); for these, the coefficient of rank correlation is +.4003.

Another indication of the lack of much commonality between NASA and DOD employment at contractor plants is in the attrition and employment absorption patterns at plants providing data to the joint NASA-DOD Economic Information System. In June 1967, *NASA employment* in the aircraft and the missiles industries was 30 percent below that of June 1966. *Total employment* in June 1967 at these plants engaged on space projects in the aircraft industry was less than 1 percent higher than the level existing a year earlier. This compares with a 10 percent increase for the entire industry. In the missiles industry, total employment at NASA plants in June 1967 was about 7 percent below the June 1966 level. For the industry, employment was about 1 percent higher than the

June 1966 level.[48] With the tight labor market during the period, a somewhat larger increase in total employment at NASA plants might have been expected, especially in the aircraft industry.

In part, the lack of strong commonality between NASA and DOD employment at contractor plants has resulted from a widening of the national R & D base, i.e., the establishment of NASA's manned space flight field centers in the South. As indicated, about one-fourth of total NASA employment in industry was located in the five southern states that border the Gulf of Mexico. The share of space employment in the mid-1960's of the five states is at least twice as much as their share of DOD employment, and perhaps even three times as much.[49] Plants of NASA contractors in three of the five southern states are engaged in work exclusively for NASA. These contractors support projects at the Michoud Assembly Facility, the Mississippi Test Facility, and the Manned Spacecraft Center in Houston.

The Department of Defense has a large number of military and civil service personnel employed in Huntsville, Alabama; but DOD contractor employment is small—less than one-tenth the size of NASA's contractor work force. Segregation of research by source of funding, however, is not solely a phenomenon of geographic isolation. Although there is a large DOD contractor work force in the Cape Kennedy area (in excess of 20,000 in the mid-1960's), few of those are employed at plants engaged in work for both NASA and DOD. Chrysler, Douglas Aircraft (now McDonnell Douglas), General Dynamics, International Business Machines, and North American Aviation (now North American Rockwell) are all large defense contractors, but in June 1966, there was no defense employment at their plants in the Cape Kennedy area.

Another force that affected the structure of the aerospace industry helps to explain the lack of strong association between NASA and DOD employment. That is the governmental strategy to maintain a broad mobilization base and to establish multiple supply sources to preclude domination by a single contractor of any major area of technology. As Stekler indicates, however, the policy of spreading contracts among firms has never been clearly outlined as a goal of public policy.[50]

GOVERNMENT PATENT POLICY, ECONOMIC CONCENTRATION, AND MONOPOLY POWER

Although it is a controverted issue within the federal government, the Congress, and the academic community, the effect of government patent policy on the concentration of economic power has aroused much less public attention than other aspects of the possible role played by the federal government in contributing to monopoly or to any inequitable distribution of federal procurements. The government patent policy controversy is over the disposition of patent rights that arise from government-financed R & D. Simply conceived, the government has two choices: it can retain ownership rights (title policy), or it can waive titles and permit contractors to secure ownership rights (license policy).[51]

The Economics of Government Ownership of Patent Inventions[52]

The economic argument in favor of the patent system is that it advances technology by granting exclusive or temporary monopoly rights. In turn, advances in science and technology can stimulate economic growth. The economic argument against the patent system is that the exclusive use of inventions can preclude widespread diffusion of the technology and can hinder the maintenance of competition. The economy derives social benefits from economic growth, whereas social costs can be incurred when monopoly power increases. The philosophy underlying the patent system, of course, is that social benefits gained from advances in science and technology outweigh the social costs of exclusive rights.

Government retention of ownership rights strikes the heart of the patent system. With NASA as the only fairly significant exception, it has been the policy of the federal government not to permit exclusive use of the patented inventions it owns. Again with the exception of NASA and then only under limited circumstances, it is the policy of the government to grant revocable, non-exclusive, royalty-free licenses upon request. Typically, that kind of license agreement has not meant much because the government does little or nothing to enforce it. In addition, the government has brought only one infringement suit against private industry for the use of government-owned patented inventions without licenses.[53]

Government Retention of Patent Rights

Advocates of a title policy contend that government ownership of patents resulting from government-financed R & D is necessary to maintain free access to technology developed at public expense and to prevent further economic concentration. Inseparably inter-related with the prevention of monopoly is the avoidance of potential adverse effects that might occur when the public is left to the mercy of a single firm owning a revolutionary invention originating from government-financed research. Although patented inventions are usually small components of complete products and processes, it is conceivable that a revolutionary invention might embody all the necessary technology for something as important as a weather control system or a cure for cancer.

The problem of monopoly and economic concentration arising from market imperfections is compounded, so advocates of the title policy argue, when titles to patented inventions are retained by contractors.

Most of the dollar volume of R & D contracts is awarded to the largest firms, and it is possible that a license policy would increase industrial concentration. Advocates of the title policy also believe that free availability of patents originating from government-financed research promotes the widest possible use and thus contributes as much as possible to economic growth.

Proponents of government ownership have arguments to supplement those on the prevention of monopoly and economic growth. They believe that economic injustice occurs when the public is forced to make a double payment for a patented invention. This can happen, so it is argued, when titles to the patented inventions remain in private hands. The public initially pays for the research and development from which the inventions originate, and then there is a second payment in the form of monopolistic prices when the public buys the finished product. This, to be sure, assumes that the patented inventions are developed and also commercialized. Protagonists for the title policy also believe that if the government retains patent rights, it can recover all or part of its research and development expenditures through royalty-bearing licenses. The recovery of these expenditures is deemed to be equitable because it shifts the burden of paying for research and development from the general taxpayer to the purchasers of the final product who benefit from the inventions.[54]

Contractor Retention of Patent Rights

Those arguing that the government should not retain patent rights believe that the patent system does, indeed, foster the disclosure of new technology and acts as a stimulant for invention and for innovation. The introduction of new products and the use of new processes have at least three stages—development of new scientific knowledge, application of basic research, and commercial exploitation. Exploitation or innovation can result from invention and the development of new knowledge. Defenders of the patent system contend that even if exclusive rights do not operate as an incentive to encourage the disclosure of new technology or to stimulate invention, temporary monopoly rights are necessary to promote the utilization of inventions. The exclusive rights compensate for future uncertainties or uninsurable risks.

In addition to believing that private ownership of patented inventions promotes commercial utilization, advocates of a license policy think that retention of patent rights by contractors promotes efficiency in procurement. If the value of patent rights is a significant factor for contractors, then acquisition of titles by business firms can provide an incentive for superior research and development work. It can also act as an inducement to get the most efficient private enterprises to accept government research and development contracts. Greater efficiency can result in either: (1) more, and presumably better, research and development with the same utilization of resources; or (2) the same amount of research activity using fewer resources.

NASA'S PATENT POLICY

By law, the National Aeronautics and Space Administration is required to take titles to inventions made under NASA contracts, but the Administrator of NASA may waive part of the rights of the federal government if he determines that such action is in the public interest.[55] Although the atmosphere in 1958 was one of urgency, with the possibilities of space research resulting in unprecedented discoveries, it was probably the intent of Congress to provide in the Space Act a balance between the need of the government and the preservation of incentives in the private sector of the economy. The Atomic Energy Commission is also required by law to take titles to inventions resulting from its research. The Atomic Energy Act of 1954 contains provisions for the exclusive licensing of inventions for specific non-atomic energy purposes; the Act does not provide a waiver policy. No similar law applies to the Department of Defense, and DOD usually permits contractors to retain titles to inventions resulting from its research.[56]

NASA's Waiver Policy

NASA has interpreted its right to waive titles to its patented inventions cautiously. The agency established an Inventions and Contributions Board (ICB) to study each petition for waiver. NASA requires its contractors to explain how an invention fits into the business plans of the company, and also to explain the steps the company will take to exploit the invention commercially. If the Inventions and Contributions Board recommends that a waiver be granted, the petition is sent to NASA's General Counsel for concurrence and then to the Administrator for final approval. Even after this deliberate process, NASA retains some control over its inventions in the form of "march-in-rights." These provisions give NASA a means to compel contractors to issue licenses to other firms. The "march-in-rights" can be exercised if: (1) within three years of the patent issue, the contractor does not use the waived invention and make it available to the public at a reasonable price, (2) the invention is essential for public health, and (3) the invention is required for public use by government regulations.[57]

NASA grants two kinds of waivers. The first waives NASA's right to individual inventions. Between October 1958 and December 1968, NASA granted 471 waivers to individual inventions and denied waivers to 98 inventions. The principal reason for denial of waivers is that petitioners fail to show NASA how inventions have commercial potential.

The second kind of waiver NASA grants is a blanket waiver, also known as a class waiver and as an advance waiver. These waivers, most of which have been granted since July 1965, cover all inventions resulting under a contract. Between July 1965 and December 1968, NASA granted 214 blanket waivers. This is a surprisingly small number compared with the thousands of contracts NASA let during the period. By the end of 1968, only 105 inventions had resulted from the 214 contracts containing advance waiver provisions. NASA's conservative waiver policy is reflected in the fact that the total number of inventions under waivers comprise only 3.8 percent of the 14,884 invention disclosures made by contractors between the inception of the Space Agency and the end of 1968.

NASA's Licensing Policy

NASA also grants two kinds of licenses—non-exclusive and exclusive. There is no monopolization in the use of NASA inventions under non-exclusive license because such licenses are available to all upon request. On December 31, 1968, NASA had 780 patented inventions available for non-exclusive licenses, and at that time, 117 non-exclusive licenses had been issued on 65 different patents. The number of non-exclusively licensed patents is about 8 percent of the total number available for licensing. If NASA does not grant a non-exclusive license within two years of the issue of a patent, the agency can grant an exclusive license. At the end of 1968, NASA had licensed only 4 percent of the 372 patents available for exclusive licenses. Under provisions of NASA's exclusive licensing policy, business firms must consent to spend a specified sum of money on the development and commercialization of the invention. At the request of the company, NASA terminated an exclusive license agreement with Union Carbide. The company could not justify the $20,000 it had agreed to spend annually for the development of the invention after a substitute became available.[58]

Examination of NASA's patent files shows that companies licensed to use NASA's inventions are somewhat different from contractors who have been granted waivers. Almost all of NASA's licensees are small businesses scattered over a wide geographic area. The principal product lines of the companies are as diverse as are the geographic locations of the firms. A few of the product lines of the companies include: consulting engineering, residential real estate, fire and burglary alarm devices, manufacture of marine equipment, ink and epoxy compounds, and fishing tackle accessories.[59] Statistics on waivers show that 27 percent of contractors holding waivers are small business; 14 percent of them are nonprofit organizations; 23 percent are aircraft companies; and 36 percent of them are large companies, with major product lines outside of the aircraft and parts industry.[60]

CONCENTRATION OF PATENTS IN INDUSTRY

Before looking at the concentration of patents resulting from federally sponsored research, it is useful to look at the data on the concentration of patents in American industry. The accumulation of patents, and their concentration, by corporations has been viewed by many economists as an evil much worse than other kinds of concentration because of the monopoly right that the patent grant confers—a monopolization of monopoly power.[61] At present, there is nothing illegal about the accumulation of large numbers of patents. To discourage large holdings of patents, however, two economists have recommended that a cumulative tax be levied on patents. As proposed, the amount of the tax would increase with the number of patents held by a company. The tax would also be levied at a higher rate over each year of the life of a patent.[62]

The data displayed in Table 8–8 show the concentration of patents assigned to all United States corporations in three different 17-year periods. The 17-year period is selected because that is the legal life of a patent. The last year in each period gives the number of unexpired patents at a point in time. The research and development data are for the earliest year for which the National Science Foundation published information on concentration ratios for total research performed in industry. Mainly because of the long period in which patent applications pend in the Patent Office, the lag between research performance and resulting patents is about five years. As the data show, concentration in-

TABLE 8–8: CONCENTRATION OF PATENTS AND RESEARCH AND DEVELOPMENT FUNDS IN INDUSTRY, SELECTED PERIODS (*PERCENTAGE DISTRIBUTION*)

NUMBER OF COMPANIES	PATENTS 1921–1938	PATENTS 1939–1955	PATENTS 1946–1962	TOTAL RESEARCH AND DEVELOPMENT, 1958
First 4	9	11	9	20
First 8	14	17	16	33
First 20	20	25	26	54
First 40	25	33	34	68
First 100	35	43	46	81
Total number	320,121	317,726	373,513	$8,389,000,000
Total percent	100	100	100	100

Source: Donald S. Watson and Mary A. Holman, "Concentration of Patent Ownership in Corporations," *Journal of Industrial Economics* (Spring 1970).

creased between 1938 and 1955, but not between 1955 and 1962. It is remarkable that the large increase in R & D during the postwar period, especially since the end of the Korean War, has had such a small effect on the concentration of patents. The data clearly show that patent concentration is almost 50 percent less than the concentration of research funds in industry.[63]

Concentration of Patents Resulting from Government-Financed R & D

Concentration ratios for industry-owned patents resulting from government-financed research are displayed in Table 8–9. The concentration of federally financed research is also shown in the table. The data indicate about the same for federally financed as for privately financed patents, i.e., their concentration is less than that of the research funds going to corporations. The difference between patent and R & D concentration, however, is not as great in the federal sector. The data in the table separate patents going to all private organizations from those going to the largest R & D contractors. To be among the largest R & D contractors, a business firm had to be on the list of the first 100 Department of De-

TABLE 8–9: CONCENTRATION OF PATENTS RESULTING FROM FEDERALLY FINANCED R & D IN INDUSTRY, SELECTED PERIODS (*PERCENTAGE DISTRIBUTION*)

NUMBER OF COMPANIES	LICENSE POLICY PATENTS OWNED BY BUSINESS FIRMS			GOVERNMENT-FINANCED R & D PERFORMED IN INDUSTRY
	1946–1955	1956–1962	1946–1962	1959
First 4				
Largest contractors	45	23	31	[a]
All contractors	29	18	23	29
First 8				
Largest contractors	59	40	45	[a]
All contractors	38	31	32	41
First 20				
Largest contractors	80	69	72	[a]
All contractors	52	54	52	68
First 40				
Largest contractors	92	87	87	[a]
All contractors	59	68	63	84
First 100				
Largest contractors	100	99	99	[a]
All contractors	64	78	71	94
Size of universe				
Largest contractors	5,407	7,664	13,071	[a]
All contractors	8,417	9,700	18,117	$5,635,000,000

[a] Not available.

Source: Adopted from Donald S. Watson and Mary A. Holman, "Concentration of Patents from Government-Financed Research in Industry," *Review of Economics and Statistics*, XLIX, 3 (August 1967), p. 377.

fense R & D contractors in any fiscal year between 1955 and 1962; or have had contract awards from either the Atomic Energy Commission or NASA in excess of $2 million. The number of firms that qualified under these criteria is 177.[64]

A glance at the table shows that the concentration of patents among the nation's largest R & D contractors decreased between 1946–1955 and 1955–1962. The same is true of all contractors up to the first 20. A few of the organizations in the "all contractor group" are patent-conscious nonprofit research organizations and universities. The concentration for the first 40 and the first 100 increased in the "all contractor group." This results from the fact that there were a larger number of contractors in the later period. In addition, each had a greater number of patents. For example, the one-hundredth firm owned one license policy patent in the 1946–1955 period, compared with six patents in the 1956–1962 period. Participation of a larger number of contractors in the later period was the principal reason for the decrease in patent concentration among the 177 largest R & D contractors.

Concentration of NASA's Waived Inventions

As Table 8–10 shows, 20 contractors accounted for 56 percent of all inventions waived by the Space Agency between its inception and December 1967. Two of the companies are small businesses. A comparison of Tables 8–10 and 8–3 shows that for the first 20 companies, waivers are much less concentrated than NASA's prime contract awards, and only slightly more concentrated than total NASA employment. Hidden within the concentration ratios is the fact that only half of the companies listed in Table 8–10 were among the 20 largest in terms of total NASA employment in June 1966. Seven of the companies holding five or more waivers had fewer than ten employees engaged on space projects at that time. Conspicuously absent from the list of companies granted at least five waivers to individual inventions are Boeing Aircraft, Chrysler, Grumman Aircraft Engineering, Hughes Aircraft, and General Motors. These companies were among 10 companies with the largest NASA labor forces in June 1966. By the end of 1967, however, Hughes Aircraft had petitioned for nine blanket waivers and had been granted eight of them. Other companies granted five or more blanket waivers by the end of 1967 include: Ampex (6), Honeywell (8), Texas Instruments (7), and United Aircraft Corporation (7).

One could view the fact that only 20 companies owned 56 percent of all waivers granted by NASA as extremely startling. Yet, the number of waivers granted is so small (only a few hundred) that these inventions have had a negligible effect on enlargement of the patent portfolios of the companies holding waivers. That is with the exception of the small business. For example, the General Electric Company is the largest private

TABLE 8-10: WAIVERS GRANTED TO BUSINESS FIRMS ON INDIVIDUAL INVENTIONS, OCTOBER 1958-DECEMBER 1967

	NUMBER	PERCENT
North American Rockwell	32	8.5
General Electric	21	5.6
Westinghouse Electric	17	4.5
TRW, Inc.	15	4.0
Subtotal 1st 4 companies	85	22.6
AVCO Inc.	13	3.5
United Aircraft	11	2.9
G.C.A., Inc.[a]	10	2.7
McDonnell Aircraft[b]	9	2.4
Subtotal 1st 8 companies	128	34.1
Sperry Rand	9	2.4
International Business Machines	9	2.4
Honeywell	8	2.1
Whittaker Corp.[a]	8	2.1
Collins Radio	7	1.9
Electro Optical Systems[a]	7	1.9
Varian Associates	7	1.9
Douglas Aircraft[b]	6	1.6
Bendix Corp.	6	1.6
Monsanto Research Corp.	6	1.6
Ampex Corp.	5	1.3
Electric Storage Battery	5	1.3
Total above	211	56.2
Other	164	43.8
Total granted	375	100.0

[a] Small business.

[b] Merged April 1967.

Source: National Aeronautics and Space Administration, *Semi-Annual Reports to Congress.*

patent holder in the United States. That company owns more than 12,000 unexpired patents. The 21 waivers granted by NASA have had virtually no impact on the size of General Electric's patent portfolio. Moreover, NASA's waivers have not increased patent concentration in industry. The combined patent holdings of those companies granted NASA waivers amounts to between 75,000 and 80,000 unexpired patents.[65] It is clear that NASA's several hundred waivers are a negligible fraction of that total. Besides that, because many of the waivers to inventions granted by NASA will not be allowed by the United States Patent Office, the patent applications will never issue into patents.

MONOPOLY POWER AND NASA'S PATENT POLICY

The possibilities for increased monopoly power resulting from NASA's patent policy depend, to a large extent, upon the commercial value of the inventions going to industry. Several empirical studies show that not only is the rate of commercial use of inventions resulting from government-financed R & D low, but the commercial value of the inventions is small.[66] Table 8–11 displays the results of five different research studies on the rate of commercialization of government R & D patented inventions. As the fourth column of the table shows, the studies provide information about: (1) the commercial use of license policy patents, mainly resulting from R & D funded by the Department of Defense; (2) the commercial use of government-owned, or title policy, patents; and (3) rights to inventions waived by NASA. A rate of commercial use of government R & D inventions of between 7 percent and 15 percent compares with a rate of commercial use of privately financed and privately owned patented inventions of about 50 percent.[67] It also compares with a rate of government use of its own inventions of between 65 and 75 percent. It has been estimated that the Atomic Energy Commission and the Department of Defense use about 75 percent of the government-owned patented inventions that they administer.[68] It has also been estimated that the government uses about 65 percent of the inventions resulting from NASA's R & D.[69] The somewhat lower rate of government use of NASA's inventions reflects the agency's policy of filing patent applications on some inventions that appear to hold commercial potential, but little promise for government use. That policy, in turn, reflects the goals outlined in the Space Act of 1958.

In addition to the low rate of actual commercial use, the commercial value of government R & D patented inventions is slight. A research group employed by Harbridge House, Incorporated found that only 2.7 percent of the inventions included in their sample of 2,100 patents played a critical role in the development of the commercial products.[70] By the end of 1965, the latest year for which information is available, gross development expenses exceeded gross sales revenue for all but three inventions waived by NASA.[71] Each of the five studies outlined in Table 8–11 gives the same reason for the slight commercial potential of government R & D inventions. In the aggregate, the inventions lack market potential because they incorporate highly specialized and esoteric technology.

Why then do companies go to the expense of securing patents on inventions resulting from contract research? Corporate officials state that patent rights are sometimes secured for such diverse reasons as: a means to recognize employee inventiveness, a way to accumulate a reservoir of patents to cross-license with patents owned by other firms, a means to enhance the image of a company, and a way to gain competitive advantage in securing future government contracts.

Because NASA's inventions hold slight commercial potential and because their commercial value is low, the possibilities for monopolistic

TABLE 8–11: RATE OF COMMERCIAL UTILIZATION OF PATENTED INVENTIONS RESULTING FROM GOVERNMENT-FINANCED RESEARCH AND DEVELOPMENT

RESEARCH GROUP MAKING STUDY	DATE OF STUDY	RATE OF COMMERCIAL USE (PERCENT)	OWNERSHIP
Watson, Bright, and Burns[a]	1960	13	License policy patents resulting from DOD research
Senate Subcommittee on Patents, Trademarks, and Copyrights[b]	1961	7	License policy patents resulting from DOD research
Holman[c]	1963	10–15	Title policy patents resulting from research of all federal agencies
Watson and Holman[d]	1966	7	Rights on title policy patents waived by NASA
Harbridge House[e]	1968	12	Combined license and title policy patented inventions resulting from research of all federal agencies

Sources:

[a] D. S. Watson, H. F. Bright, and A. E. Burns, "Federal Patent Policies in Contracts for Research and Development," *Patent, Trademark, and Copyright Journal,* 4, 4 (Winter 1960), p. 342.

[b] U.S. Congress, Senate, Subcommittee on Patents, Trademarks, and Copyrights of the Committee on the Judiciary, *Preliminary Report: Patent Practices of the Department of Defense,* 87th Cong., 1st. Sess., 1961, p. 35.

[c] Mary A. Holman, "The Utilization of Government-Owned Patented Inventions," *Patent, Trademark, and Copyright Journal,* 7, 2 (Summer 1963), p. 155.

[d] Donald S. Watson and Mary A. Holman, *An Evaluation of the Patent Policies of the National Aeronautics and Space Administration,* A report prepared for NASA and published by the U.S. House Committee on Science and Astronautics, 89th Cong., 2d Sess. (Washington, D.C.: U.S. Government Printing Office, 1966).

[e] Harbridge House, Inc., *Government Patent Policy Study,* Vol. 4: *Effect of Government Patent Policy on Commercial Utilization and Business Competition* (Boston, Mass.: Harbridge House, 1968), pp. 1–2.

abuse or misuse are also slight. Although the likelihood of misuse is faint, NASA, under its 1966 waiver regulations, can void a waiver if the use of the patent is held to have been in violation of the antitrust laws.[72] In addition, the government, under general welfare clauses, has the authority to control the use of revolutionary inventions. Governmental authority to control inventions that vitally affect public health, safety, and welfare, of course, extends to privately developed and owned patented inventions.

SUMMARY

Although the maintenance of competition is an objective of national economic policy, some economists and public officials charge that federal procurement, especially for research and development, contributes to geographic and industrial concentration and also to the monopolistic control of American industry.

As a result of external and internal economies that increase efficiency in the conduct of research, five major R & D complexes have developed within the United States during the last several decades. The areas include Boston-Cambridge, New York-Northern New Jersey, Washington, D.C., San Francisco-Palo Alto, and Los Angeles-Long Beach. The sense-of-the-Congress in the mid-1960's, or at least of those Congressional committees primarily responsible for the nation's research and development activities, was that the geographic distribution of federal R & D funds should be broadened wherever feasible. Although not proved or even spelled out by those recommending that the geographic concentration of federal R & D be reduced, the only economic justification for such a policy is that the geographic distribution of federal R & D funds existing in the mid-1960's was inefficient because it over-estimated the scientific capabilities of those areas of great R & D concentration, and it under-estimated the scientific capabilities of those areas that were not engaged in much R & D effort. In 1965, President Johnson called upon the heads of all executive agencies and departments to administer R & D programs so as to increase the number of academic institutions capable of performing high-quality research. In authorization acts of the mid-1960's, Congress specified that it was in the national interest for NASA to expand the geographic distribution of its research and development effort.

It was in the early 1960's, however, when NASA actually began to widen the national R & D base by establishing its manned space flight field centers and its production and test facilities in the five Gulf Coast States. The Space Agency did expand the geographic distribution of the federal R & D effort. For example, in June 1967, the five Gulf Coast States comprised about 26 percent of total NASA civil service and contractor employment, compared with about 11 percent of the industrial and civil service work force of the Department of Defense. Because of NASA's logistics requirement that research, production, and test facilities be in close proximity to one another and also be located on year-around navigable water, it is probable that the Space Agency's expansion outside the established R & D agglomerations and into the South was not accomplished at the expense of economic efficiency.

Because it includes comprehensive information about the employment created by subcontracting activity, the joint NASA-DOD Economic Information System provides for the first time a means to analyze the structure and the concentration of government-created employment in American industry. When concentration is measured by employment rather than by prime contract awards, concentration ratios drop by at least one-third. For the first four firms, the concentration ratio drops from 51 percent (prime contract awards data) to 27 percent (employment data), and for the first 100 firms the ratio drops from 93 percent (prime contracts) to 60 percent (employment data). Although more concentrated than total employment created by NASA, employment of scientific personnel is also much less concentrated than NASA's prime contract awards.

Because NASA's systems method of procurement has resulted in an inter-industry flow of subcontracting activity, prime and subcontract employment created by the Space Program has not been concentrated in the same firms. Although NASA and the Department of Defense rely on many of the same companies for the production of goods and services, the commonality of contractors and plants between NASA and DOD employment is not strong. This lack of strong association between NASA and DOD employment reflects the fact that the aerospace industry is a captive industry, with almost 70 percent of total sales being sales to the federal government. Data appear to confirm the belief that it is governmental strategy to spread contracts among business firms to maintain a broad mobilization base and to establish multiple supply sources to preclude domination by a single contractor of any major area of technology. In addition, data indicate that it is the strategy of industry to segregate work by source of funding, i.e., to have some plants engaged in commercial work, other plants engaged on defense projects, and still other plants engaged in space research and development. This means that the link between NASA prime and subcontractors and also between NASA and DOD contractors is not as much of a closed network of technologically specialized contractors and establishments as some have argued.

A final issue in the possible role played by the federal government in contributing to monopoly and economic concentration pertains to the private ownership of patented inventions resulting from government-financed R & D. Although by law NASA is required to take titles to inventions arising from its contract research, NASA's Administrator can waive the rights of the government if it is in the public interest. NASA has employed an extremely cautious waiver policy. Between October, 1958 and December, 1968, NASA waived titles to only 471 individual inventions, or less than 4 percent of all inventions made by NASA contractors. Between July, 1965 and December, 1968, NASA granted 214 blanket waivers, i.e., waivers that cover all inventions resulting under a contract. Compared with the thousands of prime and subcontracts let during the three-and-a-half–year period, 214 blanket waivers is a surprisingly small number. Under its licensing policy, NASA had granted exclusive licenses on only 4 percent of its portfolio of patents (372) available for exclusive licenses by the end of 1968.

The likelihood for monopolistic abuse under NASA's waiver and exclusive licensing policies is not high because the commercial potential of the inventions is small and the commercial value is low. Also, NASA's few hundred waivers have had a negligible effect on the concentration of patents in industry. The combined patent holdings of those companies granted NASA waivers amounts to between 75,000 and 80,000 patents.

NASA also grants non-exclusive licenses on its patents. By the end of 1968, NASA had granted 117 non-exclusive licenses on 65 different patents. There is no monopoly element in NASA's non-exclusive licensing policy because the Space Agency grants royalty-free licenses upon request.

NOTES

1. Walter Adams and Horace M. Gray, *Monopoly in America: The Government as Promoter* (New York: Macmillan, 1955), especially Chapter 5, "Procurement for Defense," pp. 101–16. The Adams and Gray study examines a large number of government policies that are alleged to contribute to monopoly. Other researchers have been concerned with the effect of government activity in particular industries. A typical case study is that made by Stanley E. Boyle, "Competition in the Synthetic Rubber Industry," *Economic Journal* (April 1961), pp. 151–69.

2. Mark S. Massel, "Government Purchasing and Research and Development," in *Competition and Monopoly: Legal and Economic Issues* (Garden City, N.Y.: Doubleday, 1964), pp. 78–82.

3. U.S. Congress, Senate, *Equitable Distribution of R & D Funds by Government Agencies, Hearings,* before the Senate Committee on Government Operations, 3 parts, 89th Cong., 2d Sess., 1966 and 1967.

U.S. Congress, Senate, *Economic Concentration,* Hearings, before the Subcommittee on Antitrust and Monopoly of the Senate Committee on the Judiciary, 5 parts, 89th Cong., 1st Sess., May 18–June 17, 1965, especially Part 3, *Concentration, Invention, and Innovation.*

U.S. Congress, Senate, *The Role and Effect of Technology in the Nation's Economy: A Review of the Effect of Government Research and Development on Economic Growth, Hearings,* before the Senate Select Committee on Small Business, 6 parts, 88th Cong., 1st Sess., May 20–June 6, 1963 and June 20–December 17, 1963.

U.S. Congress, Senate, *Economic Aspects of Government Patent Policies, Hearings,* before the Senate Select Committee on Small Business, 88th Cong., 1st Sess., March 7–14, 1963.

U.S. Congress, House, *Study of the Allocation of R & D Funds: The Decision-Making Process, Hearings,* before the House Committee on Government Operations, 89th Cong., 2d Sess., January 7–11, 1966.

U.S. Congress, House, *Government and Science: Distribution of Federal Research Funds, Hearings,* before the House Committee on Science and Astronautics, 88th Cong., 2d Sess., May 5–June 4, 1964.

U.S. Congress, House, *Investigation of Expenditures for Federally Conducted Research*

Programs, Hearings, before the House Select Committee on Government Research, 3 parts, 88th Cong., 1st and 2d Sess., November 18, 1963–January 22, 1964.

4. U.S. Congress, Senate, Subcommittee on Government Research of the Committee on Government Operations, *A Bibliography: An Inventory of Congressional Concern with Research and Development,* 90th Cong., 2d Sess., 1968.

5. For a discussion of the dissatisfaction in Congress with the then existing distribution of R & D funds, see John M. Brazzel, *Regional Equity and the Geographic Distribution of Federal R & D Funds* (Columbia, Mo.: School of Business and Public Administration, University of Missouri, September 1967).

6. U.S. Congress, House, Subcommittee on Science, Research, and Development of the Committee on Science and Astronautics, *Geographic Distribution of Federal Research and Development Funds, Report,* 88th Cong., 2d Sess., February 1965, p. 53.

7. *Ibid.,* pp. 7–17.

8. U.S. Congress, House, Subcommittee on Government Research, *Equitable Distribution of R & D Funds,* Part 2, p. 305.

9. U.S. Congress, House, Subcommittee on Science, Research, and Development of the Committee on Science and Astronautics, *Second Progress Report: Inquiries, Legislation, and Policy Studies—Re: Science and Technology,* 89th Cong., 2d Sess., 1966, p. 8.

10. *Ibid.*

11. The unique requirement of NASA to ship heavy equipment by water between its several facilities is in part an exception to this point.

12. Charles T. Stewart, *A Study of Changing Regional Patterns of Research and Development and Science-Based Technology in Influencing Regional Development* (Washington, D.C.: The George Washington University, Department of Economics, 1964), pp. III-36-38.

13. *Ibid.,* pp. III-37-39. Also see Charles T. Stewart, "Regional Allocation of Public Funds for Economic Development," *Land Economics,* XLIII, 4 (November 1967), pp. 421–33.

14. U.S. Congress, Senate, Testimony of Albert Shapero, "Impact of Federal Support for Technology on Regional Economic Development," *Equitable Distribution of R & D Funds by Government Agencies, Hearings,* Part 2, pp. 379–89. Also Albert Shapero, Richard P. Howell, and James R. Tombaugh, *The Structure and Dynamics of the Defense R & D Industry: The Los Angeles and Boston Complexes* (Menlo Park, Calif.: Stanford Research Institute, November 1965).

15. Brazzel, *Regional Equity and the Geographic Distribution of Federal R & D Funds,* pp. 13–15.

16. Harry G. Johnson, "Federal Support of Basic Research: Some Economic Issues," *Basic Research and National Goals,* A Report to the Committee on Science and Astronautics by the National Academy of Sciences, Washington, D.C., March 1965, p. 140.

17. "The Pentagon Short that Misfired," *Business Week,* April 8, 1967, p. 38.

18. John T. Burgess, *An Economic Analysis of Research and Development Procurement: The NASA Subcontract Program* (St. Louis, Mo.: Department of Economics, Washington University, 1967), pp. 134–48.

19. *Ibid.,* p. 138. The results of another study on NASA subcontracting flows produced findings similar to those reported by Burgess. See Marvin Hoffenberg, *Analysis of NASA Postcard Subcontract Data* (Los Angeles, Calif.: Institute of Government and Public Affairs, University of California, 1966).

20. U.S. Congress, Senate, Statement by William H. Miernyk, *Equitable Distribution of R & D Funds by Government Agencies, Hearings,* pp. 731–39.

21. A possible exception might be the dressed chicken industry, a recent phenomenon resulting from automation. Dressed chickens are one of the few agricultural commodities not produced under one or another kind of government subsidy.

22. Donald S. Watson, *Economic Policy* (New York: Houghton Mifflin, 1960), pp. 200–202.

23. Antitrust Law Section of the American Bar Association, *Antitrust Developments: 1955–1968, A Supplement to the Report of the Attorney General's National Committee to Study the Antitrust Laws* (Washington, D.C.: American Bar Association, 1968), pp. 24–28.

24. Watson, *Economic Policy*, p. 209.

25. *Report of the Attorney General's National Committee to Study the Antitrust Laws*, Stanley N. Barnes and S. Chesterfield Oppenheim, Co-Chairmen (Washington, D.C.: U.S. Government Printing Office, 1955), pp. 339–40.

26. *Technological Innovation: Its Environment and Management*, a panel report by an advisory committee to the Secretary of Commerce, Robert A. Charpie, Chairman (Washington, D.C.: U.S. Government Printing Office, 1967), pp. 47–52. Also see the American Bar Association, "The Role of the Federal Trade Commission as a Guide to the Businessman," *Antitrust Developments: 1955–1968*, pp. 339–40.

27. For a discussion of the contribution of antitrust policy to the maintenance of a competitive environment in the United States, see Theodore J. Kreps, *An Evaluation of Antitrust Policy: Its Relation to Economic Growth, Full Employment, and Prices*, Study Paper No. 22 of the Joint Economic Committee, 86th Cong., 2d Sess., 1960.

28. Merton J. Peck and Frederic M. Scherer, *The Weapons Acquisition Process: An Economic Analysis* (Cambridge, Mass.: Graduate School of Business Administration, Harvard University, 1962), Part II, "The Structure and Dynamics of the Weapons Industry," pp. 67–277; Frederic M. Scherer, *The Weapons Acquisition Process: Economic Incentives* (Cambridge, Mass.: Graduate School of Business Administration, Harvard University, 1964), Part I, "Competitive Incentives, pp. 19–130; Herman O. Stekler, *The Structure and Performance of the Aerospace Industry* (Berkeley, Calif.: University of California Press, 1965), Chap. 4, "Structure of the Industry: the Nature and Development of Competitive Practices," pp. 58–77; Murray L. Weidenbaum, "Concentration and Competition in the Military Market," *Quarterly Review of Economics and Business* (Spring 1967), pp. 7–18.

29. Scherer, *The Weapons Acquisition Process: Economic Incentives*, pp. 2 and 49.

30. Stekler, *The Structure and Performance of the Aerospace Industry*, pp. 56–57 and 152–53.

31. Weidenbaum, "Concentration and Competition in the Military Market," pp. 7–18.

32. U.S. Congress, Senate, Testimony of M. A. Adelman, *Economic Concentration, Hearings*, Part I, p. 231.

33. U.S. Congress, Senate, Testimony of Willard F. Mueller, *Economic Concentration, Hearings*, Part I, pp. 113–14.

34. Adelman, *Economic Concentration, Hearings*, Part I, pp. 231–33. N. R. Collins and L. E. Preston, "The Size and Structure of the Largest Industrial Firms, 1909–1958," *American Economic Review*, LI (December 1961), pp. 968–1011. Sylvia S. Small, "Concentration of Industrial Employment," *Monthly Labor Review*, 87, 9, (September 1964), pp. 997–1001. William G. Shephard, "Trends of Concentration in American Manufacturing Industries, 1947–1958," *Review of Economics and Statistics* (May 1964), pp. 200–212.

35. A number of individuals have argued that the general concentration of economic power in American industry is strengthened as a result of concentration in the research sector because of the enormous benefits accruing to firms engaged in research and development. This, in turn, so it is alleged, puts firms in an impregnable bastion from which they can gouge the American people. See, for example, Richard J. Barber, "Seeds of Industrial Concentration," *The Politics of Research* (Washington, D.C.: Public Affairs Press, 1966), pp. 71–90, especially Barber's conclusion on page 90. Daniel Hamburg's position is equally strong, but his argument is less emotional. Daniel Hamburg, "Size of Firm, Industrial Con-

centration, Research, and Innovation," *Essays on the Economics of Research and Development* (New York: Random House, 1966), pp. 35–68.

36. James S. Worley, "The Changing Direction of Research and Development Employment Among Firms," in *The Rate and Direction of Inventive Activity* (Princeton: Princeton University Press, 1962), pp. 233–51.

37. For 1964, value added was about 43 percent of total contractor costs; it was about 41 percent in 1965.

38. Because of the technical orientation of in-house personnel, NASA's scientific work force in industry is smaller (78 percent of all scientists and engineers) than total employment in industry (85 percent of the agency's work force). Data on NASA's scientific personnel come from the joint NASA-DOD Economic Information System, surveys of service support contractors, and from value added data provided by the Bureau of Labor Statistics. See National Aeronautics and Space Administration, *Manpower Information Digest* (Washington, D.C.: NASA, January 1968). (mimeographed)

39. For an estimate of DOD employment in industry, see Vernon M. Buehler, "Economic Impact of Defense Programs," U.S. Bureau of the Budget, *Statistical Reporter* (July 1967), p. 2.

40. National Science Foundation, *Basic Research, Applied Research, and Development in Industry,* NSF 67–12, p. 37; and Worley, *Rate and Direction of Inventive Activity,* pp. 234 and 236.

41. A cumulative distribution function, alpha can be derived from the equation:

$$\log N(y) = \log A - \alpha \log Y$$
$$N(y) = Ay^{-\alpha}$$

Where: N is the number of companies with NASA employment (also computed for prime contract awards) exceeding y; y is the size of employment (prime contract awards); and A and α are parameters (constants) of the distribution.

When plotted on double logarithmic paper, the Pareto distribution yields a straight line—the steeper the line, the more nearly equal the distribution. For a discussion of the merits and limitations of the Pareto slope coefficient as a measure of the concentration of income, see Mary Jean Bowman, "A Graphical Analysis of Personal Income Distribution in the United States," *American Economic Review,* XXXV, 4 (September 1945), pp. 607–28.

42. It is possible to obtain these additional dimensions because, unlike conventional concentration ratios, it is not necessary to know the entire universe to compute alpha. The joint NASA-DOD Economic Information System requests contractors to identify the percentage of government employment under prime and subcontracts.

43. Because many NASA suppliers have establishments in more than one industry, the distribution given in Table 8–5 is based on the primary product line of the parent company. Although no company reports a primary product line in the guided missiles industry (SIC 1925) to the Securities and Exchange Commission, over half of the establishments of the aircraft companies (SIC 372) supplying space systems to NASA are classed as guided missiles and space vehicles plants.

44. The fact that subsystems firms frequently produce sophisticated hardware, which requires considerable R & D, does not support the contention of the Attorney General's Report of 1956 that there is little subcontracting of R & D. The report states that most subcontract effort consists of the purchase of raw materials and standardized components. See U.S. Attorney General, Report Pursuant to Section 708 (e) of the Defense Production Act of 1950, as amended November 1956, p. 37. Section 708 (e) of that act directs the Attorney General to make surveys to determine any factors that might tend to eliminate competition, create or strengthen monopolies, injure small business, or otherwise promote undue concentration of economic power.

45. See Marvin Hoffenberg, *Analysis of NASA Postcard Subcontract Data,* pp. III 9–11; also, John T. Burgess, *An Economic Analysis of Research and Development Procurement: The NASA Subcontract Program,* pp. 140–41.

46. The inter-industry pattern of subcontracting is more discernible if the cut-off point drops to the 25 largest contractors that have both prime and subcontract employment—rho is +.2527. Rho is a low +.1375 for the 25 firms with the largest NASA work forces (3 report no employment under subcontract). Eleven of these companies have primary product lines in the aircraft industry, 5 in the electronics industry, and 9 are in other industries.

47. For an excellent discussion of this transition, see Herman O. Stekler, *The Structure and Performance of the Aerospace Industry,* pp. 96–118.

48. Data from the Bureau of Labor Statistics show that employment in the aircraft industry was 743,500 in June 1966 and 820,330 in June 1967. In the guided missiles industry, employment totaled 161,600 in June 1966, compared with 162,700 a year later.

49. Mary A. Holman and Ronald M. Konkel, "Manned Space Flight and Employment in the South," *Monthly Labor Review* (March 1968), pp. 30–36.

50. Stekler, *The Structure and Performance of the Aerospace Industry,* pp. 56–58, 128–29, and 199–201. Also see, Peck and Scherer, *The Weapons Acquisition Process: An Economic Analysis,* pp. 376–77.

51. Before the sharp rise in contract research, the argument was whether or not the government should or should not retain titles to inventions arising from the work of federal employees. That argument was resolved with Executive Order 10096, dated January 23, 1950. That Executive Order provides that the government shall obtain the entire right, title, and interest in all inventions made by government employees during working hours or with government money, and facilities. In most instances when the government does not believe the invention is sufficiently meritorious to warrant the cost of government-financed patent prosecution, the government permits the inventor to apply for a patent application. The inventor, of course, bears the cost of legal and Patent Office fees.

52. Much of the material on the economics of government patent policy has been adapted from previous works of the author, which include: "The Utilization of Government-Owned Patented Inventions," *Patent, Trademark, and Copyright Journal,* 7, 2 and 3 (Summer and Fall 1963), pp. 109–61 and pp. 321–75; "Inventions from Government-Financed R & D: A New Resource?" *Land Economics* (August 1965); and "The Economics of Government of Patented Inventions," in *Nurturing New Ideas: Legal Rights and Economic Roles,* ed. by L. James Harris (Washington, D.C.: National Affairs Press, 1969).

53. The federal government lost its case against the small business firm, Tektronix, Incorporated. The court judged that although the license to use a patent is a formal waiver of the right to sue, the government had long followed an avowed policy of not prosecuting infringers of its patents.

54. U.S. Congress, Senate, *Government Patent Policy, Hearings,* before the Subcommittee on Patents, Trademarks, and Copyrights of the Senate Committee on the Judiciary, Part II, 87th Cong., 1st Sess., 1961, p. 574.

55. National Aeronautics and Space Act, Section 305 (a) and Section 305 (f) (1958).

56. In an attempt to provide more uniformity in the patent policies of federal agencies, President Kennedy issued a Presidential Memorandum on Government Patent Policy on October 10, 1963. The Memorandum, however, provides no guides as to the proportions of titles that should go to contractors and to the government. The various R & D agencies have interpreted the criteria set forth in the Memorandum differently. It is not heroic to state that the federal agencies essentially continue to follow the same patent practices they followed before the Memorandum.

57. Donald S. Watson and Mary A. Holman, *An Evaluation of the Patent Policies of the National Aeronautics and Space Administration,* a report prepared for NASA and published

by the House Committee on Science and Astronautics, 89th Cong., 2d Sess. (Washington, D.C.: U.S. Government Printing Office, 1966), pp. 47–52.

58. *Ibid.*, p. 31.

59. *Ibid.*, p. 32.

60. The data are for the period between October 1958 and June 1967. NASA publishes the names of companies granted waivers in each *Semi-Annual Report to Congress*.

61. For a discussion of the opinions held by economists on the concentration of patents, see Donald S. Watson and Mary A. Holman, "Concentration of Patent Ownership in Corporations," *Journal of Industrial Economics* (Spring 1970).

62. Carl Kaysen and Donald F. Turner, *Antitrust Policy: An Economic and Legal Analysis* (Cambridge, Mass.: Harvard University Press, 1959), p. 176.

63. National Science Foundation data show that there has been remarkably little change in R & D concentration between 1958 and 1966. Variation during the period does not fluctuate by more than a percentage point. That fluctuation is down as well as up. National Science Foundation, *Research and Development in Industry*, NSF Report 68–20, p. 40.

64. Donald S. Watson and Mary A. Holman, "Concentration of Patents from Government-Financed Research in Industry," *Review of Economics and Statistics*, XLIX, 3 (August 1967), pp. 375–81.

65. Watson and Holman, *An Evaluation of the Patent Policies of NASA*, p. 81.

66. A sixth study does not estimate the rate of use, but finds that the commercial use of NASA-waived inventions has not been significant. Six years after the inception of NASA, only 10 inventions, under NASA waivers, had been commercialized. Robert A. Solo, "Patent Policy for Government-Sponsored R & D," *Patent, Trademark, and Copyright Journal*, 10, 2 (Summer 1966), p. 175.

67. Joseph Rossman and Barkev S. Sanders, "The Patent Utilization Study," *Patent, Trademark, and Copyright Journal*, 1, 1 (June 1957), pp. 74–111.

68. Holman, "The Utilization of Government-Owned Patented Inventions," pp. 359–64.

69. Watson and Holman, *An Evaluation of the Patent Policies of NASA*, p. 45.

70. Harbridge House, Incorporated, *Government Patent Policy Study*, Vol. 4, "Effect of Government Patent Policy on Commercial Utilization and Business Competition" (Boston: Harbridge House, 1968).

71. Watson and Holman, *An Evaluation of the Patent Policies of NASA*, p. 74.

72. National Aeronautics and Space Administration, 1966, *Waiver Regulation* (1245.109 (7) (b)).

9 Performance: Economic Growth

Economic growth is commonly defined as an increase in the productive capacity of an economy. In the United States and throughout the world, economic growth generally is deemed desirable and usually is a national goal.[1] For a nation to enjoy higher standards of living, the total increase in real goods and services must rise at a faster rate than the increase in population. In industrialized societies, economic growth depends principally upon (1) a quantitative increase in the factors of production, generated mainly by capital formation; and (2) a qualitative improvement in the utilization of a given amount of resources, propelled by knowledge and technological change.

THE THEORY OF ECONOMIC GROWTH

Chapter 6 presented an analysis of the economic growth of the communities surrounding several space centers. This section looks at the contribution of the space program, and of science and technology in general, to economic growth at the national level. Although it may be superfluous to say so, the matter of economic growth is extremely complex, and fraught with a wide diversity of opinions. The economic literature abounds with case studies of the growth of firms and industries; of the uses of technical reports, inventions, and patents; and of the transfer of people and skills. There is also a wide literature containing mathematical and econometric growth models.[2] And, of course, there is a large body of thought and empirical investigations on the social and political aspects of economic development. The fundamental difficulty is that of interpreting and measuring cause and effect relationships in the process of economic growth and of using estimated results for forecasting future growth.

Most economists would agree that one key variable in analyzing economic growth is the level of output. One functional set of relationships used for estimating the level of output in an economic system is:

$Y = f(K, N, L, S, U)$

Where: Y is the amount of output;

K is the services of the plant equipment (capital stock) in the economic system;

N is the rate of use of natural resources;

L is the employed labor force;

S is the fund and use of knowledge existing in the society; and

U is the cultural and institutional environment of the society.[3]

Clearly, none of the variables are homogeneous and most of them are inter-related. The value of the services of the plant and equipment existing in an economic system will vary by firm and by industry, as well as by accounting procedure. The same is true of the value of the heterogeneous kinds of natural resources and of the dissimilar kinds of labor skills and rates of individual efficiency. A multi-component model, which measures the value of capital, natural resources, and labor by kind and by industry, reduces but does not eliminate the problems arising from the evaluation of incommensurate inputs.

As Adelman so succinctly states, the variables S and U are heuristic devices used for conceptual purposes. They stand for forces that are undoubtedly vital to economic growth, but whose quantification at best is possible only in an ordinal way.[4] The variable S, the fund of scientific, technical, and managerial knowledge, and the variable U, the institutional complex, are also multi-dimensional.

The matter of interdependence among the variables, as well as their multi-components, creates the most glaring difficulties in estimating rates of change of values. For example, changes in the value of services of capital affect the labor and the natural resource variables. Changes in the socio-cultural environment certainly influence the fund of available knowledge, as well as the three physical inputs. Similarly, changes in the fund of technical knowledge affect the institutional arrangement of a society and the productivity of capital, labor, and the natural resources.

NASA, Congress, and the public have sought quantitative answers as to how the space program has changed and will change the variables that determine economic growth. Although more attention and publicity has been focused on NASA in relation to the knowledge variable (*S*), there can be no doubt that the space program has, in some way, changed each of the other variables, including the socio-cultural environment variable (*U*). Given the present state of the art, social scientists cannot provide the quantitative information that is desired about NASA in relation to national economic growth. Nevertheless, much can be said about economic growth, science and technology, government spending for research and development, and the Space Agency.

MEASURES OF ECONOMIC GROWTH

The most commonly used yardsticks of economic growth measure growth that has actually materialized. Forecasts of economic growth usually have some relationship to the trend of the immediately preceding period.

The Partial Productivity Index

A familiar yardstick used to measure economic growth is the increase in output per man-hour. The index is a partial productivity index because output is related to only one input. The output per man-hour productivity index does not measure changes in output that result from changes in other inputs. For example, an increase in the productivity of labor can result from the substitution of capital for labor in response to changes in the relative prices of the inputs, rather than from technological change.[5]

The consensus among economists is that changes in productivity do not measure the rate of technological change, although technological change affects the productivity index. Table 9–1 shows the average annual rate of change in output per man-hour in selected periods, i.e., the change in the amount of labor time required to produce a unit of output. The data necessarily reflect changes in many factors, including the rate of capacity utilization, managerial skill, labor skill, capital investment per worker, as well as changes in technology. Since the end of World War II (1947–1966), productivity in the private sector of the economy has risen at an

TABLE 9–1: AVERAGE ANNUAL RATE OF CHANGE IN OUTPUT PER MAN-HOUR OF WORK, 1947–1968

PERIOD	TOTAL PRIVATE	FARM SECTOR	TOTAL NONFARM	NONFARM SECTOR MANUFACTURING	NON-MANUFACTURING
1947–1966	3.3	6.0	2.8	3.0	2.7
1950–1955	3.1	5.5	2.5	2.7	2.3
1955–1960	2.2	5.5	1.7	1.6	1.8
1960–1965	3.8	5.9	3.4	4.2	3.0
1965–1968	2.9	5.1	2.6	1.6	3.0

Source: Computed from data in Council of Economic Advisers, *Economic Report of the President*, January, 1969, p. 266.

average annual rate of 3.3 percent. Although not shown in the data displayed in the table, this 3.3 percent growth rate compares with a pre-World War II (1900 to 1928) growth rate of 1.9 percent per annum.[6] The specific causes for the marked increase in productivity during the post-World War II period are not completely clear. Kendrick and others, however, suggest that the productivity gains could have resulted from the spread of the scientific management movement, the expansion of college- and graduate-level training in business administration, and from the increase in research and development activities.[7]

As is well known, output per man-hour in the agricultural sector made spectacular gains during the last two decades. Farm productivity rose at an average annual rate of 6.0 percent compared with an average annual rate of 2.8 percent in the nonfarm sector. The long-term trend shows that productivity gains have been somewhat greater in manufacturing industries than in nonmanufacturing industries. Most of the gain in manufacturing productivity occurred between 1960 and 1965. The data in Table 9–1 clearly show that productivity gains since 1965 have been below the long-term trend and markedly less than the 1960 to 1965 period. This is especially evident in the manufacturing sector of the economy. Aggregate measures of productivity change reflect short-term changes in the general economic environment. As an economic system operates at higher and higher levels to meet increased production demands, less efficient resources are usually drawn into the production process. The use of less efficient resources, as the economic system approaches full employment, of course, dampens productivity gains. This is precisely what happened after the 5-year period of sustained growth prior to the buildup for the Vietnam War, and the even more dramatic claim on resources after the buildup in mid-1965.

Total Productivity Indexes

Several economists have devised total productivity indexes. Unfortunately, because the inputs that are measured differ, as do the time periods used, the indexes yield somewhat different results.[8] An index that measures total productivity relates changes in output to capital and to labor inputs. The most widely used total productivity index is that compiled by John W. Kendrick. Kendrick's index shows that productivity rose from 1.3 percent per annum between 1900 and 1928 to 2.2 percent during the 1947 to 1957 decade. In the manufacturing sector, the largest gains in productivity took place in the production of tobacco products, chemicals, fabricated metals, and transportation equipment. Manufacture of lumber products, furniture, leather goods, food and beverages, and apparel experienced the smallest gains in productivity.[9]

Another total productivity index, compiled by Edward F. Denison, uses a more complicated production function, i.e., the relationship between productive inputs and output. In addition to changes in the number of man-hours worked and to changes in the amount of capital, Denison's index shows changes in total output resulting from increases in scientific, technological, and managerial knowledge; from increases in economic efficiency; and from economies of large-scale production.[10] The results of Denison's work indicate that gains in output per unit of total input amounted to about 0.4 percent annually between 1909 and 1929, and to about 1.1 percent per annum between 1947 and 1958.[11] Although the numbers are not comparable, both the Denison and Kendrick measures of productivity show a substantial increase in the growth rate during the post-World War II period, compared with that during the early decades of the twentieth century.

Table 9–2 presents Denison's estimate of the relative contribution of the different sources of growth to measured changes in output. The total increase in output during each of the three periods is equal to 100 percent. Although the proportion has been declining, a little over one-half of the increase in measured output is attributable to an increase in labor and capital. A rising fraction of measured growth, but less than one-half of the total, is attributable to an increase in output per unit of input, i.e., an increase in efficiency, according to Denison.

TABLE 9–2: PERCENTAGE ALLOCATION OF SOURCES OF ECONOMIC GROWTH IN THE UNITED STATES, 1929–1962, SELECTED PERIODS

SOURCES OF GROWTH	1929–1957	1950–1955	1955–1962
Total increase in output	100	100	100
Increase in inputs	68	60	56
Labor	54	34	32
Education	23	11	17
Land	0	0	0
Capital	15	26	24
Increase in output per unit of input	32	40	44
Advance in knowledge	20	20	25
Economies from growth of national and regional markets	10	11	10

Sources: (1929–1962) Edward F. Denison, *The Sources of Economic Growth in the United States and the Alternatives Before Us,* Supplementary Paper No. 13 of the Committee for Economic Development (New York: Committee for Economic Development, 1962); and (1950–1962) Edward F. Denison, *Why Growth Rates Differ: Postwar Experience of Nine Western Countries* (Washington, D.C.: Brookings Institution, 1967).

Of most concern here is Denison's estimate of the contribution to growth made by the improvement in the quality of education and by advances in knowledge. Denison distinguishes between (1) changes in output arising from an improvement in the quality of inputs, using a given stock of knowledge; and (2) changes in output arising from an advance in the state of the arts, using a given quantity of resources. Education is the prime mover of the first, whereas advances in knowledge cause the second.[12]

According to Denison, improved education resulted in 11 of the 34 percentage points allocated to the labor input in the 1950 to 1955 period and to 17 of the 32 percentage points allocated in the 1955 to 1962 period. Denison used a weighted index relating changes in the number of years of schooling and changes in income to changes in the size of the labor force. Denison construes the notion of an advance in knowledge broadly; it includes management techniques, as well as technical knowledge. Economic growth arising from advances in knowledge is no less elusive to Denison than to other economists. His estimate of the contribution of knowledge to economic growth is a residual, obtained after estimating the other components of economic growth. On that basis, he concludes that between 20 and 25 percent of the total increase in output in the United States has resulted from advances in knowledge.

Per Capita Gross National Product

Another commonly used measure of economic growth is the increase, over time, of the per capita real gross national product. This measure takes into account the relationship between changes in the level of real output and changes in the size of the population, but it does not measure the way in which output is distributed within the various sectors of the economy or among individuals in the society. Historically, the real per capita gross national product has risen substantially during wars and other periods of national emergency. However, because production is geared to the public sector, standards of living do not rise appreciably at these times. Between 1940 and 1945, the real per capita gross national product rose at the remarkable average annual rate of 8 percent. In 1944, at the peak of that war effort, slightly more than 40 percent of all goods and services were produced for national defense purposes and were not available to the civilian population.

The long-term trend data on per capita real gross national product show the same tendencies that the productivity series show. Between 1910 and 1968 the per capita real gross national product rose at an average annual rate of 1.7 percent. The per capita real (1958 dollars) gross national product was $1,300 in 1910, and it was $3,500 in 1968. The growth rate, however, has been substantially higher since World War II. This is true even if the depression years of the 1930's are excluded from the pre-World War II series. Between 1910 and 1929 the average annual rate of increase in the per capita gross national product was 1.3 percent. This compares with an average annual rate of 2.3 percent since World War II (1947 to 1968). The latter growth rate includes the period of damped economic activity between 1955 and 1960 when the per capita real gross national product rose at the dismal rate of only 0.3 percent per annum.

Some economists are dismayed by the fact that total productivity indexes and other measures of economic growth show such mild rates of increase in response to the enormous acceleration in the amount of research and development that has taken place in the United States during the last thirty years.[13] Yet, other economists attribute virtually all long-term economic growth to the growth of technological knowledge—"the increase in knowledge about useful goods and how to make them."[14]

MEASURES OF SCIENTIFIC OUTPUT

The main problem in any attempt to ascertain the effect of NASA on economic growth is to devise a statistical measure of scientific output. During the last decade, the National Science Foundation has gathered voluminous information about the economic characteristics of research and development on the input side. The output side of the R & D process, however, eludes statistical proof and sophisticated measurement. The search for a way or ways in which to measure the ouptut of research and development is closely tied to cost-effectiveness analysis and other methods to determine the optimum allocation of available resources.

Because many of the results of scientific activities are purely intellectual in nature, they do not take the form of physical values that can be measured. Fritz Machlup, an economist, distinguishes between the parts of scientific output that are intangible and not subject to measurement, and those parts that are measurable.[15] By measurable output, Machlup has in mind the means by which information is transferred. Such instruments of information transfer include blueprints, memoranda, patents, technical reports, and new products and processes. Nonmeasurable outputs of the research process consist of ideas, hypotheses, and theories. Both the measurable and the nonmeasurable outputs are the direct results of research and development. In addition, there is indirect output, i.e., the effects of advancing science and technology, which take the form of increased sales; increased profits; changes in the rate of return on business investment; and the growth of firms, industries, and of the economy as a whole. The productivity indexes discussed above embody, indirectly, all of the different kinds of scientific output.[16]

Publications as A Measure of Scientific Output

Because published papers are, in Machlup's concept, a direct output of R & D, a number of researchers have made counts of publications in an attempt to measure scientific output. Another reason for the belief the publications can reflect scientific output is that virtually all important research findings, especially basic research, eventually get published.[17] Most organizations that employ scientists and engineers encourage, or at least tolerate, the publication of research results. The reason for this varies from increasing the prestige of the organization to a fear that scientists will go to an environment where publication is permitted.

The results of publications counts show that no constant relationship exists between scientific inputs and publications as a measure of the scientific output. One reason for this is the fact that quality standards established by scholarly journals vary over time and by field of scientific endeavor. Publication acceptance standards also differ within a given academic field. Another reason for the failure of publications counts to measure scientific output is that the choice of the means of counting affects the measurement of criterion. For example, a count of citations to articles would yield one result; a count of abstracts of articles would yield another result, and a count of the number of words contained in a publication would give still another result. Furthermore, many prolific authors publish essentially the same article or research finding in a number of different journals. The lag between the time when research is conducted and the time when results are published is yet another reason why counts of publications cannot be used as a satisfactory measure of scientific output.[18]

Patent Activity as a Measure of Scientific or Technological Output

Like publications, patents pass some kind of recognition or acceptability test for describing an invention that does, in fact, contribute something new. Therefore, patent counts have been considered as a possible measure of scientific output. By and large, research results have proved that patents are not a good measure of scientific output. The main reason for this is that there is no way to know (1) whether the ratio of patents to inventions in a given scientific field remains constant over time, and (2) whether the importance of an "average invention" in a given field remains constant over time.

Critics of the patent system contend that business firms tend to patent only those inventions that could probably not be kept as trade secrets. If this is true, patents represent only a fraction, and a changing fraction, of the total number of patentable inventions. To be sure, there are other causes for not patenting, including the cost of attorneys' fees, Patent Office fees, and the possible cost of future litigation fees to uphold the validity of patents.

Research indicates that there is wide variation among business firms and also among federal government agencies in their propensities to file patent applications. Between 1945 and 1963, for example, the Department of Agriculture filed patent applications on 80 percent of its invention disclosures. This compares with 21 percent for the Atomic Energy Commission, 11 percent for the Air Force, 26 percent for the Army, and 40 percent for the Navy during the same period. NASA's propensity to patent was 20 percent between 1959 and 1963. The patent departments of the six agencies apply different criteria for the selection of inventions for patent applications. The only criterion used by the Department of Agriculture is patentability. The Atomic Energy Commission and NASA include commercial potential as one criterion; the military agencies do not. There is also a relationship between the propensity to patent and the number of invention disclosures reported to a federal agency. The Department of Agriculture administers less than one-tenth the number of invention disclosures handled by the military agencies. In addition, there is an inverse relationship between the propensity to file patent applications and the rate of commercial use of patented inventions. Economic growth, which originates from inventions, is realized only when producing units actually use inventions.[19]

New Products and Processes as a Measure of Scientific Output

Another possible direct measure of the results of the R & D process is an identification and a count of new products and processes. Appealing though this method may appear, new products and processes, it must be admitted, can be far removed from the research and development process. This is true despite the fact that an increasing share of corporate profits comes from products that were not on the market a decade or so ago. Guy Suits, among others, correctly argues that science seldom produces new products and processes; rather, science creates an opportunity for the development of new products and processes.[20]

Actual innovation that depends on profit expectations can be induced or it can be autonomous. Producers can be forced by competitive market pressures and falling profits to employ cost-reducing processes and to introduce product improvements to stay in business; or they can assiduously seek new techniques and markets to reap the profits of possible successful innovation. If the innovation is required for survival, the primary business management decision is whether or not to continue to re-invest capital in the firm. The range of alternative investment opportunities is wide for firms actively seeking opportunities to invest in innovation, but non-innovational investment possibilities probably are even larger. Executive decision-makers probably make at least a rough calculation of the expected yield from contemplated innovational investment compared with the productivity from other possible investment outlets.

More important than that, counts of new products and processes cannot be a satisfactory index of the output of research and development because most innovations incorporate a number of component parts and processes. In most instances, it is virtually impossible to identify given research endeavors with the resulting components contained in new products and processes. This identification problem is compounded because of time lags between the research and development process and the appearance of new products.

The time interval between invention and innovation has also been the subject of several studies. For eleven inventions in the petroleum industry, for example, the average time lag amounted to 11 years. For any given product or process, of course, the time lag varies substantially. The lag between invention and innovation for freon refrigerants was only one year, compared with a lag of 79 years for the fluorescent lamp. The following gives the estimated time interval between invention and innovation for a dozen well-known innovations: DDT (3 years), Streptomycin (5 years), power steering (6 years), ball-point pen (6 years), nylon (11 years), radar (13 years), xerography (13 years), crease-resistant fabrics (14 years), jet engine (14 years), television (22 years), zipper (27 years), and the gyrocompass (56 years).[21]

NASA AND MEASURES OF ECONOMIC GROWTH

Data are not readily available on the numbers of scientific publications and technical reports resulting from research sponsored by the Space Agency. The foregoing should make it clear that making such an estimate would not be easy, and most probably the costs of undertaking such a study would exceed the benefits. Although they are not a measure of the scientific output of NASA, data are available on the number of invention disclosures resulting from NASA's research. The data do, however, reveal information about the inventive process in NASA, especially when they are compared with similar information for other federal government agencies. Table 9–3 displays data on the number and the average cost of invention disclosures and patent applications resulting from NASA's research and the research of seven other government agencies. The data on the average cost of each invention disclosure and patent application indicate relative rates of disclosure.

The data show that, on the average, the dollar volume of research and development yields fewer inventions from NASA than from other government agencies. There are several reasons why. First, the R & D data published by the National Science Foundation probably overestimate the amount of research actually done by NASA, compared with other government agencies. During the 1962 to 1965 period, construction of plant and equipment loomed large for NASA; so did, and does, the

TABLE 9-3: DISTRIBUTION OF INVENTION DISCLOSURES AND PATENT APPLICATIONS BY GOVERNMENT AGENCY, FISCAL 1963-1969

	INVENTION DISCLOSURES		PATENT APPLICATIONS	
AGENCY	NUMBER	AVERAGE COST[a] (THOUSANDS)	NUMBER	AVERAGE COST[b] (THOUSANDS)
Agriculture	1,800	$ 600	800	$ 1,300
Atomic Energy Commission	8,700	$ 800	1,800	$ 3,500
Health, Education, and Welfare	1,700	$2,600	200	$18,100
Interior	800	$ 800	300	$ 1,900
National Aeronautics and Space Administration	15,000	$1,300	1,500	$ 9,700
Army	12,200	$ 700	5,900	$ 1,300
Navy	14,700	$ 600	6,800	$ 1,400
Air Force	14,400	$1,500	5,100	$ 4,100
Total[c]	71,400	$1,100	22,600	$ 3,000

[a] Total funds for research lagged two years.

[b] Total funds for research lagged three years.

[c] Includes other federal agencies, such as the Department of Commerce and the Tennessee Valley Authority.

Source: Patent Advisory Panel, Federal Council for Science and Technology; and National Science Foundation, *Federal Funds for Research, Development, and Other Scientific Activities*, Vol. XVI, NSF 67-19.

development and test part of the R & D process. Few inventions can come from the construction of buildings, launch facilities, other specialized structures, and from elaborate testing activities. The data in Table 9-3 show that the average cost of each invention disclosure coming out of research sponsored by the Department of Health, Education, and Welfare was about twice as much as that for NASA. The explanation for the high cost of HEW's inventions is the exact opposite of that for NASA. Most of HEW's research is basic research and, by conventional definition, the expectation is that basic research yields few if any inventions.

Another probable cause of NASA's lower rate of disclosure, compared with the Army, Navy, and AEC, is that most of NASA's hardware must be made as reliable as possible to be "man-rated." A small increment from one level of reliability to another is likely to increase costs markedly, and it is unlikely to produce many inventions. This is true because the research emphasis must be on using proved methods and devices rather than on creating new ones.

Another probable cause for NASA's lower rate of invention disclosure is a combination of the agency's patent policy and the number of its contracts. As indicated in Chapter 8, NASA follows the title policy;

it takes ownership rights to contractor inventions. A prevailing image of NASA in industry is that it waives titles to its inventions only grudgingly.[22] A title policy can work well in getting invention disclosures only if the agency has few contractors. When a government agency puts direct pressure on contractors to disclose more, more inventions will be disclosed. This of course says nothing about the quality of the additional disclosures. With thousands of prime contractors and tier upon tier of subcontractors, as is the case for NASA, the cost of monitoring invention disclosures becomes prohibitively high. The federal agency with the oldest title policy, the Department of Agriculture, sponsors little contract research and does not have to monitor the reporting of large numbers of invention disclosures from industry. The Atomic Energy Commission has the most vigorous title policy. But in sharp contrast with NASA, the Atomic Energy Commission has relatively few contractors. Because AEC gets 90 percent of its invention disclosures from 14 contractors, stricter monitoring of invention disclosures would be less difficult and less costly for AEC than it would be for NASA.

It is impossible to determine how much of NASA's research has been incorporated into new products and processes. Furthermore, it should be clear that even if the number of new products and processes incorporating NASA's technology were known, the information would be, at best, an indirect and imperfect measure of the agency's contribution to economic growth. However, consistent data on the use of NASA inventions that are covered by patent applications and issued patents are available. Table 8–11 in Chapter 8 presented data showing that the rate of commercial use of inventions resulting from government-financed research, including NASA's, is low (7 to 15 percent), compared with the rate of government use of these inventions (65 to 75 percent), and the rate of commercial use of inventions resulting from privately financed research (50 percent). The reasons for the relatively low rate of commercial use of government R & D inventions are insufficient market demand for the products, components, and processes embodied in the inventions, and that the technology is too sophisticated and expensive.

The commercial use or non-use of government R & D inventions and technology by private business creates a serious dilemma for the federal agencies with large amounts of contract research. The agencies, principally the Atomic Energy Commission, the Department of Defense, NASA, and the Department of Health, Education, and Welfare, are damned if the rate of commercial use is high, and they are damned if the rate is low. If the rate of commercial use is high, or seemingly high, the agencies are assailed for contributing to monopoly and economic concentration; they are condemned for enhancing the powers of the so-called "military-industrial complex." If the commercial use of the results of billions of dollars of research is not clearly apparent, and if the public is not seen to directly benefit, the agencies are assailed for not sufficiently promoting economic growth and social progress.

TECHNOLOGY TRANSFER

It is much easier to perceive of the concept of technological transfer than to understand the process by which it takes place. Technology transfer has been defined as the acquisition, development, and use of technological information in a context different from that in which it originated.[23] The ultimate result of the transfer of technology is innovation. The effective flow of technical information is an important ingredient, if not the principal factor, in the process of technology transfer.

Experts in the field of technology transfer distinguish between vertical and horizontal information flows. Vertical transfer takes place when information flows within a given institutional environment: pure science leads to applied research; applied research to developmental research; and eventually and if successful, development becomes hardware, or even in some cases "software." In contrast, horizontal transfer takes place among institutions or between sectors of the economy.[24] It is believed that the science-oriented industries witness vertical transfer, whereas horizontal transfer occurs mostly in the non-science—based industries.

It is possible, if not usual, to have successful and complete information transfer without the economic conditions that cause innovation. Technological information transfer must be distinguished from entrepreneurial skill and initiative, because it is entrepreneurial initiative, combined with market forces, that gives rise to innovation, i.e., the use of new technology.

Experience of the Department of Defense

One of the most extensive studies of technological transfer was Project Hindsight, undertaken by the Department of Defense. The project, well over a 50—man-year effort, was an attempt to analyze the past history of the development of 17 weapons systems (1945 to 1963) to identify the most important factors in the growth and application of defense technology.[25] A team of five to ten scientists and engineers from DOD's in-house laboratories analyzed the technology embodied in the subsystems and components of each of the 17 weapons systems. The purpose was to trace the evolution and to identify the relative importance of component devices and materials to the total weapon system. The results of Project Hindsight were disappointing to many. Despite the size of the research effort and the number of weapons systems examined, there were no monumental or even unusual findings. The principal lesson gained from tracing Department of Defense technology transfer was that advancing technology is made up of many pedestrian advances. Although each of the "inconsequential improvements" took time and money, it was the aggregate effect that was substantial.

Findings of the National Planning Association

Under a contract with the National Science Foundation, the National Planning Association (NPA) undertook an investigation of the sources and the impacts of externally generated stimuli to technological innovation. A secondary objective of the research was to determine the extent of commercial spill-over from government-financed research. The study incorporated an empirical investigation of 567 different innovations in the computer, housing, and railroad industries. The results of the NPA study confirmed the major finding of Project Hindsight, i.e., that most innovations are minor. Major technological advance comes from an army of technical employees, each of whom has the daily job of gradually improving the technology of the organization for which he works.[26]

The NPA study indicates that firms do surprisingly little search and information retrieval for the acquisition of technical information. Two-thirds of the innovators queried by NPA stated that they had applied technical information that was widely diffused and generally was known by those working in the field. According to Myers, the project director, innovation is not so much a matter of access to information as it is the ability to perceive an application for the available information.[27] Of the innovations studied, about 10 percent resulted in technology transfer from the government to the private sector of the economic system. Twenty-six of the 56 innovations based on government-financed information were in the computer industry, 16 in the housing industry, and 14 were in the railroad industry.[28]

The Nasa Technical Information System

As the newest R & D-oriented agency, NASA, and its technology utilization program, has had the benefit of the experience of other government agencies, especially the Atomic Energy Commission and the Department of Defense. Shortly after it was established, the Atomic Energy Commission initiated information identification and dissemination programs for the civilian application of nuclear and non-nuclear technology developed by the Commission. Most of the Atomic Energy Commission's activities with private industry are undertaken by the Office of Industrial Cooperation at the Argonne and Oak Ridge National Laboratories. Within the Department of Commerce, the government maintains a central Clearinghouse for Federal Scientific and Technical Information. The Clearinghouse provides, at relatively low cost, unclassified information about the results of government-sponsored research undertaken by all agencies in the Executive Branch. The Clearinghouse receives about 22,000 technical documents each year.[29] The Small Business Administration, of course, provides special technical assistance to small businesses throughout the country.

Shortly after its inception, NASA established its own Office of Technology Utilization. That office has two complementary programs—one to provide technical information to the aerospace community and the other to stimulate the use of NASA-developed technology by non-aerospace users. In College Park, Maryland, NASA has what is called the Scientific and Technical Information Facility. The Facility is a computer storage system that provides technical information to NASA employees and employees of government contractors working in the aerospace industry. Inputs into the information system come from essentially the same groups as those using the system. In the late 1960's, documents were being introduced into the system at the rate of about 70,000 each year. Under contract with NASA, the American Institute of Aeronautics and Astronautics reviews and indexes books, journal articles, and conference proceedings, and publishes abstracts in its *International Aerospace Abstracts* and in its *Scientific and Technical Aerospace Report.*

NASA established eight Regional Dissemination Centers (RDC's) to foster the use of its technology by non-aerospace firms. Each center is affiliated with a university or a nonprofit research institute. In 1968, the locations and affiliations were: University of Connecticut, Storrs, Connecticut; University of Pittsburgh, Pittsburgh, Pennsylvania; North Carolina Science and Technology Research Center, Triangle Park, North Carolina; Indiana University, Bloomington, Indiana; Wayne State University, Detroit, Michigan; Midwest Research Institute, Kansas City, Missouri; University of New Mexico, Albuquerque, New Mexico; and the University of Southern California, Los Angeles, California. The services of each of the centers are available to private industry for a fee, which ranges from about $140 to $15,000, depending on the kind and extent of the services rendered. The principal service provided by information specialists working at these centers is that of putting technical information together in useful packages. In essence, the information packages contain bits and pieces of all relevant information in the computer system about a given technological problem. Apart from supplying information about possible product and process innovations, the two most important functions performed by the RDC's for industrial clients are (1) providing technical information so as to minimize the duplication of research efforts, and (2) providing information that keeps scientific and professional personnel knowledgeable about current R & D findings.

In addition to these specialized services, NASA's Technology Utilization Offices at the various field installations put together "Tech Briefs" and "Technological Surveys." The NASA Tech Briefs, available from the Clearinghouse for Federal Scientific and Technical Information for about 15 cents each, are single-page descriptions of individual inventions, devices, methods, and concepts. In contrast, NASA Technological Surveys put together NASA contributions to an entire area of technology. The cost of a survey varies from about 15 cents to 10 dollars.

A somewhat newer program is designed to provide information about medical advances gained from space research. NASA has organized three Biomedical Application Teams that provide information about instruments and mechanisms which might be helpful to physicians and surgeons. The Biomedical Applications Teams are located in Durham, North Carolina; Kansas City, Missouri; and San Antonio, Texas.

The Use of NASA Information Systems

The most recent study on the use of NASA's technical information systems comes from the Denver Research Institute's Project for the Analysis of Technology Transfer. The Denver Research Institute team sent questionnaires to 11,013 individuals and business firms requesting NASA technical support information. The requests had been made to NASA field installations between June 1966 and October 1968. Completed questionnaires were received from 54 percent of those surveyed.

The results of the study reveal that requests for NASA-generated information, by subject area, vary according to the size of the organization requesting information. Small companies, with fewer than 500 employees, requested more information about mechanical devices (27 percent of all requests), than about materials (22 percent) and technology in the physical sciences (5 percent). In contrast, larger companies, with more than 5,000 employees, asked for more information about materials (30 percent) and technology in the physical sciences (20 percent), than information about mechanical devices (16 percent). Undoubtedly, this reflects the less esoteric nature of the technology embodied in mechanical devices and, therefore, the possibility of wider use by small business firms. NASA reports containing information about electronics comprised 41 percent of total requests, and computer programs comprised about 2 percent. In these fields, there was little discernible difference between the small and large companies in the number of requests.

The questionnaires sent out by the Denver Research Institute team also asked those requesting information about NASA-generated technology to indicate the usefulness of that information. Three percent of the respondents stated that the information contained in NASA reports was of no value, but 42 percent stated that the information increased their knowledge about the state-of-the-art. In terms of applicability to the firm or to the individual, 5 percent of those responding stated that the information was of no value; 32 percent believed that the information was of limited value; and 8 percent believed the information to be of great value. As is true of other studies, the technical information included in the Denver survey resulted in only a few new commercial products or processes—less than 1 percent of the total. The time lag between the request of the Denver team about use of the information, however, was extremely short, ranging from about 1 month to 18 months.[30] The time lag was too short to expect a rate of commercial use higher than 1 or 2 percent.

NASA TECHNOLOGY AND FUTURE INNOVATIONS

Those evaluating the role played by NASA and other R & D agencies in contributing to economic growth must face the fact that some things are unknown and unknowable. Among the unknowns are the magnitude and importance of innovations that will materialize in the future as a result of new technology developed by the Space Agency. To date, it can be said that none of the identified inventions coming out of NASA programs can be called major. In addition, all those inventions that have been identified as being in commercial use have not been of much commercial value to the innovators.[31] A major innovation, however, is almost always recognized as such after its success has been widely demonstrated. History shows that many years elapse between an invention and its use in products and processes.

In 1879, Mr. Keates remarked in the House of Commons, about the utility of electric lighting, which was used widely by the turn of the century: "I do not think that there is the slightest chance of its [electricity] competing, in any general way, with gas. There are defects about the electric light which, unless some essential change takes place, must entirely prevent its application to ordinary lighting purposes."[32]

TECHNOLOGICAL FORECASTING

Technological forecasting, which itself has become an academic subject in graduate schools of business and engineering, attempts to analyze the rate and kind of technological change, especially the use and diffusion of new products and processes. As Mansfield suggests, technological forecasts can be put into three groups: one kind of technological forecast is based on expert opinion, on an individual or group basis; another comes from the extrapolation of statistical trends; and the third takes the form of stochastic models. Mansfield believes that technological forecasts made by experts tend to be subject to large errors, especially when experts believe that something is not possible, rather than when they forecast that something *is* possible.[33]

The most important forecast, based on expert opinion, made for NASA technology was the two-year summer study conducted at Woods Hole, Massachusetts. The principal conclusions reached by the experts working on that study, as discussed in Chapter 3, were that the benefits to be derived from earth resources satellites are far larger than those initially expected by the group and also far larger than the costs of achieving the benefits. The panel of experts recommended that Congress maintain sizable budget appropriations for the development and use of this space-generated technology.

Extrapolation of statistical trends is the second kind of technological forecast. Historical data based on the empirical research of many individuals indicate that the rate and diffusion of government-generated technology into the commercial sector will be accomplished only after a substantial time lag and will take the form of incorporating rather insignificant components and parts into new products and processes. For good reason, Mansfield believes that the accuracy of technological forecasts based on extrapolation of statistical trends tends to be inversely related to the time peered into in the future.

It is the third kind of technological forecast that is coming into vogue in graduate schools. These forecasts are based on statistical models that attempt to explain and to measure the variables that cause technological change.[34] A variant of such a model, as applied to space technology, is the RAND model that was discussed in Chapter 3. The success of these models is yet unknown; in fact, they are an infant innovation. To be useful, the information put into the models must be significantly better than that which has been available about the factors that spur technological change.[35]

FUTURE APPLICATIONS AND TECHNOLOGICAL BENEFITS GENERATED FROM NASA RESEARCH

The following is a brief discussion of some of the contributions that the space program has made, might have made, or might make to technological advance and social progress. The list of examples is by no means complete.[36] The discussion purposely refrains from analyzing the examples in terms of such essential economic criteria as market demand, competitiveness with alternatives, and profitability. To do so, at this point in time, might lead to pessimistic, if not cynical, conclusions.

The Fuel Cell

Although the idea of a fuel cell is over 100 years old, the space program has given a boost to its development and economic practicability. By means of chemical reaction, a fuel cell produces electrical power directly from fuel, by-passing the use of conventional generators. Some of its advantages include: (1) efficiency or a greater amount of energy per unit of fuel consumed than steam, diesel, or gasoline generators; (2) the absence of movable parts that reduces the possibility of breakdown and increases reliability; and (3) nonpollutant waste because its by-products are water and carbon dioxide. For use in the Gemini and Apollo spacecrafts, NASA was the first federal agency to specify fuel cells as a source of power. Some believe that NASA's research, development, and actual use of the fuel cell increased the chances of its widespread use.[37] The most dramatic future uses of the fuel cell will probably be for complete urban public utility systems and for electrically propelled vehicles.

Medical Technology

The necessity of providing man with a safe environment in space has resulted in the development of a number of devices, instruments, and techniques in medicine. Many are the outcome of studies about the effects of the terrestrial environment on the human body. NASA has also advanced the state of the art in the field of cardiovascular monitoring. While in space, of course, the circulatory and respiratory systems of the astronauts are continually monitored. The treatment of certain kinds of heart diseases and other medical problems (severe burns) requires continuous monitoring of vital parameters in special coronary and intensive care wards. Space medicine has contributed to the development of an ultra-small, enclosed thermocouple that is a component part of a probe used in cryogenic brain and eye surgery. The thermocouple measures and records minute changes in temperature and provides vital information for cryogenic surgery. NASA research in bacteria control, developed for sterilization of the spacecraft, has application in the sterilization of surgery rooms, the manufacture of germ-free food and pharmaceutical products, and also in research projects requiring the isolation of bacteria.[38]

Hayes International Corporation, a NASA contractor, has produced a sight switch control device for the motorization of wheelchairs. The unit can be used by paraplegics for self-mobility. The device, worn in the shape of a pair of eyeglasses, controls the movement of a wheelchair. The sight switch was developed for use by astronauts during launch when high-gravity forces reduce physical maneuverability.

A final example of space technology in the medical field is a walking machine. Unlike the other examples, this device is not a product of space-medicine research. Rather, it was developed as an instrument carrier for use in the unmanned exploration of the lunar surface. The walking machine, which can be adapted for use by disabled people, moves on reciprocating legs. The vehicle permits travel over terrain inaccessible to invalids in ordinary wheelchairs.

Communications Satellites

Echo I, NASA's first communications satellite, was launched in August 1960. It proved the possibility of using orbiting satellites for international communications. Although NASA still is engaged in advanced research on satellite communications systems, the quasi-public Communications Satellite Corporation (COMSAT) has principal responsibility for the development and provision of commercial communications service. In addition to the television services of the communications satellite that are well known to the public, the communications satellite system provides telephone and telegraph services. It is estimated that the telephone capacity across the Atlantic has increased by about 50 percent since the inception of the satellite system.[39] Future advances in the use of communications satellites appear to be greatest in educational television services in under-developed countries, data exchange between widely separated computer systems, and the transmission of other kinds of live television for social and educational purposes (visual surgical techniques, for example).

Weather Satellites

The first United States meteorological satellite, Tiros I, was also launched in 1960. As in the case of the communications satellites, NASA has relinquished much responsibility for the operation of the weather satellite system. Under a 1964 agreement, the Weather Bureau is responsible for the acquisition, analysis, and dissemination of weather data via satellite. NASA's continuing role concerns advanced research and the launch operations for the satellites. The weather satellite system has improved weather forecasting and has also provided an advance warning system on the paths of hurricanes and other storms. Many meteorologists are hopeful that weather satellites will eventually provide sufficient information to make accurate five-day weather forecasts. At present, accurate forecasts can be made only for a maximum of two days.

Computer Technology

NASA's technical requirements for computer development and use, no doubt, have contributed to the rapid growth of this new industry. NASA uses computers to design, develop, test, launch, and track all of the space systems it produces and puts into orbit. The Apollo Mission Simulator and the Lunar Module Simulator, used for astronaut and flight control personnel training, are computer-based systems. NASA has purchased or leased computer equipment from virtually all competing companies in the computer industry. The names of the companies include: Burroughs Corporation, Control Data, Decision Control Corporation, General Electric, Honeywell, International Business Machines, Packard-Bell Electronics, Radio Corporation of America, Raytheon, Scientific Data Systems, and Sperry Rand. Some of NASA's newest computer equipment is capable of making over 16 million computations a second. A computation rate of that speed is about 15 times higher than the capability of computers designed several years ago. The principal contributions of the space program to computer technology have been in the miniaturization of components, increased reliability, and greater speed. Micro-electronic circuitry is essential for the exploration of space because payloads must be limited in size and weight. In a joint effort between the University of Illinois and the Burroughs Corporation, a computer is being developed that will be capable of making one billion computations a second. The Space Agency has plans to employ that computer.[40]

Computer software developed by NASA, such as tapes, card decks, and run instructions, is readily available to other potential users. In July 1966, NASA made an arrangement with the University of Georgia to collect, evaluate, document, and disseminate information about the agency's computer software products. Since June 1966, the Department of Defense and the Atomic Energy Commission have joined the University of Georgia Computer Software Management and Information Center. One of the benefits of the joint venture is reduced duplication in the development of computer software.

Natural Resources

Remote sensors from space can be used for the discovery, inventory, evaluation, development, and conservation of natural resources. For many earth resource programs, remote sensing and photography from space have a number of advantages over ground or aerial observations, including: shorter data acquisition time, synoptic views of regions, greater independence of turbulent weather, observation of inaccessible areas, and continuity of observation. Photographs taken from 100 miles in space have provided as much, if not more, detailed imagery as those taken at 40,000 feet. Space photos have shown such details as airfields and road construction projects. On one manned space flight, the crew photographed 80 percent of Peru in only three minutes.[41]

Many, including the participants in the Woods Hole Space Summer Study, believe that NASA's earth resources program is one of the agency's most exciting programs and one with the greatest promise to the entire world. Even the United States Treasury is interested in the possibility of using spaceborne remote sensing equipment—to identify the location of narcotic plants and thus improve control over the traffic in narcotics.[42]

Education

The post-Sputnik period created an educational environment designed to upgrade the quality of education, especially in mathematics and the physical sciences. Some, no doubt, would argue that the impetus for improvement in education is too strong and that youth protests are a form of reacting to pressures and standards of excellence with which they cannot cope. Nevertheless, all levels of education have been influenced by the nation's space program. The returns from NASA's educational programs consist of factors that can contribute to economic growth. To enrich curriculum at the elementary and secondary levels, NASA provides materials upon request about aeronautics and space projects. NASA also provides lecture-demonstration teams that describe the nature of space exploration to school children, as well as to civic and other adult groups.

NASA support at the university level, of course, has been much greater than that at the elementary and secondary levels. In 1961, NASA initiated a predoctoral scholarship program. A principal purpose of the program was to increase the nation's pool of scientific talent, and, thereby, avoid any possible manpower shortage resulting from a rapid expansion of the space program. Between 1962 and 1968, NASA provided 5,414 different scholarships. About 85 percent of the Ph.D's awarded were in engineering and the physical sciences and 15 percent in the life, behavioral, and social sciences. Some colleges and universities in each of the 50 states and the District of Columbia, at one time or another, received NASA fellowship support. Under its Sustaining University Program, NASA financed the construction of 35 laboratories and other kinds of research facilities. Most of NASA's funds for the construction of facilities went to academic institutions heavily entrenched in space research. By mid-1968, the University of California, Massachusetts Institute of Technology, University of Minnesota, Case Western Reserve University, and Stanford University had received more than $2 million each from NASA for the construction of facilities.[43]

Another area of substantial NASA support has been in research. The magnitude and objectives of NASA's support of research in the social sciences were discussed in Chapter 5. Excluding the quasi-governmental Jet Propulsion Laboratory at the California Institute of Technology, about 5 percent of NASA's prime contract awards have gone to educational and nonprofit institutions. About two-thirds of NASA's grants and contracts with universities have been for specific projects. The project approach serves the dual purpose of enlisting the expertise of academicians to solve space-related problems and of providing university faculties with an opportunity to expand knowledge through research and to impart the acquired information to students.

Transportation and Air Travel

A world-wide, satellite-based navigation system holds the promise of becoming an effective way to control sea and air traffic routing. The function of such a system would be to provide traffic control agencies, aircraft, and ships with weather reports, weather forecasts, position points and paths, and safety information.

Research undertaken by NASA on vertical take-off and landing (VTOL) and short take-off and landing (STOL) transports is designed to solve the urban problem of transporting cargo and people in the congested megalopolis (San Francisco-Los Angeles, Boston-New York-Washington, and Buffalo-Chicago). It has been predicted that VTOL and STOL service will materialize by the end of the twentieth century. For long-haul flights, NASA is working with the Air Force on the development of a hypersonic transport that could travel at four to six times the speed of sound. In the late 1960's, NASA was also supporting the Department of Transportation in the development of the supersonic transport, designed for long-haul flights at twice the speed of sound.

George E. Mueller, Associate Administrator for Manned Space Flight at the time of the first manned lunar landing, and others at NASA are firmly convinced that a so-called "space-shuttle system" will provide efficient transportation in the future. According to Mueller, the principal breakthrough required for a space shuttle is the development of a competitively cost, re-usable launch vehicle. He conceives of a space shuttle system as similar to commercial air transport methods in current use. The shuttle would take off vertically from a small pad and return from orbit by gliding, almost noiselessly, to a runway landing. Mueller believes that as a global transport for point-to-point traffic, the space shuttle would provide safety and comfort standards equivalent to those of the large jet transports in current use. Travel time from New York to Tokyo might be reduced to 47 minutes.[44]

Space Sciences

NASA's contribution to pure science transcends immediate and direct applications. No one can predict the ultimate value to mankind of new knowledge and of greater understanding of the universe. Although most of the actual work is applied research, development, and test, the greatest contribution to basic knowledge has been the manned lunar landing as a scientific entity in and of itself. As put by George E. Mueller:

For a million and half years there have been manlike creatures on this planet; and during all these eons, man has been confined to the thin membrane of his earth's atmosphere. Now . . . man has broken these confines. For some, the meaning of this achievement is seen only dimly and superficially. The purposes of space exploration and discovery are no clearer to many men in this age than they were in the days of Galileo.[45]

As indicated in Chapter 2, most fundamental research undertaken by the Space Agency is administered by NASA's Office of Space Sciences and Applications. Research in the agency's Physics and Astronomy Program is designed to increase fundamental knowledge about the space environment and the physical nature of the earth. Some of the projects in the Physics and Astronomy Program include the Solar and Astronomical Observatories, the Pioneer and Explorer inter-planetary satellites, and Sounding Rockets. Under the Lunar and Planetary Exploration Program, the Surveyor, Lunar Orbiter, Mariner, and Ranger projects help to provide new information about the solar system by means of automated spacecraft. At present, NASA's Bioscience Program is designed to search for signs of extraterrestrial life (Exobiology) and to determine the effects of the space environment on terrestrial life (Environmental and Behavioral Biology).

NASA's budget identifies "basic research" as one program in the Office of Advanced Research and Technology. That basic research program is divided into four broad disciplines: fluid physics, electrophysics, materials research, and applied mathematics. Fluid physics encompasses research on propulsion systems, power generation, and the dynamics of atmospheric entry. The goal of the electrophysics program is to gain knowledge about atomic, molecular, solid state, and nuclear physics where electric fields are important in the exchange and release of energy. As indicated, materials research is the third broad research field. The principal role of NASA's materials research project is to determine ways in which high temperature, ionization radiation, micrometeorite impact, corrosive fluids, high vacuum, and electromagnetic fields affect materials used for space devices. The fourth area is what NASA calls its applied mathematics program. The objective of that program is to provide rapid and indisputable mathematical means to determine and to predict cause and effect relations for the solution of engineering and scientific problems.[46]

SUMMARY

Because the principal objectives of the space program are the exploration of space and the expansion of man's understanding of the universe, the primary benefits of the program must be expressed in terms of the fulfillment of these objectives. Nevertheless, NASA, the Congress, and the public have tried to find out how much the space program contributes to economic growth.

Failure to provide the answer is not the fault of those directing the course of the space program. To date, the answer to that question and to similar questions for other R & D-oriented federal programs eludes statistical proof. Unfortunately, economists have not yet found a satisfactory way to measure the contribution of research and development to economic growth. This is true because scientific activities are intellectual in nature and most of the results of the R & D process do not take the form of physical values that can be measured. Some economists, do, however, distinguish between scientific outputs that are intangible and not subject to measurement and those outputs that are measurable. The nonmeasurable outputs consist of ideas, hypotheses, and theories, whereas the measurable outputs are the means by which information is transferred, i.e., technical reports, publications, patents, and even new products and processes. The results of research show that counts of publications, patent applications, patents, and new products and processes are, by and large, poor measures of scientific output and do not reflect the contribution of science and technology to economic growth.

Economists have indirect measures of the value of technological advance and economic growth. The most common yardsticks used to measure changes in economic welfare show a marked increase in the rate of change in the post-World War II period. The surge of research and development activities in the United States, of course, has been during the last two and a half decades. For example, between 1910 and 1968 the per capita real gross national product rose at an average annual rate of 1.7 percent. Prior to 1929, the average annual rate of increase was about 1.3 percent. This compares with a rate of 2.3 percent since World War II. The index that measures output per man-hour shows a rate of increase of 1.9 percent before 1929 and a rate of change of about 3.3 percent per annum since 1947. Although productivity change does not measure the rate of technological change, technological change affects productivity by increasing efficiency. Research indicates that technological advance does not occur in a dramatic fashion. Rather, major technological advance takes place slowly, incorporating many pedestrian improvements. It is the exception rather than the rule when an innovation results in immediate and widely diffused benefits.

The matter of the commercial use by private business of new technology generated by NASA and other federal agencies with large amounts of contract research creates a serious dilemma for those agencies and for the public in general. If commercial benefits appear to be high, the agencies are assailed for contributing to monopoly and economic control; they are condemned for enhancing the powers of the so-called "military industrial complex." If use of the results of billions of dollars of research is not clearly apparent and if the struggling farmer in Appalachia and the underprivileged Negro in urban ghettos are not seen directly to benefit from the scientific research undertaken by these agencies, the agencies are assailed for hampering economic growth and social progress.

The greatest contributions of the space program to economic growth and economic welfare might not be realized for decades or even for generations. Although they cannot be measured, some of the contributions that the space program has made, might have made, or might make to technological advance, social progress, and economic growth include: acceleration in the development of the fuel cell; instruments for cardiovascular monitoring; devices and techniques for cryogenic surgery; improved television, telephone, and telegraph services via communications satellites; new weather forecasting techniques via meteorological satellites; micro-miniaturization of electronic components; increased reliability and speed of computers; earth resources information provided by earth-orbiting satellites; and a space-shuttle transportation system. To be sure, NASA's greatest contribution to advanced technology has been the manned lunar landing as a scientific entity in and of itself.

NOTES

1. In his book on the costs of economic growth, the British economist Ezra J. Mishan argues that the continued pursuit of economic growth by Western countries is more likely to reduce rather than to increase social welfare. For example, Mishan is convinced that the invention of the private automobile is one of the greatest disasters to the human race because of air pollution and destruction of the countryside with suburbanization. Ezra J. Mishan, *The Costs of Economic Growth* (New York: Frederick A. Praeger, 1967).

2. For the most complete survey and analysis of the economic literature and the theoretical and empirical progress made by economists in the field of economic growth, see Edwin Mansfield, *Industrial Research and Technological Innovation: An Econometric Analysis* (New York: W. W. Norton, 1968); and Edwin Mansfield, *The Economics of Technological Change* (New York: W. W. Norton, 1968).

3. The production function used above is adapted from the model described by Irma Adelman, *Theories of Economic Growth and Development* (Stanford, Calif.: Stanford University Press, 1961), pp. 8–24.

4. *Ibid.*, p. 11.

5. Edwin Mansfield, "Technological Change: Measurement, Determinants, and Diffusion," *The Employment Impact of Technological Change*, Appendix Vol. II, Report of the National Commission on Technology, Automation, and Economic Progress (Washington, D.C.: U.S. Government Printing Office, 1966), pp. II 100–102.

6. U.S. Department of Commerce, Bureau of the Census, *Long Term Economic Growth: 1860–1965* (Washington, D.C.: U.S. Government Printing Office, 1965), pp. 188–89.

7. Mansfield, "Technological Change: Measurement, Determinants, and Diffusion," p. II 101; John W. Kendrick, *Productivity Trends in the United States* (Princeton: Princeton University Press, 1961).

8. The May 1969 issue (Part II) of the U.S. Department of Commerce's *Survey of Current Business* is devoted to an analysis of the major conceptual problems in the measurement of productivity. The article highlights the different measurement procedures, as well as differences in the findings about the relative importance of factors contributing to productivity, used by Edward F. Denison and by Dale W. Jorgenson and Zvi Griliches for estimating changes in productivity in the United States. See U.S. Department of Commerce, Office of Business Economics, "Some Major Issues in Productivity Analysis," *Survey of Current Business,* 49, 5, Part II (May 1969).

9. Kendrick, *Productivity Trends in the United States,* pp. 136–37.

10. Edward F. Denison, *The Sources of Economic Growth in the United States and the Alternatives Before Us,* Supplementary Paper No. 13 of the Committee for Economic Development (New York: Committee for Economic Development, 1962).

11. As published in U.S. Department of Commerce, *Long Term Economic Growth,* pp. 188–89.

12. Edward F. Denison, *Why Growth Rates Differ: Postwar Experience of Nine Western Countries* (Washington, D.C.: Brookings Institution, 1967), p. 79.

13. Kenneth J. Arrow, "Explaining the Growth of Output," *Science* (May 9, 1969), p. 700.

14. Jacob Schmookler, *Invention and Economic Growth* (Cambridge, Mass.: Harvard University Press, 1966), p. 196.

15. Fritz Machlup, *The Production and Distribution of Knowledge in the United States* (Princeton, N.J.: Princeton University Press, 1962), pp. 180–81. Edward Ames makes a similar distinction in his article, "Research, Invention, Development, and Innovation," *American Economic Review,* LI, 3 (June 1961), pp. 370–81.

16. For a discussion of other indirect measures, see Sterling Scott Sutton, "Scientific Output: Methods of Measurement" (unpublished thesis submitted to The George Washington University, June 1964).

17. Consensus of a group of experts in the field of scientific output measures at a symposium on the Exploration of Methods to Measure Scientific Output sponsored by the National Science Foundation, July 18–19, 1963, Washington, D.C.

18. Three of the best works on and criticisms of publications counts as a measure of scientific output are William Shockley, "Individual Variations of Productivity in Research Laboratories," *Science,* CXXII (November 4, 1955), p. 879; Derek J. de Solla Price, *Science Since Babylon* (New Haven, Conn.: Yale University Press, 1961), pp. 100–105; and Helen S. Milton and Ellis A. Johnson, *Sponsorship of Research: A Survey of the Scientific Literature, 1920–1960* (Bethesda, Md.: Operations Research Office, The Johns Hopkins University, September 1961).

19. Donald S. Watson and Mary A. Holman, "The Federal Government's Propensity to Patent," *Patent, Trademark, and Copyright Journal,* 10, 1 (Spring 1966), pp. 61–74.

20. C. Guy Suits, "Organization of Technical Activities at the General Electric Company," *Organizing the Research Function for Profit,* Research and Development Series, No. 1 (New York: American Management Association, 1956), p. 8.

21. John L. Enos, "Invention and Innovation in the Petroleum Refining Industry," *The Rate and Distribution of Inventive Activity* (Princeton, N.J.: Princeton University Press, 1962), pp. 307–8.

22. Donald S. Watson and Mary A. Holman, *An Evaluation of the Patent Policies of the National Aeronautics and Space Administration,* Report of the U.S. House Committee on Science and Astronautics, 89th Cong., 2d Sess., 1966, pp. 20–23.

23. Richard S. Rosenbloom, *Technology Transfer—Process and Policy: An Analysis*

of the Utilization of Technological By-products of Military and Space R & D, Special Report No. 62 (Washington, D.C.: National Planning Association, 1965), pp. 8–10.

24. Harvey Brooks, "National Science Policy and Technology Transfer," *Proceedings of a Conference on Technology Transfer and Innovation* (Washington, D.C.: U.S. Government Printing Office, 1967), pp. 53–55. The Conference was held in Washington, D.C. on May 15–17, 1966, under the auspices of the National Planning Association and the National Science Foundation.

25. Raymond S. Isenson, "Technological Forecasting: Lessons from Project Hindsight," in *Technological Forecasting for Industry and Government*, ed. by James R. Bright (Englewood Cliffs, N.J.: Prentice-Hall, 1968), pp. 35–54.

26. Sumner Myers, *Technology Transfer and Industrial Innovation* (Washington, D.C.: National Planning Association, 1967).

27. *Ibid.*, p. V–12.

28. *Ibid.*, p. V–24.

29. U.S. Congress, Senate, Subcommittee on Science and Technology of the Select Committee on Small Business, *Report: Policy Planning for Technology Transfer*, 90th Cong., 1st Sess., 1967, pp. 15–24.

30. Dean C. Coddington, et al., *Project for the Analysis of Technology Transfer: First Annual Report* (Denver, Colo.: Industrial Economics Division, Denver Research Institute, University of Denver, November 1968). John S. Gilmore, et al., *The Channels of Technology Acquisition in Commercial Firms, and the NASA Dissemination Program* (Denver, Colo.: Industrial Economics Division, Denver Research Institute, University of Denver, June 1967).

31. Watson and Holman, *An Evaluation of the Patent Policies of NASA*, p. 87.

32. Nancy T. Gamarra, *Selection of Statements on Erroneous Predictions and Negative Comments Concerning Exploration, Territorial Expansion, and Scientific and Technological Development*, Prepared at the request of the Senate Committee on Aeronautics and Space Sciences (Washington, D.C.: The Library of Congress, Legislative Reference Service, 1967), p. 31.

33. Mansfield, *The Economics of Technological Change*, pp. 38–40.

34. *Ibid.*

35. For examples of these kinds of models outside the area of space technology, see James R. Bright, ed., *Technological Forecasting for Industry and Government: Methods and Applications* (Englewood Cliffs, N.J.: Prentice-Hall, 1968), and Murray Brown, *On the Theory and Measurement of Technological Change* (London: Cambridge University Press, 1966).

36. David Cochran, *The Benefits of the National Space Program and Their Application and Understanding by the American Public*, 2 vols. (Menlo Park, Calif.: Stanford Research Institute, 1967 and 1968). John G. Meitner, *Some Major Impacts of the National Space Program*, 6 vols. (Menlo Park, Calif.: Stanford Research Institute, 1968). "Space By-products: Moon-Shot Technology Spawns Ideas for Home, Business, and Medicine," *Wall Street Journal*, July 9, 1969, p. 1 and p. 13.

37. "The Case for Space," *Forbes*, July 1, 1968, pp. 72–74.

38. U.S. Congress, House, Subcommittee on NASA Oversight of the Committee on Science and Astronautics, *Staff Study: The National Space Program—Its Values and Benefits*, 90th Cong., 1st Sess., 1967, pp. 41–46.

39. *Ibid.*, p. 24.

40. "The Case for Space," *Forbes*, p. 56.

41. David Cochran, *The Benefits of the National Space Program and Their Appreciation and Understanding by the American Public*, *op. cit.*, Vol. II, pp. 67–76.

42. U.S. Congress, House, Committee on Science and Astronautics, *The National Space Program—Its Values and Benefits*, p. 33.

43. National Aeronautics and Space Administration, *A Study of NASA University Programs* (NASA-SP-185) (Washington, D.C.: NASA, 1968).

44. George E. Mueller, "The New Future for Manned Spacecraft Developments," *Astronautics and Aeronautics* (March, 1969), pp. 24–32.

45. George E. Mueller, "Manned Space Flight: Programs, Progress, and Prospects," in *Scientific Progress and Human Values*, ed. by Edward and Elizabeth Hutchings (New York: American Elsevier Publishing Co., 1967), p. 59.

46. U.S. Congress, House, Testimony of Herman H. Kurzweg, *1968 NASA Authorization, Hearings*, before the Subcommittee on Advanced Research and Technology of the House Committee on Science and Astronautics, Part 4, 90th Cong., 1st Sess., March and April, 1967, pp. 201–227.

10 The Future Profile of the Space Program

After spending many billions of dollars annually during the decade of the 1960's, the United States successfully completed in July 1969 the largest research and development project it had ever undertaken. This is true despite the fact that Apollo flights are scheduled throughout the early 1970's. In no way can public enthusiasm over post-Apollo flights be compared with that aroused by the July 1969 first manned lunar landing flight. There can be little doubt that a number of key NASA personnel believed that Apollo 11 completed the most important phase of the present space program. Within several months after the epic July flight, a number of key NASA officials separated from the civilian space program, including the NASA Associate Administrator for Manned Space Flight, George E. Mueller, the Apollo Program Director, Samuel C. Phillips, and Astronaut Michael Collins.

With the goal of the manned lunar landing accomplished, the question, of course, is what, if any, space objectives should be pursued by the nation in the future? Using a basic principle of economics, i.e., maximization of social benefit and minimization of social cost, it would be ideal for this book to conclude with a profile of a national space program to be undertaken during the decade of the 1970's and beyond. But to provide an outline of a future space program based on economic criteria would require measurement of all social wants—both private and public. It should be clear from preceding chapters, especially Chapter 3, that social scientists have not been able to devise an effective way to measure the social benefits and costs of public activities with such disparate objectives as the elimination of poverty, reduction of air and water pollution, maintenance of a strong national defense, prevention of crime, and space research and development.

This chapter assumes the more limited task of using information in the preceding chapters to analyze the future of the space program in terms of the multiplicity of concepts covered by the term "forecasting." The future of the space program will be reviewed in relation to five kinds of forecasts. Irving H. Siegel calls the forecasts the "Five P's," which are: (1) prediction, (2) projection, (3) program or policy, (4) propaganda, and (5) poetry or primitive projections.[1]

PREDICTING THE FUTURE OF THE SPACE PROGRAM

As used here, a forecast based on prediction means that all future events can be anticipated without qualification. As Siegel states, all forecasters would like to make predictions, but they must usually settle for something much less because prediction requires the forecaster to correctly take into account and to appraise all of the pertinent "ifs" and "buts" of future events. This, of course, is impossible for a forecast of the future of the space program, or for that matter, almost any other private or public undertaking. Complete and correct knowledge about the relevant variables, as well as their inter-relationships, is simply not available. Siegel conjectures that if and when future socio-economic events can be easily and accurately foretold, man would probably no longer anticipate the future because man would have become part of either a regimented or a stationary state.[2]

PROJECTIONS OF THE FUTURE SPACE EFFORT

The second kind of forecast, which is projection, is based on "grounded possibilities," i.e., the alternatives that seem to be logical in the context of historical experience. With projection, the forecaster does not take it for granted that all the events postulated will actually materialize.

Selected for discussion here are five possible funding levels for the space program between 1969 and 1980. Three of the five possible funding levels were outlined by a panel of experts, President Nixon's Space Task Group, to define the possible scope of spending for the exploration of space during the decade of the seventies.[3] These projections are displayed in Table 10–1. The other two possible funding levels establish a high-low range based on projections of spending for space witnessed during the 1960's.

Table 10–1 also provides an estimate of the gross national product for selected years. The GNP estimates come from work undertaken by the Center of Economic Projections of the National Planning Association.[4] Under the guidance of Gerhard Colm, the National Planning Association has pioneered the development of economic projection techniques for the past two decades. An increase in the gross national

TABLE 10–1: SPENDING FOR SPACE IN RELATION TO THE GROSS NATIONAL PRODUCT, 1966–1980 (*BILLIONS OF DOLLARS*)

		NASA PROGRAM ALTERNATIVES					
	GROSS	OPTION I		OPTION II		OPTION III	
YEAR	NATIONAL PRODUCT	DOLLARS	PERCENT OF GNP	DOLLARS	PERCENT OF GNP	DOLLARS	PERCENT OF GNP
1966[a]	$ 750	$5.9	0.79	$5.9	0.79	$5.9	0.79
1969[b]	$ 943	$4.0	0.42	$4.0	0.42	$4.0	0.42
1971[c]	$1,027	$4.3	0.42	$4.0	0.39	$4.0	0.39
1973[c]	$1,111	$5.8	0.52	$4.3	0.39	$4.3	0.39
1975[c]	$1,215	$7.7	0.63	$5.5	0.45	$5.5	0.45
1980[c]	$1,522	$9.4	0.62	$7.7	0.51	$5.5	0.36

[a] Current dollars.
[b] Current dollars, third quarter of calendar year.
[c] 1969 dollars.

Sources: National Planning Association, *National Economic Projections,* Report No. 68-N-1 (Washington, D.C.: National Planning Association, 1969), p. 128. U.S. President, Space Task Group Report to the President, *The Post-Apollo Space Program: Directions for the Future,* September 1969, p. 22.

product, in constant 1969 dollars, from $943 billion in 1969 to $1,522 billion in 1980 assumes that output will increase during the decade of the seventies at an average annual rate of 4.4 percent. This compares with a slightly higher actual average annual rate of increase of 4.7 percent between 1960 and 1967, and a much lower rate of increase of 2.2 percent per annum between 1955 and 1960. Clearly, the National Planning Association's estimate assumes that policy makers will not permit the economy to revert to its dismal performance during the mid- and late-1950's.

A Phase-Out of the Space Program

Many Americans seem to think that spending for space exploration during the decade of the 1960's took a much larger share of the gross national product than the actual allocation of resources to the program. As Chapter 5 indicated, resource allocation for the space program, at its peak, never amounted to as much as 1 percent of the GNP. Similarly, because of the three-year lag between the peak of the space effort and the first manned lunar landing, many do not realize how sharply the nation's resource allocation to space exploration was reduced between 1966 and 1969.

If spending for the space program were to continue to decrease from the 1969 level of about $4.0 billion, at the rate of $640 million a year, the space program would be terminated by the end of 1975. A reduction of $640 million annually, over a 5-year period, is based on actual spending patterns between 1966 and 1969.

Projections Made by the President's Space Task Group

The Space Task Group developed a set of options for spending for space exploration between what the group called, "a maximum pace—limited not by funds, but by technology," and a low-level program without a Manned Space Flight Program after the completion of the Apollo and Apollo Applications Programs.[5]

The three spending options shown in Table 10–1 include the same major manned and unmanned space missions. The various missions, along with a tentative schedule for the accomplishment of each event, are outlined in Table 10–2. Option I illustrates what the Space Task Group "believes" would be a dramatic increase in funding for space exploration. If undertaken, and if successful, Option I might result in a manned expedition to Mars in the mid-1980's; establishment of an earth space station in earth orbit in the mid-1970's; and an earth-to-orbit space shuttle transportation system also in the mid-1970's. With Option I,

TABLE 10–2: THE TIME SCHEDULE FOR FUTURE SPACE MISSIONS

	COMPARATIVE PROGRAM ACCOMPLISHMENTS			
FUTURE SPACE MILESTONES	MAXIMUM PACE	PROGRAM OPTION I	PROGRAMS OPTIONS II & III	LOW LEVEL
Manned systems				
Space station (earth orbit)	1975	1976	1977	–
50-Man space base (earth orbit)	1980	1980	1984	–
100-Man space base (earth orbit)	1985	1985	1989	–
Lunar orbiting station	1976	1978	1981	–
Lunar surface base	1978	1980	1983	–
Initial Mars expedition	1981	1983	II-1986	–
			III-Open	–
Space transportation system				
Earth-to-orbit	1975	1976	1977	–
Nuclear orbit transfer stage	1978	1978	1981	–
Space tug	1976	1978	1981	–
Scientific				
Large orbiting observatory	1979	1979	1980	–
High energy astronautics capability	1973	1973	1981	1973
Out-of-ecliptic survey	1975	1975	1978	1975
Mars—High resolution mapping	1977	1977	1981	1977
Venus—Atmospheric probes	1976	1976	Mid-80's	1976
Multiple outer planet "tours"	1977–79	1977–79	1977–79	1977–79
Asteroid belt survey	1975	1975	1981	1975
Applications				
Earliest operational earth resource system	1975	1975	1976	1975
Demonstration of direct broadcast	1978	1978	Mid-80's	1978
Demonstration of navigation/traffic control	1974	1974	1976	1974

Source: U.S. President, Space Task Group Report to the President, *The Post-Apollo Space Program: Directions for the Future*, September 1969, p. 20.

spending for space would continue at a relatively low level (of less than $5.0 billion in 1969 constant dollars) until 1973. By 1980, spending for the space program would amount to about $9.4 billion.

As outlined by the Space Task Group, the only difference between Option II and Option III is that in Option III, funding of a manned mission to Mars would be deferred indefinitely, or at least until the period beyond 1986. In the early 1970's (1970 through 1972), there is little difference between Option I, compared with Options II and III; under each, spending for space exploration is below $5.0 billion. The difference between the three possibilities does not amount to much until 1973, when spending under Option I would rise to $5.8 billion and spending under Options II and III would remain under the $5.0 billion mark. With Option III, the nation would allocate no more than $5.5 billion to space exploration between 1969 and 1980. With the Mars expedition included in Option II, the Space Task Group envisions spending for space rising to about $7.7 billion by 1980.

Future Space Spending in Relation to the 1966 Peak

The fifth projection relates future spending on a civilian space program to the relative share of resources allocated to space exploration in the mid-1960's. If the gross national product of the United States does rise by 4.4 percent annually, reaching $1,522,000,000 by 1980, each of the projections discussed would be substantially less than the relative share of the GNP that actually was allocated to space exploration in the 1960's.

As Table 10–1 shows, NASA programs comprised 0.79 percent of the GNP in 1966. To account for the same relative share of a $1,522,000,000 GNP in 1980, would require that spending for space rise to about $12.0 billion by 1980. In constant 1969 dollars, this amounts to a space program about one-fourth larger than the Space Task Group's accelerated plan under Option I. A program using the same share of national resources and output in 1980 as used in 1966 would be almost 60 percent greater than Option II, and more than twice the size of the projected space program under Option III.

The Allocation of Manpower to Future Space Programs

The nation's future space effort can also be projected in terms of employment. The data in Table 10–3, which present space employment in relation to a projection of total employment, tell essentially the same story as the GNP projections. The estimates of total civilian employment in the decade of the seventies were also made by the Center for Economic Projections of the National Planning Association. The projection of total civilian employment amounting to 94 million in 1980 assumes that civilian employment will rise by 1.8 percent each year; a rate that is identical with the 1960 to 1967 period.[6]

At the peak of the space program in 1966, NASA's combined civil service and contractor work force was about 400,000, and, as indicated in Chapter 5, amounted to about one-half of 1 percent of total civilian employment. If the space program in the future would use the same relative share of total national employment, spending would have to be sufficiently large to create about 500,000 jobs. This is based on a value-added per man-year cost rate of about $21,000 in 1969 constant dollars. None of the options presented by the Space Task Group approximates a future labor force on the space program of 500,000. A labor force of that size on the space program is about 15 percent larger than that under Option I in 1980, 40 percent larger than Option II, and 95 percent larger than Option III.

TABLE 10–3: SPACE EMPLOYMENT IN RELATION TO TOTAL CIVILIAN EMPLOYMENT, 1968–1980 (*THOUSANDS*)

		NASA PROGRAM ALTERNATIVES					
	TOTAL CIVILIAN	I		II		III	
YEAR	EMPLOYMENT	NO.	PERCENT	NO.	PERCENT	NO.	PERCENT
1966[a]	74,000	396	0.54	396	0.54	396	0.54
1969[b]	78,000	190	0.24	190	0.24	190	0.24
1971[c]	81,000	202	0.25	188	0.23	188	0.23
1973[c]	83,000	278	0.33	202	0.24	202	0.24
1975[c]	86,000	366	0.43	259	0.30	259	0.30
1980[c]	94,000	447	0.48	363	0.39	261	0.28

[a] June 1966.
[b] September 1969.
[c] NASA employment estimated from a value added per man-year cost-rate of $21,100 in 1969 dollars.

Source: National Planning Association, *National Economic Projections,* Report No. 68-N-1 (Washington, D.C.: National Planning Association, 1969), p. 128. U.S. President, Space Task Group Report to the President, *The Post-Apollo Space Program: Direction for the Future,* September 1969, p. 22.

The projected employment data displayed in Table 10–3 also show that under Options II and III future employment on space programs would not attain the 400,000 level actually reached in 1966. This is not true if spending reached the amount outlined under Option I. Although not shown in the table, with Option I, civilian space employment would reach the 400,000 mark in late 1976 or early 1977.

THE FUTURE SPACE EFFORT AND PROGRAM OR POLICY FORECASTING

The next kind of forecast to examine in relation to the future role to be played by the space program is the program or policy forecast. This variety of forecasting reflects the will of a purposeful planner or administrator who charts the course of a deliberate policy toward a predetermined goal.[7]

The Present Image of the Space Program

From the vantage point of the end of the Apollo period, as defined by the first successful manned lunar landing, there is little evidence to indicate that a policy-oriented forecast is feasible. The space program appears to have adopted the image of hundreds of other public policies—it has no clearly defined course of action to be pursued in the future. As Chapter 2 indicated, it was the lunar landing decision that brought solidarity to the space program in the early 1960's. The goal-oriented nature of the space program was the key factor that set NASA apart from other federal agencies and departments. Also, as discussed in Chapter 3, NASA selected, within approximately one year from the Kennedy decision to go to the moon, the means by which the goal could be accomplished.

Contrast NASA's unique status in the 1960's of having a goal to achieve and a clearly defined way to do it with any number of other public policies, i.e., defense, elimination of poverty, transportation congestion, and crime. For example, it is not far from the truth to say that an agreed-upon and workable definition of poverty has yet to be devised. In addition, the thousands of people working on the poverty problem have yet to find the most effective way or combination of ways to alleviate or to eliminate poverty—income maintenance via welfare payments, versus the negative income tax, versus medical, educational, and training programs, versus almost half a dozen different food distribution programs. Moreover, there is no agreed-upon information that shows whether, and if so by how much, anti-poverty programs actually reduce the incidence of poverty, or for that matter, how much they might enhance poverty and possibly crime.

The so-called clash between the scientific community and engineering groups over the United States space goals in the post-Apollo period makes NASA much more akin to other federal agencies and departments in 1969–1970 than to the NASA of the mid-1960's. It is argued that engineers are enthusiastic over the possibility of building larger spacecraft and more powerful boosters to permit the space program to conquer other planets, whereas the scientific community is urging NASA to devote increased attention to flights that will provide more information about the moon's origin and evolution.[8] If NASA shifts emphasis to solely scientific priorities, it probably will do so at the expense of public enthusiasm. As Chapter 3 on Planning, Programming, and Budgeting in NASA indicated, a highly scientific-oriented space program is likely to have relatively little public appeal.

Space Programs Outlined by the President's Space Task Group

The report of the President's Space Task Group can, at best, be described as a disappointing document. Although program options are outlined, the program alternatives and their possible costs are given only in indefinite terms. Like NASA's Ten-Year Plan made in 1960, the report of the Space Task Group lacks a goal orientation.[9] The reason for the Space Task Group's inability to look to the future is clear. Although not written as a caveat to the recommendations made to the President, the report itself contains the explanation:

By its very nature, the exploration and exploitation of space is a costly undertaking and must compete for funds with other national and individual enterprises. Now that the national goal of the manned lunar landing has been achieved, discussion of future space goals has produced increasing pressures for reexamination of, and possible changes in, our national priorities.[10]

Chapter 1 discussed the fact that in the United States there does not exist a quantifiable hierarchy of national goals, and also the fact that national objectives cannot be changed in any rational way, unless something more about the structure of national goals is known than vague banalities about hundreds of possible competing public objectives. Although proposals have been made for the establishment of a National Goals Commission, the United States has nothing, other than the Congress itself, that approximates a national goals steering committee.[11] As also indicated in Chapter 1, the rate and direction of public policies are determined through complicated interactions in the market place, the political process, and the judicial system.

The longer NASA delays in proposing a single space goal, or possibly several goals, that capture the imagination of the public and of the Congress, the greater will be the danger that resource allocation to space research and development will revert to the pre-manned space flight patterns of the 1959–1960 period. That is, of course, until another Soviet space accomplishment, comparable to the launching of Sputnik I in October 1957, once again shocks the nation into reacting by defining and accomplishing space objectives.

An Image of Leadership

A discussion of policy forecasting and the future space effort would not be complete without mentioning James E. Webb. As Administrator of NASA between 1961 and 1968, he was, both fondly by some and disdainfully by others, called a great huckster—possibly the best ever in the Congressional arena. Rudy Abramson said, upon Webb's retirement, that James E. Webb could convince Congress to allocate $5 billion a year to NASA because it was desirable, humanitarian, romantic, moral, and mandatory to explore space. Furthermore, after getting money for the space program, Webb made many Congressmen feel a little guilty because they had not done enough for the space program.[12]

THE PROPAGANDA FORECAST

A fourth kind of forecast is the propaganda forecast. *Webster's Seventh New Collegiate Dictionary* defines propaganda as, "the spreading of ideas, information, or rumor for the purpose of helping or injuring an institution or a cause." Although many think of propaganda as a disdainful political or private device, its use should not necessarily be viewed as something dishonest or untrue. There is "white" propaganda and there is "black" propaganda; the definition given above clearly includes both. This section examines the future role of the space program, as well as federal spending for alternative programs, in terms of their contribution to national objectives. The analysis, of course, is subject to controversy and disagreement because of the inability of social scientists to derive precise measures of the costs and the benefits of most federal programs and policies.

The Space Program and Economic Growth

As discussed in Chapter 9, economic growth depends upon an increase in the quantity and the quality of resources. It is education and technological change that cause a qualitative improvement in a given amount of resources. Although an indirect measure of the value of technological change, productivity indexes show a marked increase in the post-World War II period—an average annual rate of increase of about 1.3 percent per annum prior to 1929, compared with 3.3 percent between 1947 and 1966. An important characteristic of the post-World War II period is the surge of research and development activity, including education and training because a substantial amount of research money has gone to improve the quality of scientific and technical manpower.

Federal spending for R & D rose markedly between 1950 and 1967, from $1.1 billion to $17.2 billion. As Table 10–4 shows, the federal government spent about $500 million less on R & D in 1969 than it did two years earlier. In 1967, federal R & D purchases comprised 2.2 percent of the GNP and only 1.8 percent in 1969. The data in Table 10–4 clearly show that the decrease in R & D spending for the exploration of space was only partially offset by increases in other kinds of research projects. In 1965, R & D expenditures for the nation's space effort comprised 35 percent of total federal research compared with 23 percent in 1969. At least in the immediate future, federal R & D spending is likely to stabilize unless something unforeseen happens.

Because an ever-expanding body of scientific knowledge is essential for the continued development of an industrialized economy and because

TABLE 10–4: FEDERAL FUNDS FOR RESEARCH AND DEVELOPMENT AND R & D PLANT (*BILLIONS OF DOLLARS*)

YEAR	TOTAL	NASA	DOD	AEC	HEW
1950	1.1	0.1	0.7	0.2	a
1958	4.9	0.1	3.5	0.8	0.2
1959	7.1	0.3	5.2	0.9	0.3
1960	8.1	0.5	5.8	1.0	0.4
1961	9.6	0.9	6.7	1.1	0.5
1962	11.1	1.7	6.8	1.3	0.6
1963	13.7	3.4	7.4	1.4	0.7
1964	15.3	4.8	7.4	1.5	0.8
1965	15.7	5.5	6.9	1.5	1.1
1966	16.2	5.3	7.1	1.4	1.1
1967	17.2	5.0	8.1	1.5	1.2
1968	16.5	4.5	7.9	1.5	1.3
1969	16.7	3.9	8.1	1.8	1.4

[a] Less than $50 million

Source: National Science Foundation, *Federal Funds for Research, Development, and Other Scientific Activities*, Vol. XVIII.

private enterprise will not provide for it adequately, the expansion of science and technology becomes an economic function of society as a whole.[13] Two forces propel the government to undertake research and development work: (1) military and defense requirements; and (2) meritorious or vital civilian needs. If left alone, the price and resource allocating mechanism of the private market system would not supply, or would supply "insufficient amounts" of, the "needed" research and development effort that satisfy these requirements. This is true because private producers do not find it sufficiently profitable to engage in the necessary amount of R & D work in such important technological fields as space, atomic energy, agriculture, saline water, and medicine. For satisfactorily high rates of long-term economic growth, the United States simply cannot afford not to invest in this form of social overhead capital.

Reactivity of the United States Space Program to Soviet Accomplishments

It was not until the actual manned lunar landing in July 1969 that the United States took the lead from the Soviet Union in accomplishing a space-related event that captured the imagination, sympathy, and respect of the world. As discussed in Chapter 1, a chronology of space events by the United States and by the Soviet Union clearly shows that the USSR was the first to: (1) launch an artificial satellite, (2) experiment with living animals aboard a spacecraft, (3) launch and have a safe return of a manned spacecraft, (4) put a multi-passenger space vehicle into orbit, and (5) perform manned extra-vehicular activities.

Estimates based on the content and accomplishments of the Soviet space program suggest that USSR resource allocation to space-related research and development must have been at least double the relative share of that in the United States in the mid-1960's. Without a clearly defined space objective for the decade of the 1970's, and without an increase in spending for space above a level that sustains two or three manned lunar expeditions each year, the United States probably will relinquish its newly acquired world prestige in space to the Soviet Union. The reputable British periodical, *The Economist,* predicts that the Soviet Union wants to have its space program appear to be marking time while waiting for the development of a new rocket. After the rocket becomes available, the USSR will make a quantum jump by constructing a sizable space station.[14]

Diverting Resources from Space to Eliminate Poverty

The presence of Ralph C. Abernathy, members of the Southern Christian Leadership Conference, and the train of mule-drawn wagons at Cape Kennedy for the launch of the first spacecraft to take man to the moon, portrays the belief that if the United States can successfully put a man on another world, it can, with the same amount of money, eliminate poverty. Furthermore, it portrays the belief that resources devoted to the exploration of space should be diverted to "more important" social problems existing on Planet Earth. In fact, it is unlikely that spending for space has diverted funds from the poor. Between 1964 and 1968, federal aid to reduce poverty rose from an annual expenditure of $13 billion to $24 billion.[15] During that period, spending for the exploration of space dropped by about $300 million. See Table 10–4. In fact, in 1969, federal income-in-kind programs alone amounted to more than $10 billion dollars. Income-in-kind anti-poverty programs include food distribution programs; public housing and rent supplements; and health service programs, such as medicare, maternity and infant care, special programs for Indians, and neighborhood health centers.[16]

As indicated earlier in this chapter, many policies designed to cure social problems lack both defined objectives and the means to accomplish dimly stated problems such as poverty, crime, and urban congestion. In his book on the Office of Economic Opportunity and the anti-poverty programs of the Great Society Program of the Johnson Administration, Sar A. Levitan gives a number of examples of both the high costs and low accomplishments of our current piecemeal approach to the alleviation of poverty. For example, in fiscal 1967, the average annual cost per enrollee in the Job Corps ranged from $7,400 for those at conservation centers to $9,600 for females attending women's urban centers.[17] In February 1967, Louis Harris and Associates conducted a survey of wages earned by "no shows" for the Job Corps, i.e., accepted applicants who did not go to a Job Corps training center. A comparison of the results of that survey with wages earned by "graduates" of the Job Corps reveals that ex-corps members earned only 9 cents an hour more than the so-called "no-shows."[18]

Levitan reaches the conclusion that programs sponsored during the 1960's under the Economic Opportunity Act did not contribute appreciably to the reduction of poverty. According to Levitan, the EOA programs were designed to remedy the causes of poverty rather than to mitigate its symptoms. This simply means that the EOA programs failed to do much, if anything, for those who could not benefit from the self-help approach. It also means that poverty in the United States is permanent, according to Levitan, unless the nation begins to spend in the vicinity of $40 billion (1968 dollars) annually for income maintenance programs and for income-in-kind goods and services for the poor.[19] That is, about ten times the size of the space effort in the late 1960's and about five times the amount proposed by the President's Space Task Group to be spent in 1980 under Option I.

If, as Levitan and many others argue, the poor do not benefit much from self-help assistance, it means that national poverty programs should be viewed as pure welfare expenditures rather than investments that ultimately result in productivity gains. It also means that the problem of extreme poverty is becoming an ever-increasing problem of maintaining income and consumption levels of families and individuals who have little prospect of providing self-maintenance, i.e., poverty is increasingly being witnessed by the aged, the physically and mentally handicapped, and by members of households without employable male heads. To divert resources from the space program and from other activities that increase productivity makes no sense in the long run because it is self-defeating. Productivity gains contribute to economic growth, economic growth contributes to higher levels of national income, which, in turn, provides a wider tax base to support larger income and consumption maintenance poverty programs.[20]

THE SPACE PROGRAM AND THE POETIC FORECAST

To conclude this book with a look at the future of the space program based on the poetic forecast is to conjecture about living and working in the year 2000 and beyond. Under the heading of the forecast based on poetry or primitive projection come the visions and speculations made by scientists, historians, and journalists. The poetic forecast is dateless, it is the next logical step, or perhaps, the logical extreme.

The October 1969 issue of *The Futurist,* published by the World Future Society, was devoted to man's future in space. The World Future

Society is a nonprofit educational and scientific association, established in 1966, to study alternative futures. The October 1969 issue of *The Futurist* is the secondary source for the following poetic projections of man's future in space made by well-known people.

Addressing a group of students invited to the White House, President Richard Nixon said, after the Apollo 11 flight: "In the year 2000 we on this earth will have visited new worlds where there will be a form of life."

Earl Hubbard, artist and philosopher, believes that, "the peaceful means of solving the population problem is a combination of voluntary birth control and finding new worlds." According to Hubbard, "Our gateway to the future is the National Aeronautics and Space Administration. It should change its name to the Agency for New Worlds."

Eugene B. Konecci, a physiologist teaching at the University of Texas, anticipates the day when space orbital hospitals, with a gravity-free and germ-free environment, will provide patients suffering from severe burns and heart ailments with medical treatment and comfort unavailable on earth.

The columnist and scientist, Joshua Lederberg, believes that the colonization of the moon provides insurance for the perpetuation of humanity, "The colonization of the moon and planets is . . . the most economical insurance we know how to buy for the perpetuation of human culture through the next century . . . Existing plans for the exploration of space, with their mix of manned and instrumented flights, embody a sensible strategy for this kind of insurance."

The science-fiction writer Arthur C. Clarke in a *Look Magazine* special on the Apollo 11 flight, envisions that a trip to the moon will cost no more than a round-the-world jet flight. According to Clarke, "The fact that the present cost is millions of times greater is the measure of our present ignorance and the primitive state of technology; the time will come, through the use of reusable boosters, orbital refueling, nuclear propulsion and other foreseeable developments, when the cost of a lunar journey may be comparable to that of round-the-world jet flights."

Many might wish to call these poetic forecasts unattainable goals of science fiction. Yet man's first step on another world must mark the beginning rather than the end of the exploration and the exploitation of space.

NOTES

1. Irving H. Siegel, "Technological Change and Long-Run Forecasting," *Journal of Business,* XXVI, 3 (July 1953), pp. 141–50.

2. *Ibid.,* p. 144.

3. The Space Task Group membership included, Spiro T. Agnew, Chairman; Robert C. Seamans, Secretary of the Air Force; Thomas O. Paine, NASA Administrator; and Lee A.

Dubridge, Science Adviser to the President. Alexis Johnson, Under Secretary of State for Political Affairs; Glenn T. Seaborg, Chairman, AEC; and Robert P. Mayo, Director, Bureau of the Budget, were participating observers.

4. Ahmad Al-Samarrie, *National Economic Projection Series-Report No. 68-N-1* (Washington, D.C.: National Planning Association, 1969).

5. U.S. President, *Space Task Group Report to the President, The Post-Apollo Space Program: Directions for the Future,* September 1969, p. 20.

6. Ahmad Al-Samarrie, *National Economic Projections Series—Report No. 68-N-1,* p. 129.

7. Irving Siegel, "Technological Change and Long-Run Forecasting," p. 144.

8. "Rift in NASA," *Wall Street Journal,* November 14, 1969, p. 1 and p. 15.

9. U.S. President, *Space Task Group Report to the President,* pp. 20 and 21.

10. *Ibid.,* p. 3.

11. The old Eisenhower concept of a National Goals Commission has been revived by the Nixon Administration. In early 1970, a small national goals group, established in the Executive Office, began to examine the matter of national goals and criteria for selecting goals.

12. "NASA's Boss a Great Huckster," *Washington Post,* September 25, 1968, p. K-4.

13. Theodore Suranyi-Unger, Jr., "The Role of Knowledge in Invention and Economic Development," *American Journal of Economics and Sociology,* 22 (October 1963), pp. 463–72.

14. "Where Have All the Cosmonauts Gone?", *The Economist,* July 26, 1969, p. 14.

15. Sar A. Levitan, *The Great Society's Poor Law: A New Approach to Poverty,* (Baltimore, Md.: The Johns Hopkins Press, 1969), p. 10.

16. U.S. President, *Economic Report of the President, 1969,* p. 167.

17. Sar A. Levitan, *The Great Society's Poor Law,* pp. 286–87.

18. *Ibid.,* p. 304.

19. *Ibid.,* pp. 309, 317.

20. Charles T. Stewart, "Economic and Social Benefits of the Manned Space Flight Program," The George Washington University, November, 1969. (unpublished paper)

11 Epilog

The profile of the United States space program has not changed much since the Space Task Group made its report to President Nixon in September 1969. By mid-1973, commitments had been made to undertake (1) a space shuttle program and (2) a joint United States-Soviet rendezvous and docking mission of the Apollo and the Soyuz spacecrafts. Although these projects have become new items in the NASA budget, combined spending for the projects is not anticipated to be more than $565 million in fiscal 1974.

THE SPACE PROGRAM AND NATIONAL GOALS AND PRIORITIES

It is clear that the chief economic concern of President Nixon is to cope with the dual menace of high rates of unemployment and rapidly rising prices. In August 1971, President Nixon presented his "New Economic Policy" to the nation. On the domestic front, the policy called for: reinstitution of the investment tax credit, reduction of the federal excise tax on automobiles, and an increase in the standard personal income tax deduction. These policy recommendations were designed to increase consumer and investment spending and thereby stimulate the economy. The proposed decrease in government spending and the ceiling placed on the level of federal employment, which have dampening effects on the economy, were no doubt incorporated into the New Economic Policy to enhance acceptance by the House Ways and Means Committee and the Congress in general.

The New Economic Policy also called for an "incomes policy," or price and wage controls to reduce inflation. As a result of the Phase II price and wage controls, the rate of increase in the consumer price index dropped from 5.9 percent in 1969 to 3.3 percent in 1972. With the system of almost complete voluntary compliance under Phase III, the rate of inflation again began to soar, with the consumer price index rising at an average annual rate of about 8.0 percent during the first half of 1973. It remains to be seen whether or not Phase IV, which began in August 1973, will curb inflation without creating large distortions taking the form of shortages in important industries and sectors of the economy.

President Nixon's policy recommendations were designed to stimulate the economy through increases in total consumer spending and increases in business investment spending. It is still uncertain whether the policies are strong enough to substantially narrow the gap between the actual and the potential gross national product. The gap between what the economy could have produced and what it actually produced rose to a high of $55 billion (average annual rate in 1958 prices) in the last half of 1970. By the end of 1972, the gap still amounted to $26 billion. As indicated in Chapter 1, the largest previous level of unused resources for any one year since the mid-1950's occurred in 1961 when the gap between the potential GNP and actual GNP amounted to $43 billion. In terms of current purchasing power, a GNP gap of about $50 billion in 1958 constant dollars amounts to approximately $65 billion in 1972 current dollars or more than four times as much as average annual spending for all R & D programs of about $15 billion in the early 1970's.

It is doubtful that the New Economic Policy will have much of a direct or even an indirect effect on the level of research and development undertaken in this country in the foreseeable future. The only reference to research and development in the August 1971 statement was that President Nixon ". . . directed the Secretary of the Treasury to present to the Congress in January new tax proposals for stimulating research and development of new industries and technologies to help provide 20 million new jobs that America needs for the young people who will be coming into the job market in the next decade."

Spending for Research and Development

By the late 1960's, the research revolution, inaugurated primarily by President Kennedy, began to lose substantial momentum, and by the beginning of 1970 there was an R & D depression in the United States.[1] Two broad national philosophies can be cited as contributing to the R & D depression of the early 1970's. The first was and is the rapidly increasing demand for R & D results rather than merely R & D activity. The second is the increasingly popular concern about technology assessment and the ecological and environmental effects of science and technology. In many parts of society, the second concern is beginning to reach hysterical proportions, and it is likely that environmental interests will become a major challenge to the progress of research and development during the 1970's. The National Science Foundation has shifted its R & D mission away from fostering basic research into solving earth-oriented problems. In the spring of 1970, Congressman Joe L. Evans told William McElroy, Director of the National Science Foundation, that "we want the National Science Foundation to tell us how to solve the problems of the environment."[2] Few will deny that such serious national problems as poverty, disease, pollution, and urban congestion require solutions. These problems must and will be solved. The solutions to those problems arising from rapid population growth and urbanization will require massive amounts of federal funds. The bulk of federal support will take the form of expenditures for income maintenance; the provision of low-income housing; the construction of a number of different kinds of rapid transit facilities; and the construction of efficient and adequate solid, liquid, and nuclear waste treatment facilities; and not investment in high technology research and development.

The skepticism about the benefits to be derived from research and development is amply reflected in the decrease in federal spending for R & D. Between 1967 and 1971, federal outlays for R & D dropped from $16.5 billion to $15.4 billion. If a conservative estimate of the increase in the costs of R & D of 6 percent per annum because of inflation is applied to the 1971 spending level, the purchasing power of federal R & D spending in 1971 was about 25 percent below expenditures made in 1967.[3]

The data displayed in Table 11–1 clearly show the change in national priorities, with the exception of national defense, between the late 1960's and the early 1970's. Spending for national defense rose to a high of $81.2 billion in 1969, with the peak approximately in August of that year, and then began to decline, being about 3 percent less than the peak in 1972. Of the ten major programs listed in Table 11–1, space research and technology was the only national goal to experience a decrease in federal support. Federal spending for each of the other national goals, with the exception of defense and natural resources, rose by more than 33 percent during the three-year period. Among the largest increases were a rise of $3.5 billion for housing and community development; $47.3 billion for health, labor, and welfare; and $9.7 billion for general government.

The Aerospace Industry and Major NASA Locations

The shift in national priorities away from spending for space and for research and development in general is reflected in the sharp decline in employment in the aerospace industry. Between 1968 and 1972, employ-

TABLE 11-1: FEDERAL SPENDING FOR SELECTED NATIONAL GOALS, 1967 AND 1972 (*BILLIONS OF DOLLARS*)

NATIONAL GOAL	1967	1972	PERCENT CHANGE
National defense	$74.5	$75.3	5.6
Space research and technology	4.9	3.4	–30.6
General government	14.8	24.5	65.5
International affairs	2.7	3.6	33.3
Education	3.9	6.6	69.2
Health, labor, and welfare	40.5	87.8	116.8
Veterans benefits and services	7.1	12.6	77.5
Commerce, transportation, and housing	8.4	15.6	85.7
Transportation	5.7	7.8	36.8
Housing and community development	1.0	4.5	350.0
Agricultural resources	4.2	5.8	38.1
Natural resources	2.6	3.4	30.8
Revenue sharing	–	2.6	–
Total	$163.6	$244.6	49.5

Source: United States Department of Commerce, *Survey of Current Business*, Volumes 51 and 53, Nos. 7, July 1971 and July 1973.

ment in the aerospace industry dropped from 1,108,300 to 591,400, which means that in 1972, aerospace employment was almost 50 percent below the 1968 annual average.[4]

Table 11–2 presents information about NASA employment in selected localities. Total NASA employment dropped from a peak of about 400,000 in 1966 to 137,000 in June 1971. About 40 percent of the entire decrease in NASA employment took the form of reduced contractor work forces in Los Angeles. Because of the mission of the Manned Spacecraft Center, there has been a relatively small decrease in NASA employment in Houston. On the basis of 1972 unemployment patterns, Brevard County, Florida; Pearl River County, Mississippi; New Orleans, Louisiana; and Los Angeles, California are on the United States Department of Labor's list of areas with substantial unemployment, and are eligible for a number of different kinds of federal assistance for redevelopment under the Public Works and Economic Development Act.[5] The unemployment rate in the Anaheim–Santa Ana–Garden Grove sector of the Los Angeles metropolitan area was over 7 percent in June 1971, but it dropped to 4.7 percent by December 1972.

Of the principal localities supporting the space program, Houston and Huntsville have made the best adjustment in absorbing decreases in space employment. As indicated in Chapter 6, Houston has enjoyed startlingly rapid growth during the postwar period, a growth broadly based among all economic sectors. In addition, employment cutbacks resulting from decreased spending for space amounted to only 16 percent

TABLE 11-2: ESTIMATE OF NASA AND NASA CONTRACTOR EMPLOYMENT IN SELECTED LOCALITIES

LOCATION	PEAK EMPLOYMENT YEAR	PEAK EMPLOYMENT NUMBER	EMPLOYMENT JUNE 1971	PERCENTAGE DECREASE
Kennedy Space Center, Brevard County, Fla.	1968	25,700	15,000	41.6
Manned Spacecraft Center, Houston, Tex.	1968	12,800	10,800	15.6
Marshall Space Flight Center, Huntsville, Ala.	1964	23,000	10,100	56.1
Michoud Assembly Facility, New Orleans, La.	1964	13,400	2,000	85.1
Mississippi Test Facility Hancock County, Miss.	1965	5,900	1,411	76.1
Los Angeles, Calif.	1966	140,000 to 158,000	30,000 to 35,000[a]	75.0 to 81.0
Total space employment	1966	400,000	137,000	65.7

[a] Estimated from data in the Joint NASA/DOD Economic Information System.

Source: National Aeronautics and Space Administration.

in Houston compared with 85 percent in New Orleans. Space employment in Houston remained fairly stable through the scheduled Apollo launches. Sharp decreases in space employment in Houston will occur if manned space flights are discontinued after the Skylab series. Huntsville is no longer the small stagnant textile town that it was before the infusion of government R & D activities. Mainly as a result of positive action on the part of the business and industrial groups, Huntsville unexpectedly appears to have developed an economic base that is likely to sustain economic growth. After the initial outflow of trained technical personnel, the Huntsville Industrial Committee and the Chamber of Commerce tried to lure new business into the area. As a result of these efforts, four new industries, creating an estimated 3,000 new jobs, located in Huntsville.[6]

Brevard County, Florida, and Hancock County, Mississippi, have been profoundly affected by reductions in the space program. Unemployment in the area surrounding the Kennedy Space Center rose above 7 percent in the early 1970's. As of April 1971, about 4,000 FHA homes had been repossessed. Increased tourism, forthcoming in part from Disney World, located about 50 miles from Cocoa Beach, is the most promising prospect for rejuvenating economic prosperity in Brevard County. NASA, in cooperation with other government agencies, is attempting to compensate for the lack of an economic base in the rural area surrounding the Mississippi Test Facility by diversifying the base of the facility itself. NASA is working with the Environmental Protection Agency, the Coast Guard, the Department of the Interior, and the National Oceanographic and Atmospheric Agency to expand the use of the facility into the areas of natural resources development and environmental protection. NASA still hopes the test facility can be used to test the proposed Space Shuttle Program.

Unemployment Resulting from Reductions in NASA Spending

Under a NASA contract, the Battelle Memorial Institute conducted a study of the rate and duration of unemployment resulting from cutbacks in NASA programs. Battelle sent questionnaires to 5,000 randomly sampled individuals from a list of 27,171 people laid off by NASA's seven largest contractors between June 1968 and September 1970. For a survey of this kind, Battelle received the rather high rate of response of 57 percent. The Battelle survey indicated that the unemployment rate among those surveyed was 32 percent in the spring of 1970, 40 percent in Los Angeles, and 47 percent in the age group fifty and older. Among those subsequently finding employment, the average duration of unemployment was 31 weeks. Only 18 percent of those laid off found alternative permanent jobs in the aerospace industry.[7]

The National Science Foundation conducted a similar study in the spring of 1971. The results of the NSF study differ from those of the Battelle study because the NSF sought responses from the entire scientific community. Battelle surveyed all workers, semi-skilled, managerial, and scientific, known to have been laid off by NASA contractors. The NSF study, covering about 50 percent of all scientists, found the unemployment rate to be only about 2.6 percent in the spring of 1971. The NSF study did, however, confirm Battelle's finding that the duration of unemployment for those losing jobs had been long. The average length of the period of unemployment reported to the NSF was over 7 months.[8]

THE STATUS OF THE U.S. SPACE PROGRAM

Since the historic landing of Apollo 11 on the moon in July 1969, the United States has attempted six manned lunar flights—Apollo flights 12 through 17. Apollo 12 followed the first successful manned lunar landing by only four months. With a precision soft landing on the moon in mid-November 1969, the astronauts aboard Apollo 12 put long-life lunar surface sensors on the moon. The astronauts also collected and brought back to earth parts from the unmanned Surveyor 3 spacecraft that had landed on the surface of the moon and had dug soil with an automatic shovel two and a half years earlier—in April 1967.

Apollo 13 was launched on April 11, 1970. The mission, which was the third attempt by the United States at a manned lunar landing, was aborted because of loss of pressure in the service module and the failure of two fuel cells. After circling the moon, the spacecraft made an emergency return to earth. Some of the superstitious attribute the failure of this mission to the "unlucky" number assigned to the spacecraft. Others believe that it was caused by the lack of attentiveness and anxiety of most on the NASA ground support team. Of course, it is a matter of speculation as to whether large reductions in funds and personnel could have caused or contributed to the failure of Apollo 13 to complete its mission. It is a fact, however, that the largest single cutback in spending for space was between fiscal 1968 and 1969 when funding dropped by almost $600 million, causing a large number of lay-offs of NASA employees and employees of NASA contractors who had worked on the Manned Space Flight Program.

Apollo 14 was launched on January 31, 1971. After a successful mission, astronauts Alan B. Shepard, Stuart A. Roosa, and Edgar D. Mitchell were picked up in the Pacific Ocean on February 9, 1971. On July 26, 1971, Apollo 15 was launched with two "space firsts" for the United States—exploration of the moon in the lunar roving vehicle and extra-vehicular activity to collect cameras and other equipment from the command module during the return trip to earth. The accomplishments of Apollo 16 were more advanced than those of its predecessor flight, but less than those scored on the last manned mission to the moon in the foreseeable future. Launched on December 14, 1972, with Eugene A. Cernan, Ronald E. Evans, and Harrison (Jack) Schmitt as the crew, Apollo 17 set at least ten new records for space exploration. They included the longest number of miles traveled during a lunar flight, the longest mission in terms of time, the longest time on the moon, the longest time exploring the lunar surface, the longest single moon walk, the greatest quantity of rocks and soil brought back to earth, the swiftest speed in the lunar rover (11.1 miles per hour), the longest distance traveled in the lunar rover, and the longest time in lunar orbit.

At the end of fiscal 1972, the Manned Space Flight Program was still NASA's largest single activity, with a total expenditure of about $1.3 billion during that fiscal year. In addition, the United States has continued to improve the operational efficiency of its communications, weather, and other earth resources satellites. Between 1957 and 1970, the United States launched 77 communications satellites, 45 weather satellites, and 42 navigation and geology missions. Although NASA-developed vehicles and launch facilities are used, currently, most communication satellites are either part of a global commercial satellite system or part of a military (NATO) communications system. Other unmanned missions relate to planetary flights. On May 8, 1971, NASA launched Mariner 8 to study some of the dynamic characteristics of the planet Mars from orbit for a minimum period of three months. The intended mission was to map about 70 percent of the planet's surface, but the mission aborted when the vehicle failed. Mariner 9 was launched on May 30, 1971, to accomplish those tasks established for Mariner 8. Pioneer 10 was launched in March 1972 to provide the first measurements of the interplanetary medium beyond Mars.

Currently, the National Aeronautics and Space Administration plans only one more flight in its extended Apollo series, the rendezvous and docking of a modified Apollo spacecraft with a Soyuz-type spacecraft. In the original Apollo flight plan, two additional flights to the moon were scheduled for launch sometime in 1974. With a gap of a year or so between launches, Apollo flights 18 and 19 were scrapped because of safety considerations. Under current plans, NASA has an inventory of large launch vehicles. At the end of 1972, NASA had three Saturn V launch vehicles and seven Saturn 1-B launch vehicles. The first Skylab Workshop was launched with the Saturn V, and the Saturn 1-B will be used for the three additional Skylab Workshops and for the joint Apollo-Soyuz project. If program plans materialize, there will be an inventory of two unused Saturn V vehicles and three Saturn 1-B's.

The Saturn Inventory and Private Enterprise

In 1970, a group of citizens formed an association known as the Committee of the Future. The association includes a number of prominent and wealthy people. Millionaire toy manufacturer, Louis Marx, along with his daughter, Barbara Marx Hubbard and her artist-author-philosopher husband Earl Hubbard, are among the founders. Retired Air Force General Joseph S. Bleymaier, developer of the Titan 3-C rocket and former director of the cancelled military manned orbiting laboratory is a member of the Committee of the Future. Other committee members include: Harold W. Ritchey, chief developer of solid fuel rockets and president of the Thiokol Chemical Corporation; Richard H. Nolte, former ambassador to the United Arab Republic and Executive Director of the Institute of Current World Affairs; Ronald Winston of Harry Winston, Inc., a Manhattan diamond brokerage company; Donaldson, Lufkin, and Jenrette, the Wall Street investment broker firm; and the Department of Astronomy of Northwestern University.[9]

The group believes that an expanded space program is in harmony with other urgent national priorities because the space program can be used as an instrument to solve earth-related problems. They believe that advanced space technology, designed to sustain man on the moon, will be applied to environmental problems on earth. Further, the group believes that space-related job opportunities in frontier industries will offer both training and a challenge to change the attitudes and improve the well-being of the disadvantaged and will ultimately stimulate a commitment from the apathetic youth of today.

The main objectives of the group are to find a way to awaken public interest in the ailing space program and to explore possible ways to colonize the moon. As a first step, the Committee for the Future wants to raise money to buy one or both of the two complete Apollo/Saturn V hardware combinations that are not scheduled for use because of funding cutbacks on the space program. The *Los Angeles Times* reports that some astronauts have indicated that they would be willing to fly the proposed "people's mission to the moon."[10] The group believes that if they can raise about $250 million through public stock offerings, it would be enough to buy the two mothballed Saturn V rockets and other complementary hardware. "Market" and "economic" research conducted by the group suggests that about 100 pounds of lunar rock would be collected on each trip. Diamond expert Ronald Winston conjectures that the moon rocks would have a great deal of novelty value because of publicity. Harold W. Ritchey states that the Committee's working figure for the value of a piece of moon rock on earth is $1,500 a carat, or about $3.5 million a pound.[11] Mr. Ritchey acknowledges that many assumptions were made to reach that figure and that the plan has one or two commercial weak spots. Although tempting and fascinating, estimating the commercial value of moon rocks is beyond the scope of this book. Suffice it to say that if the Committee for the Future succeeded in their venture, they would have a monopoly on commercial moon rocks. Being at least temporarily fixed in supply, the moon rocks conceivably could bring a price of $1,500 a carat, or higher, or zero, depending on the intensity of demand. At present, the National Aeronautics and Space Administration is engaged in polite conversations and correspondence with the group but is not supporting the proposed commercial venture to the moon. Spokesmen for the Committee state that several Congressmen are interested in the idea and that the Committee will ask Congress to intervene and approve the plan.

Comparative Positions of the U.S. and Soviet Space Programs

Perhaps the greatest single difference between the United States space program and that of the USSR is that there is no evidence that the Soviet space program has peaked. United States experts familiar with the Soviet space program say that, on the basis of the numbers and kinds of launches, about 600,000 people in the USSR are working on space projects and that the program amounts to about 2 percent of the Soviet gross national product.[12]

One of the most important unknowns about the Soviet space program is the present capability of the USSR's launch vehicles. Sovietologists conjecture that the USSR is working on a space vehicle that has a 10- to 14-million-pound thrust capability. This compares with the 7.5-million-pound thrust of the Saturn V. At present, as proven by flights, the Saturn V has the best operational thrust in the world.[13]

Although the Soviet Union is not known to have placed bombs in orbit, which is against a treaty signed by both the United States and the USSR that bans weapons of mass destruction from orbit, the USSR has increased tensions by testing a fractional orbit bombardment system. Technically, these tests have not violated the treaty because there is no evidence that nuclear material was used on the test flights. In addition, the dummy warheads have flown less than one orbit and have avoided the United States.[14] The United States does not have a program for a fractional orbit bombardment system.

The Soviet Union has put two manned orbital scientific stations into orbit. According to Soviet claims, the docked combination of Soyuz IV and V in January 1969 represented a primitive space station because the combined capacity of the two spacecraft provided four rooms, with space for as many as twelve people. Five cosmonauts were aboard the two spacecraft. About 18 months later, the USSR put Soyuz VI into orbit, with cosmonauts Georgiy Timofeyevich Dobrovolskiy, Vladislau Nikolayevich Volkov, and Viktor Ivanovich Patsyev aboard. Between June 6 and 30, 1971, the three cosmonauts reportedly performed a number of successful scientific activities during their mission aboard the orbital scientific station. The Soyuz VI mission, unfortunately, resulted in a great space tragedy when the oxygen supply for the cosmonauts was cut off during re-entry into earth orbit. The three cosmonauts were found dead when the spacecraft was recovered. At least one Soviet expert postulates that a new large vehicle, if developed, could be used to lift the principal components of an orbital space station.[15] The Soviet Union might, with only one launch, put up the equivalent of our Skylab Workshop at any time.[16]

One other area of the Soviet space program warrants special attention — unmanned lunar missions. On September 12, 1970, the USSR launched Luna 16. The mission resulted in a soft landing on the moon in the Sea of Fertility, and the vehicle returned to earth on September 24, 1970, with samples of lunar rock. Luna 17 was launched on November 10, 1970. With that mission, the Soviet Union put Lunokhod-I, a lunar rover, on the moon. Each of these events was a "Space First" for the USSR, missions yet to be undertaken by the United States. Luna 18, launched September 2, 1971, was not successful because it impacted on the moon nine days after launch.

THE FUTURE OF THE U.S. SPACE PROGRAM

No international or domestic event has occurred to change the profile of the United States space program since the first lunar landing. Although a number of studies are being conducted within the Space Agency, the announced flight schedule does not go beyond 1978, with a proposed first orbital flight of the space shuttle in late 1978. NASA's budget estimates for fiscal year 1974 contain plans for the first horizontal flight of the space shuttle in late 1976. However, there are reports that NASA plans a large number of shuttle flights in the 12 years starting in 1978, with both take-offs and landings using the Cape Kennedy facilities. After the four planned Skylab launches in 1973, no manned space flights are planned until the joint United States-Soviet launch of the Apollo and Soyuz spacecrafts sometime in mid-1975. Under current plans for manned space flights, there will be a 12- to 18-month gap in the flight schedule between the last Skylab launch and the joint United States-Soviet project, and then another gap of similar length between the Apollo-Soyuz mission and the first test of the space shuttle.

Early funding requests for the shuttle program are small and they are likely to continue to be small compared with the early phases of the manned lunar landing project—below $500 million through fiscal year 1974.[17] If funding for the shuttle does not grow beyond the $500 million level, it is probable that during the decade of the 1970's, the United States space program will be devoted to the launch and operation of unmanned earth resources satellites and to aeronautical projects such as the development and operation of the inter-city short take-off and landing (STOL) and the vertical take-off and landing (VTOL) airplanes.

The 1969 Report of the President's Space Task Force still remains the best indication of future levels of spending for space objectives. The low-option plan calls for an allocation of no more than $5.0 billion to the space program during the decade of the 1970's; the high-option plan is similar until the mid-1970's, when it provides for funds rising to about $9.4 billion in the late 1970's to further manned space flight objectives. These proposed expenditures are high compared with the current space funding levels of about $3.1 billion. Yet compared with the estimated costs of other national goals, an expenditure of $5.0 to $10.0 billion is something that the economy can afford, assuming that the economy continues to grow and that the almost $50 billion gap between the actual and the potential gross national product existing in the early 1970's continues to close.

Estimates are available on the probable future costs of achieving a high-quality environment. These costs are high, probably much higher than those of a long-term viable space program. But even combined with a sizable space program, they are not so great when compared with the potential capacity of the United States economy to produce. This, of course, assumes that the fanatical fringe of the environmental protectionists are prevented from launching astronomically expensive environmental protection campaigns that have small and intangible benefits.

The Council on Environmental Quality estimated that total spending for air, water, and solid waste pollution will amount to about $105 billion between 1970 and 1975, or about $20 billion annually. Of this amount, about 23 percent will be for air pollution control, 36 percent for water pollution control, and 41 percent for solid waste management.[18] An annual spending of $20 billion a year for pollution abatement, combined with, say, $10 billion for space, would still amount to only $30 billion per year, or approximately three-fifths of the potential output of $50 billion of unemployed resources in the United States in 1971.

NOTES

1. S. Dedijer, "The R & D Depression in the United States," *Science,* Vol. 168, No. 3929, April 1970, pp. 345–346.

2. United States House Subcommittee on Science Research and Development of the Committee on Science and Astronautics, Report, *Toward a Science Policy for the United States,* 91st Congress, 2nd session, October 15, 1970.

3. National Science Foundation, *An Analysis of Federal R & D Funding by Function,* NSF 72–313, June 1972, p. 65.

4. Aerospace employment is measured broadly by SIC 1925 (complete guided missiles) and SIC 372 (aircraft and parts) and underestimates the actual decrease in aerospace employment, especially in the computer and electronics components of the industry. Employment data are from the U.S. Department of Labor, Bureau of Labor Statistics.

5. Under the Public Works and Development Act, the unemployment rate in an eligible area was, or is expected to be, 50 percent or more higher than the national average. An eligible area could have also had an average annual rate of unemployment of at least 6 percent during the preceding calendar year. United States Department of Labor, Manpower Administration, *Area Trends in Employment and Unemployment,* February 1973, pp. 24–28.

6. U.S. Congress, House, Select Committee on Small Business, Hearings, *The Impact of Federal Installations on Small Business,* 92 Congress, 1st session, June 15–16, 1971, pp. 145–147.

7. Battelle Memorial Institute, *A Survey of Aerospace Employees Affected by Reductions in NASA Contracts,* (Columbus, Ohio: Battelle Columbus Laboratories, 1971).

8. National Science Foundation, Science Resources Studies, *Highlights: Unemployment Rates for Scientists,* NSF 71–26, July 1971.

9. "There is a Plan Afoot to Sell Moon Rocks for Millions. To You," *The Wall Street Journal,* Monday, June 28, 1971.

10. "Surplus Spacecraft: Group Seeks Money for More Moon Trips," *Los Angeles Times,* Friday, April 2, 1971.

11. *The Wall Street Journal, op. cit.*

12. Charles S. Sheldon, II, *United States and Soviet Progress in Space: Some New*

Contrasts, Library of Congress, Congressional Research Service, Washington, D.C., pp. 12–13.

13. *Ibid.*

14. *Ibid.*, p. 32.

15. *Ibid.*, pp. 60–62.

16. Others believe that because the USSR is trailing so far behind the United States in space research and development, space scientists in the Soviet Union are delighted to have the opportunity to share some of the American know-how by means of the joint Apollo-Soyuz mission. See Robert G. Kaiser, "Why No Soviet Apollo?" *The Washington Post*, December 15, 1972.

17. In July 1972, NASA awarded a $2.6 billion contract to North American Rockwell for the reusable space shuttle. That award spans a six-year period, and if spent in equal annual increments, would average about $433 million during its six-year life.

18. Council on Environmental Quality, *Second Annual Report* (Washington, D.C.: U.S. Government Printing Office, 1971), pp. 144–150.

Bibliography

FEDERAL SPENDING IN RELATION TO THE ECONOMIC, SOCIAL, AND POLITICAL ENVIRONMENT OF THE POST-WORLD WAR II PERIOD

American Academy of Political and Social Science. *The Annals of the American Academy of Political and Social Science*. Vol. I: *Social Goals and Indicators for American Society* (May, 1967).

Augenstein, B. W. "Spend for Space, but . . ." *Christian Science Monitor,* August 9, 1969.

Bauer, Raymond A., ed. *Social Indicators*. Cambridge, Mass.: M.I.T. Press, 1966.

Becher, Abraham S. "Comparison of the United States and U.S.S.R. National Output: Some Rules of the Game." *World Politics,* 13, 1 (October, 1960).

Biderman, Albert D. "Social Indicators and Goals." In *Social Indicators*. Edited by Raymond A. Bauer. Cambridge, Mass.: M.I.T. Press, 1966.

Burck, Gilbert. *The Space Industry: America's Newest Giant*. Englewood Cliffs, N.J.: Prentice-Hall, 1962.

Committee for Economic Development. *Budgeting for National Objectives: Executive and Congressional Roles in Program Planning and Performance*. New York: Committee for Economic Development, 1966.

Dewhurst, J. Frederick, and Associates. *America's Needs and Resources*. New York: Twentieth Century Fund, 1955.

Diamond, Edwin. *The Rise and Fall of the Space Age*. Garden City, N.Y.: Doubleday, 1964.

Etzioni, Amitai. *The Moon-Doggle, Domestic and International Implications of the Space Race*. New York: Doubleday, 1964.

Furash, Edward F. "Business Review of the Space Effort." *Harvard Business Review,* 41, 5 (September/October, 1963).

Goldsen, Joseph M. *Public Opinion and Social Effects of Space Activity*. Santa Monica, Calif.: RAND Corporation, 1959.

Gordon, Kermit, ed. *Agenda for the Nation*. Washington, D.C.: Brookings Institution, 1968.

Harris, Seymour E. *The Economics of the Kennedy Years and a Look Ahead*. New York: Harper and Row, 1964.

Herman, Leon M. "Soviet Economic Capabilities for Scientific Research." *Soviet Space Programs, 1962–1965: Goals and Purposes, Achievements, Plans, and International Implications*. Staff Report prepared for the Senate Committee on Aeronautical and Space Sciences, December 30, 1966.

Knowles, James W. *The Potential Economic Growth of the United States*. Study Paper No. 20, Study of Employment, Growth, and Price Levels, Joint Economic Committee. Washington, D.C.: U.S. Government Printing Office, 1960.

Korol, Alexander G. *Soviet Research and Development: Its Organization, Personnel, and Funds*. Cambridge, Mass.: M.I.T. Press, 1965.

Krieghbaum, Hillier. *Science and the Mass Media*. New York: New York University Press, 1967.

Krutch, Joseph Wood. "Must Technology and Humanity Conflict?" *Bell Telephone Magazine*, 48, 2 (March/April, 1969).

Lecht, Leonard A. *The Dollar Cost of Our National Goals*. Washington, D.C.: National Planning Association, 1965.

———. *Manpower for National Goals in the 1970's*. New York: Frederick A. Praeger, 1969.

Lederman, Leonard L., and Windus, Margaret L. *An Analysis of the Allocation of Federal Budget Resources as an Indicator of National Goals and Priorities to the National Aeronautics and Space Administration*. 2 vols. Columbus, Ohio: Battelle Memorial Institute, 1969.

Moeser, John V. *The Space Program and the Urban Problem: Case Studies of the Components of National Consensus*. Washington, D.C.: Program of Policy Studies in Science and Technology, The George Washington University, 1969.

Morgenstern, Oskar. "The Worth of the Space Program." *Astronautics and Aeronautics*, February, 1968.

Morisson, Eltin E. *The American Style*. New York: Harper, 1957.

Mueller, Eva. "Public Attitudes Toward Fiscal Programs." *Quarterly Journal of Economics*, 77, 2 (May, 1963).

"National Purpose," 5 Parts. *Life* Magazine, May 23, May 30, June 6, June 13, June 20, 1960.

Norton, Hugh S. *The Role of the Economist in Government Policymaking*. Berkeley, Calif.: McCutchan Publishing Corp., 1969.

Okun, Arthur. "Potential GNP: Its Measurement and Significance." *The 1962 Proceedings of the Business and Economics Section, American Statistical Association*. Minneapolis, September 7–10, 1962.

Organization for Economic Co-operation and Development. *The Research and Development Effort in Western Europe, North America and the Soviet Union*. Paris: OECD, 1965.

———. *Reviews of National Science Policy: United States*. Paris: OECD, 1968.

Sheldon, Charles S. *Review of the Soviet Space Program.* Washington, D.C.: U.S. Government Printing Office, 1967.

Shonfield, Andrew. *Modern Capitalism.* New York and London: Oxford University Press, 1965.

Silk, Leonard. "The Impact on the American Economy." In *Outer Space: Prospects for Man and Society.* Edited by Lincoln P. Bloomfield. Englewood Cliffs, N.J.: Prentice-Hall, 1962.

Smithies, Arthur. *The Budgetary Process in the United States.* New York: McGraw-Hill, 1955.

Stein, Herbert. *The Fiscal Revolution in America.* Chicago: University of Chicago Press, 1969.

U.S. Congress. House. Committee on Science and Astronautics. *Government, Science and Public Policy: A Compilation of Papers Prepared for the Panel on Science and Technology,* 89th Cong., 2d Sess., 1966.

———. Committee on Science and Astronautics. *Report: Science Technology and Public Policy During the 89th Congress—January, 1965 through December, 1966,* 90th Cong., 1st Sess., 1967.

———. Committee on Science and Astronautics. *A Survey of Views of Leading Industrial Executives on the National Space Program,* 90th Cong., 2d Sess., 1968.

U.S. Congress. Joint Economic Committee. *Comparisons of the United States and Soviet Economies.* Washington, D.C.: U.S. Government Printing Office, 1959.

U.S. Congress. Senate. Committee on Aeronautical and Space Sciences. Staff Report; *Soviet Space Programs, 1962–1965: Goals and Purposes, Achievements, Plans, and International Implications.* Washington, D.C.: U.S. Government Printing Office, 1966.

———. Committee on Aeronautical and Space Sciences. *Hearings: Scientists' Testimony on Space Goals.* 88th Cong., 1st Sess., 1963.

———. Select Committee on Small Business. *The Role and Effect of Technology in the Nation's Economy.* 6 Parts. 88th Cong., 1st Sess., 1963.

U.S. Department of Commerce. *Long-Term Economic Growth, 1860–1965.* Washington, D.C.: U.S. Government Printing Office, 1966.

U.S. Department of Labor. "Economic Growth and Employment Opportunities." *Manpower Report of the President, 1966.* Washington, D.C.: U.S. Government Printing Office, 1966.

U.S. President. The President's Commission on National Goals. *Report, Goals for Americans: Program of Action in the Sixties.* Englewood Cliffs, N.J.: Prentice-Hall, 1960.

———. The President's Materials Policy Commission. William S. Paley, Chairman. *Resources for Freedom.* 5 vols. Washington, D.C.: U.S. Government Printing Office, 1952.

Watson, Donald S. *Economic Policy: Business and Government.* Boston: Houghton Mifflin, 1960.

Whelan, Joseph G., Beck, Leonard N., Herman, Leon M., Krivickas, Domas, and Rusis, Armins. *Soviet Space Programs, 1962–1965: Goals and Purposes,*

Achievements, Plans, and International Implications. Staff report for the Senate Committee on Aeronautical and Space Sciences, December 30, 1966.

Wiesner, Jerome B. *Where Science and Politics Meet.* New York: McGraw-Hill, 1965.

Williams, Robin M. "Individual and Group Values." *The Annals of the American Academy of Political and Social Science.* Vol. I: *Social Goals and Indicators for American Society* (May, 1967).

SPACE GOALS AND THE EVOLUTION OF A FEDERAL AGENCY

Atkins, Charles M. *NASA and the Space Science Board of the National Academy of Sciences.* NASA Historical Series, HHN-62. Washington, D.C.: National Aeronautics and Space Administration, September, 1966.

Bradley, Charles E., and McCuistion, Clayton C. *Contract Decision-Making and Incentive Fee Contracts.* Unpublished research report to NASA, December, 1965. The George Washington University, Washington, D.C.

Daddario, Emilio Q. "Congress Faces Space Policies." *Bulletin of the Atomic Scientists* (May, 1967).

Danhof, Clarence H. *Government Contracting and Technological Change.* Washington, D.C.: Brookings Institution, 1968.

Dyer, William H., and Roberts, Edward B. *The Role of Follow-on Contracts in Government-Sponsored Research and Development.* Cambridge, Mass.: Sloan School of Management, Massachusetts Institute of Technology, July, 1967. Paper No. 269–67.

Greenberg, Edward. *Relationships between R & D Contracts and Production Contracts.* St. Louis, Mo.: Department of Economics, Washington University, June, 1967. Working Paper 6707.

———. "Some Aspects of the State Distribution of Military Prime Contract Awards." *Review of Economics and Statistics,* XLVIII, 2 (May, 1966).

Hayes, George T. *The Industry-Government Aerospace Relationship.* Menlo Park, Calif.: Stanford Research Institute, 1963.

Logsdon, John M. *The Decision to Go to the Moon: Project Apollo and the National Interest.* Cambridge, Mass.: M.I.T. Press, 1970.

McKinsey and Company, Inc., Washington, New York, Chicago, London, San Francisco, and Los Angeles, Management Consultants. *An Evaluation of NASA's Contracting Policies, Organization, Performance.* Report to Dr. T. Keith Glennan, October 28, 1960.

National Academy of Sciences—National Research Council. *A Review of Space Research,* Publication 1079. Washington, D.C.: National Academy of Sciences, 1962.

———. Space Sciences Board. *Space Research: Directions for the Future.* Washington, D.C.: National Academy of Sciences, 1965.

Oates, Stephen B. "NASA's Manned Spacecraft Center at Houston, Texas." *Southwestern Historical Quarterly* (January, 1964).

Orlans, Harold. *Contracting for Atoms.* Washington, D.C.: Brookings Institution, 1967.

Peck, Merton J., and Scherer, Frederic M. *The Weapons Acquisition Process:*

An Economic Analysis. Boston: Graduate School of Business Administration, Harvard University, 1962.

Roberts, Edward B. *Marketing and Engineering Strategies for Winning R & D Contracts.* Cambridge, Mass.: Sloan School of Management, Massachusetts Institute of Technology, July, 1965. Paper No. 128–65.

———. *Questioning the Cost-Effectiveness of the R & D Procurement Process.* Cambridge. Mass.: Sloan School of Management, Massachusetts Institute of Technology, September, 1965. Paper No. 144–65.

Rosholt, Robert L. *An Administrative History of NASA, 1958–1963.* Washington, D.C.: U.S. Government Printing Office, 1966.

Swenson, Loyd S., Jr., Grimwood, James M., and Alexander, Charles C. *This New Ocean: A History of Project Mercury.* NASA Historical Series, SP-4201. Washington, D.C.: U.S. Government Printing Office, 1966.

U.S. Congress. House. Committee on Government Operations. *Organization and Management of Missile Programs,* 86th Cong., 1st Sess., February and March, 1959.

———. Committee on Government Operations. *Report: Air Force Ballistic Missile Management—Formation of the Aerospace Corporation,* 87th Cong., 1st Sess., May, 1961.

———. Committee on Government Operations. *Report: Criteria for Support Service Contracts,* 90th Cong., 2d Sess., August 5, 1968.

———. Committee on Government Operations. *Report on Government Operations in Space: An Analysis of Civilian-Military Roles and Relationships,* 89th Cong., 1st Sess., June 4, 1965.

———. Committee on Government Operations. *Support Service Contracts, Hearings,* 90th Cong., 1st Sess., June 21, 1967.

———. Committee on Government Operations. *A Cost Profile for Support Services, Hearings,* 90th Cong., 2d Sess., April, 1968.

———. Committee on Science and Astronautics. *Summary Report: Future Programs Task Group, A Report by the National Aeronautics and Space Administration to the President,* 89th Cong., 1st Sess., April, 1965.

———. Committee on Science and Astronautics, Subcommittee on NASA Oversight. *Staff Study: Future National Space Objectives,* 89th Cong., 2d Sess., 1966.

U.S. Congress. Senate. Committee on Aeronautical and Space Sciences. *NASA's Proposed Operating Plan for Fiscal Year 1969,* 90th Cong., 2d Sess., October 3, 1968.

———. Committee on Aeronautical and Space Sciences. *National Space Goals for the Post-Apollo Period, Hearings,* 89th Cong., 1st Sess., 1965.

U.S. Department of Defense, and National Aeronautics and Space Administration. *Phase I Report: Joint Report on Facilities Required at Launch Site to Support the NASA Manned Lunar Landing Program.* (Debus-Davis Report) Washington, D.C.: National Aeronautics and Space Administration, July 31, 1961.

Zangwell, Willard I. "Top Management and Selection of Major Contractors at NASA." *California Management Review,* XII, 1 (Fall, 1969).

ECONOMIC EFFICIENCY AND PLANNING, PROGRAMMING, AND BUDGETING

Banks, Robert L., and Kotz, Arnold. "The Program Budget and the Interest Rate for Public Investment." *Public Administration Review,* XXVI, 4 (December, 1966).

Baumol, William J. "On the Social Rate of Discount." *American Economic Review,* LVIII, 4 (September, 1968).

Belcher, Donald J., Hardy, Ernest E., Shelton, Ronald L., and Schepis, Eugene L. *Potential Benefits to be Derived from Applications of Remote Sensing of Agricultural, Forest, and Range Resources.* Ithaca, N.Y.: Center for Aerial Photographic Studies, Cornell University, 1967. (mimeographed)

Bendersky, David, and Smith, Paul L. *Medical Applications of NASA-Developed Science and Technology.* Kansas City, Mo.: Midwest Research Institute, March, 1966.

Campbell, H. G., and Dreyfuss, D. J. *Manned Spacecraft Cost-Estimating Relationships.* Santa Monica, Calif.: RAND Corporation, 1967. (mimeographed) RM 5317-NASA.

Doffman, Robert, ed. *Measuring Benefits of Government Investments, Studies of Government Finance.* Washington, D.C.: Brookings Institution, 1965.

Dole, S. H., Campbell, H. G., Dreyfuss, D., Gosch, W. D., Harris, E. D., Lewis, D. E., Parker, T. M., Ranftl, J. W., and String, J. *Methodologies for Analyzing the Comparative Effectiveness and Costs of Alternative Space Plans.* 2 vols. Santa Monica, Calif.: RAND Corporation, 1968. (mimeographed) RM5656-NASA.

Enke, Stephen, ed. *Defense Management.* Englewood Cliffs, N.J.: Prentice-Hall, 1967.

Enthoven, Alain C. "Economic Analysis in the Department of Defense." *American Economic Review,* LIII, 2 (May, 1963).

Frank, Charles R., Heiss, Klaus-Peter, and Baxter, Nevins D. *Cost Benefit Study of the Earth Resources Observation Satellite System: Grazing Land Management.* Princeton, N.J.: Radio Corporation of America, 1968.

Frey, H. Thomas. *Agricultural Application of Remote Sensing—The Potential from Space Platforms.* Washington, D.C.: U.S. Department of Agriculture, 1967. Information Bulletin No. 328, Economic Research Service.

Grosse, Robert N. *An Introduction to Cost-Effectiveness Analysis.* McLean, Va.: Research Analysis Corporation, July, 1965. RAC Paper-RAC-P-5.

———. *Principles of Cost Effectiveness Analysis.* Washington, D.C.: Industrial College of the Armed Forces, September, 1965. Publication No. L66–24.

Harper, E. L., Kramer, F. A., and Rouse, A. M. "Implementation and Use of PPB in Sixteen Federal Agencies." *Public Administration Review* (November/December, 1969).

Hirsch, Werner Z. "Quality of Government Services." In *Public Expenditure Decisions in the Urban Community.* Edited by Howard G. Schaller. Washington, D.C.: Resources for the Future, Inc., 1963.

Hitch, Charles J. *Decision-Making for Defense.* Berkeley, Calif.: University of California Press, 1965.

———, and McKean, Roland N. *The Economics of Defense in the Nuclear Age.* Cambridge, Mass.: Harvard University Press, 1960.

Lyden, Fremont J., and Miller, Ernest G., eds. *Planning, Programming, and Budgeting: A Systems Approach to Management.* Chicago: Markham Publishing Co., 1967.

Margolis, Milton A., and Barro, Stephen A. "The Space Program." In *Program Budgeting, Program Analysis, and the Federal Government.* Edited by David Novick. Cambridge, Mass.: Harvard University Press, 1965.

McCullough, J. D. *Cost Analysis for Planning-Programming-Budgeting; Cost-Benefit Studies.* Santa Monica, Calif.: RAND Corporation, November, 1966.

McKean, Roland N., ed. *Issues in Defense Economics.* New York: Columbia University Press, 1967.

———, ed. *Efficiency in Government Through Systems Analysis with Emphasis on Water Resources Development.* New York: John Wiley, 1965.

Meitner, John G., Endlish, Roay M., Muraca, Raffeale F., Rodovic, Milan, Samuel, Aryeh H. A., Viezee, William, and Whittick, Julia. *Priority Analysis of Manned Orbital Research Application.* 2 vols. Menlo Park, Calif.: Stanford Research Institute, 1965. (mimeographed)

Muir, Alan H. "A Study of the Economic Benefits and Application of Space Station Operations." Washington, D.C.: Planning Research Corp., 1968. (mimeographed)

———, and Summers, R. A. "The Use of Economic Benefit Analysis in Earth Resources Satellite System Planning." AIAA Paper No. 68–1077, AIAA Fifth Annual Meeting and Technical Display, Philadelphia, October 21–24, 1968.

Musgrave, Richard A. *The Theory of Public Finance, A Study in Public Economy.* New York: McGraw-Hill, 1959.

———. "Cost-Benefit Analysis and the Theory of Public Finance." *Journal of Economic Literature,* VII, 3 (September, 1969).

National Academy of Sciences—National Research Council. *Report of the Central Review Committee, Useful Applications of Earth-Oriented Satellites.* Washington, D.C.: National Academy of Sciences, 1969.

———. *Summaries of Panel Reports, Useful Applications of Earth-Oriented Satellites.* Washington, D.C.: National Academy of Sciences, 1969.

———. *Useful Applications of Earth-Oriented Satellites: Forestry-Agriculture-Geography, #1.* Washington, D.C.: National Academy of Sciences, 1969.

———. *Useful Applications of Earth-Oriented Satellites: Geology, #2.* Washington, D.C.: National Academy of Sciences, 1969.

———. *Useful Applications of Earth-Oriented Satellites: Hydrology, #3.* Washington, D.C.: National Academy of Sciences, 1969.

———. *Useful Applications of Earth-Oriented Satellites: Meteorology, #4.* Washington, D.C.: National Academy of Sciences, 1969.

———. *Useful Applications of Earth-Oriented Satellites: Oceanography, #5.* Washington, D.C.: National Academy of Sciences, 1969.

———. *Useful Applications of Earth-Oriented Satellites: Sensors and Data Systems, #6.* Washington, D.C.: National Academy of Sciences, 1969.

———. *Useful Applications of Earth-Oriented Satellites, Point-to-Point Communication, #7.* Washington, D.C.: National Academy of Sciences, 1969.

———. *Useful Applications of Earth-Oriented Satellites: Systems for Remote-Sensing Information and Distribution, #8.* Washington, D.C.: National Academy of Sciences, 1969.

———. *Useful Applications of Earth-Oriented Satellites: Point-to-Point Communications, #9.* Washington, D.C.: National Academy of Sciences, 1969.

———. *Useful Applications of Earth-Oriented Satellites: Broadcasting, #10.* Washington, D.C.: National Academy of Sciences, 1969.

———. *Useful Applications of Earth-Oriented Satellites: Navigation and Traffic Control, #11.* Washington, D.C.: National Academy of Sciences, 1969.

———. *Useful Applications of Earth-Oriented Satellites: Geodesy-Cartography, #13.* Washington, D.C.: National Academy of Sciences, 1969.

Nieburg, H. L. *In the Name of Science.* Chicago: Quadrangle Books, 1966.

Novick, David. "The Origin and History of Program Budgeting." *California Management Review* (Fall, 1968).

Root, L. Eugene. "The Return on Our Investment in Space Exploration." *Looking Ahead,* 13, 6 (September, 1965). Washington, D.C.: National Planning Association.

Russo, John A. *The Operational and Economic Impact of Weather on the Construction Industry in the United States.* Hartford, Conn.: Travelers' Research Center, 1965.

Schelling, Thomas C. *PPBS and Foreign Affairs.* Memorandum prepared at the request of the Subcommittee on National Security and International Operations of the Senate Committee on Government Operations, 90th Cong., 1st Sess., January, 1968.

Schick, Allen. "The Road to PPB: The Stages of Budget Reform." *Public Administration Review* (December, 1966).

Smith, Robert G. *Major Factors in Aerospace Planning and Decision-Making.* Washington, D.C.: Program of Policy Studies in Science and Technology, The George Washington University, May 10, 1966. (mimeographed)

———. *Planning, Programming, and Budgeting: A Technique for Federal Program Planning and Decision-Making.* Washington, D.C.: Program of Policy Studies in Science and Technology, The George Washington University, November, 1965. (mimeographed)

Snyder, William P. *Case Studies in Military Systems Analysis.* National Security Management Series. Washington, D.C.: Industrial College of the Armed Forces, 1967.

String, J., Jr. *A Model for Resource Analysis of Space Programs.* Santa Monica, Calif.: RAND Corporation, 1967. (mimeographed) RM-5347-NASA.

"Symposium: Planning, Programming, Budgeting System Reexamined: Development Analysis and Criticism." *Public Administration Review,* XXIX (March–April, 1969).

Tucker, Samuel A., ed. *A Modern Design for Defense Decision, A McNamara-Hitch-Enthoven Anthology.* National Security Management Series. Washington, D.C.: Industrial College of the Armed Forces, 1966.

U.S. Congress. House. Committee on Government Operations. *Federal Research and Development Programs: The Decision-Making Process.* 89th Cong., 2d Sess., January 7–11, 1966.

U.S. Congress. Joint Economic Committee. Subcommittee on Economy in Government. *Interest Rate Guidelines for Federal Decision-Making.* 90th Cong., 2d Sess., January 29, 1968.

Weidenbaum, Murray L. *Program Budgeting: Applying Economic Analysis to Government Expenditure Decisions,* Paper 6602. St. Louis, Mo.: Washington University, April, 1966.

______. *Program Budgeting and the Space Program,* Paper 6607. St. Louis, Mo.: Washington University, April, 1966.

MANAGEMENT TECHNIQUES AND SYSTEMS ANALYSIS

Air Transportation 1975 and Beyond: A Systems Approach. Report of the Transportation Workshop, 1967. Bernard A. Schriever and William W. Seifert, Co-chairmen. Cambridge, Mass.: M.I.T. Press, 1968.

Black, Guy. *The Application of Systems Analysis to Government Operations.* New York: Frederick A. Praeger, 1968.

Borum, Charles. "Found: A Practical Configuration Management System." *Defense Industry Bulletin,* 2, 5 (May, 1966).

Breckner, Norman V., and Noah, Joseph W. "Costing of Systems." In *Defense Management.* Edited by Stephen Enke. Englewood Cliffs, N.J.: Prentice-Hall, 1967.

Bright, James R., ed. *Technological Forecasting for Industry and Government.* Englewood Cliffs, N.J.: Prentice-Hall, 1968.

Campbell, H. G., and Dreyfuss, D. J. *Manned Spacecraft Cost-Estimating Relationships.* Santa Monica, Calif.: RAND Corporation, 1967. RM 5317-NASA.

Conrath, David W. *A Model of Sub-Organizational Decision-Making with an Application to Budgeting in NASA.* Berkeley: University of California, 1965. (mimeographed) Internal Working Paper No. 33.

Danhof, Clarence. *Government Contracting and Technical Change.* Washington, D.C.: Brookings Institution, 1968.

Dooley, Arch R. "Interpretations of PERT." *Harvard Business Review* (March/April, 1964).

Hatt, Peter K. "Management Systems Planning on New Flight Programs." *The Management of Aerospace Programs,* Vol. 12. Edited by Walter L. Johnson, Washington, D.C.: American Astronautical Society, 1967.

Hayya, Jack C. *PERT and Procurement Policy.* Los Angeles: Graduate School of Business Administration, University of California, 1966. (mimeographed) Report No. 8.

Johnson, Richard A., Kast, Fremont E., and Rosenzweig, James E. *The Theory and Management of Systems.* New York: McGraw-Hill, 1967.

Johnson, Walter L., ed. *The Management of Aerospace Programs,* Vol. 12. Proceedings of an American Astronautical Society national conference held at the University of Missouri, Columbia, Missouri, November 16–18, 1966. Washington, D.C.: American Astronautical Society, 1967.

MacDonald, Gordon J. F. "Science and Space Policy: How Does It Get Planned?" *Bulletin of the Atomic Scientists* (May, 1967).

Marschak, Thomas, Glennan, Thomas K., and Summers, Robert. *Strategy for R & D*. New York: Springer-Verlag, 1967.

Marshall, A. W., and Meckling, W. H. "Predictability of the Costs, Time and Success of Development." In *The Rate and Direction of Inventive Activity: Economic and Social Factors*. Princeton: Princeton University Press, 1962.

Miller, Robert W. *Schedule, Cost, and Profit Control with PERT: A Comprehensive Guide for Program Management*. New York: McGraw-Hill, 1963.

Nord, Ole C., and Packer, David W. *Management Science: A Primer for Managers*. Sudbury, Mass.: Management Science Associates, 1964.

Siepert, Albert F. "Creating the Management Climate for Effective Research in Government Laboratories." *The Management of Scientists*. Edited by Karl Hill. Boston: Beacon Press, 1965.

Steiner, George A. *Managerial Methods of Successful Project Managers*. Los Angeles: University of California, 1965. (mimeographed) Research Paper No. 1.

Swenson, Loyd S., Jr., Grimwood, James M., and Alexander, Charles C. *This New Ocean: A History of Project Mercury*. NASA Historical Series, SP-4201. Washington, D.C.: U.S. Government Printing Office, 1966.

U.S. Congress. House. Committee on Science and Astronautics. *Report: Ways and Means of Effecting Economies in the National Space Program,* 87th Cong., 2d Sess., 1963.

U.S. Congress. House. Select Committee on Astronautics and Space Exploration. *Report: The Next Ten Years in Space, 1959–69,* 86th Cong., 1st Sess., 1959.

U.S. Congress. Senate. Committee on Aeronautical and Space Sciences. *Report: Apollo 204 Accident,* 90th Cong., 2d Sess., January, 1968.

U.S. National Aeronautics and Space Administration, Office of Manned Space Flight. *Apollo Program Development Plan,* January, 1966.

———. Office of Technology Utilization. *Applications of Systems Analysis Models: A Survey*. Cambridge, Mass.: ABT Associates, 1968.

Vivian, Lawrence. "Can the Aerospace Industry Meet Reliability Requirements for Manned Space Flight?" *Defense Industry Bulletin,* 2, 9 (September, 1966).

von Bertalanfy, Ludwig. "Systems Everywhere." In *General Systems Theory: Foundations, Development, Applications*. New York: George Braziller, 1968.

Watson, Donald S. *Economic Policy: Business and Government*. Boston: Houghton Mifflin, 1960.

Webb, James E. *Space Age Management: The Large-Scale Approach*. New York: McGraw-Hill, 1969.

Wood, Archie L. *The Analysis of Large-Scale R & D Programs: A Case Study of Project Mercury*. Cambridge, Mass.: Sloan School of Management, Massachusetts Institute of Technology, 1965. (mimeographed) Report No. 115–65.

Wyatt, DeMarquis D. "Cost Models for Complex Programs." *The Management of Aerospace Programs*. Science and Technology Series, Vol. 12. Edited by Walter L. Johnson. Washington, D.C.: American Astronautical Society, 1967.

MEASURES OF THE SOCIAL AND ECONOMIC IMPACT OF THE SPACE PROGRAM AND OTHER FEDERAL AGENCIES

Alterman, Jack. "Interindustry Employment Requirements." *Monthly Labor Review* (July, 1965).

Battelle Memorial Institute, *A Survey of Aerospace Employees Affected by Reductions in NASA Contracts* (Columbus, Ohio: Battelle Columbus Laboratories, 1971).

Beckler, Bernard S., et al. *Methodology for Industry Impact Analysis.* Vol. I: *Methodology and Summary Results.* McLean, Va.: Research Analysis Technical Paper RAC-TP-190.

Brandwein, Robert. "The Dyna-Soar Contract Cancellation." *Business Review* (October, 1965).

Buehler, Vernon M. "Economic Impact of Defense Programs." U.S. Bureau of the Budget, *Statistical Reporter* (July, 1967).

Burgess, John T. *An Economic Analysis of Research and Development Procurement: The NASA Subcontract Program.* St. Louis, Mo.: Department of Economics, Washington University, July, 1967. Working Paper 6708.

C-E-I-R Applied Research and Management Sciences Division. *Preliminary Report: Economic Impact Analysis of Subcontracting Procurement Patterns of Major Defense Contractors.* Bethesda, Md.: C-E-I-R, September, 1966.

Chamber of Commerce of the United States. *What New Industrial Jobs Mean to a Community.* Research Study prepared by Economic Research Department, 1965.

Colm, Gerhard. *The Impact of the United States Civilian Space Program on the U.S. Economy.* A report by the National Planning Association, Washington, D.C., July, 1966. (mimeographed)

Dedijer, S. "The R & D Depression in the United States," *Science,* 168, 3929 (April 1970), pp. 345–46.

Edelstein, Robert H. *Methodology of Regional and Subregional Input-Output Studies.* Report prepared for the Economic Development Administration, U.S. Department of Commerce, Washington, D.C., June, 1968.

Glock, Charles Y. "Applied Social Research: Some Conditions Affecting Its Utilization." *Case Studies in Bringing Behavioral Science into Use.* Part I. Stanford, Calif.: Institute for Communications Research, Stanford University, 1961.

Goldman, Morris R., Marimont, Martin L., and Vaccara, Beatrice N. "The Interindustry Structure of the United States." *Survey of Current Business* (November, 1964).

Greenberg, Edward. *Employment Impacts of Defense Expenditures and Obligations.* St. Louis, Mo.: Department of Economics, Washington University, April, 1965.

Hoffenberg, Marvin. *Analysis of NASA Postcard Subcontract Data.* Los Angeles: Institute of Government and Public Affairs, University of California, 1966.

Howell, Richard P., et al. *An Exploratory Study: The Economic Impact of Defense R & D Expenditures in Terms of Value Added and Employment Gen-*

erated. Menlo Park, Calif.: Stanford Research Institute, February, 1966. SRI Project No. IMU-5452.

Igo, Donald J. *Industrial Classification and Economic Impact Analysis.* McLean, Va.: Research Analysis Corporation, May, 1966. Technical Paper RAC-TP-207.

———, Bull, Edwyn M., and Lindsay, William, Jr. *Economic Analysis, A Military Procurement Final Demand Vector.* 2 vols. McLean, Va.: Research Analysis Corporation, 1967.

Isard, Walter. *Methods of Regional Analysis: An Introduction to Regional Science.* Cambridge, Mass.: M.I.T. Press, 1960.

———, Langford, Thomas W., Jr., and Romanoff, Elishu. *Philadelphia Input-Output Study.* 2 vols. Philadelphia: Department of Regional Science, Wharton School, University of Pennsylvania, 1966.

Kirschner Associates. *Adjustments to Reduced National Defense Expenditures in New Mexico, Summary Report.* Albuquerque, N.M.: Kirschner Associates, December, 1965.

Lee, Maw Lin. *The Relationship of New Orders to Shipments of Defense Products.* St. Louis, Mo.: Washington University, September 15, 1965. Working Paper 6517.

Leontief, Wassily. *Input-Output Economics.* New York: Oxford University Press, 1966.

———. "Quantitative Input-Output Relations in the Economic System of the United States." *The Review of Economics and Statistics* (August, 1963).

Lienberg, Maurice, Hirsch, Albert A., and Popkin, Joel. "A Quarterly Econometric Model of the United States." *Survey of Current Business,* 46, 5 (May, 1966).

Miernyk, William H. *The Elements of Input-Output Analysis.* New York: Random House, 1965.

———, Bonner, Ernest R., Chapman, John H., Jr., and Snellhammer, Kenneth. *The Impact of Space and Space-Related Activities on a Local Economy.* Boulder, Colo.: University of Colorado, July, 1965.

National Planning Association, Center for Economic Projections. *Community Readjustment to Reduced Defense Spending, Case Studies of Potential Impact on Seattle-Tacoma, Baltimore and New London-Groton-Norwich.* Prepared for the U.S. Arms Control and Disarmament Agency. Washington, D.C.: National Planning Association, December, 1965.

National Science Foundation. *Highlights: Unemployment Rates for Scientists.* Science Resources Studies, NSF 71–26 (Spring 1971).

Park, Se-Hark. *Urban Employment Multipliers and Their Application to the Aerospace Industry in St. Louis.* St. Louis, Mo.: Department of Economics, Washington University, June, 1965.

Perloff, Harvey S., Dunn, Edgar S., Lampard, Eric E., and Muth, Richard F. *Regions, Resources, and Economic Growth.* Baltimore, Md.: Johns Hopkins Press, 1960.

Poole, Richard W., Tarver, James D., White, David, and Gurley, William B. *An Evaluation of Alternative Techniques for Estimating County Population in a Six-State Area.* Stillwater, Okla: Oklahoma State University, November, 1965.

Riefler, Roger F., and Downing, Paul B. "Regional Effect of Defense Effort on Employment." *Monthly Labor Review*, 91, 97 (July, 1968).

Seastone, Don, Bonner, Ernest R., Franks, Charles M., and McKormick, William. *The Impact of Space and Space-Related Activities on a Local Economy: A Case Study of Boulder, Colorado.* Part II: *The Income-Product Accounts*. Boulder, Colo.: University of Colorado, 1965.

Stewart, Charles T. *A Study of the Changing Regional Patterns of Research and Development and Science-Based Technology in Influencing Regional Development.* Report to the National Science Foundation, July 10, 1964. The George Washington University, Washington, D.C.

______. *Winston-Salem: A Case Study of the Impacts of Research-Based Manufacturing.* Report to the National Science Foundation, October, 1965. The George Washington University, Washington, D.C.

Terry, Edwin. *Linear Estimates of the Export Employment Multiplier.* St. Louis, Mo.: Washington University, November, 1964. Working Paper 6415.

Tiebout, Charles M. *The Community Economic Base Study.* New York: Committee for Economic Development, 1962.

U.S. Arms Control and Disarmament Agency. *Research Sponsored by ACDA on the Economic Impact of Defense and Disarmament in the United States.* Washington, D.C.: U.S. Arms Control and Disarmament Agency, June 30, 1968 (mimeographed)

U.S. Congress. House. Committee on Government Operations. Staff Report, *The Use of Social Research in Federal Domestic Programs,* 90th Cong., 1st Sess., April, 1967.

U.S. Congress. House. Select Committee on Small Business. Hearings, *The Impact of Federal Installations on Small Business,* 92nd Cong., 1st Sess., Vol. I, pp. 145–47.

U.S. Department of Labor Manpower Administration, *Area Trends in Employment and Unemployment,* July 1971, pp. 7–17.

U.S. National Science Foundation. *Knowledge in Action: Improving the Nation's Use of the Social Sciences.* Washington, D.C.: U.S. Government Printing Office, 1969.

Weidenbaum, Murray L. *Measures of the Impact of Defense and Space Programs.* St. Louis, Mo.: Washington University, September, 1965. Working Paper 6515.

______. *Methods of Government Assistance to Research and Development.* St. Louis, Mo.: Washington University, June, 1967. Working Paper 6705.

______. *Product Management for Defense/Space Markets.* St. Louis, Mo.: Washington University, December, 1964.

______. *The Public Nature of the Private Sector: Impacts of Space and Defense Programs.* St. Louis, Mo.: Department of Economics, Washington University, March, 1966. Working Paper 6531.

FEDERAL PROGRAMS AND THE GROWTH OF REGIONS AND LOCALITIES

Akens, David S. *Historical Origins of the George C. Marshall Space Flight Center,* MSFC Historical Monograph No. 1. Huntsville, Ala.: Office of Management Services, George C. Marshall Space Flight Center, December, 1960.

Bank of America N.T. & S. A. *Focus on Los Angeles–Long Beach Metropolitan Area: An Economic Study.* Los Angeles: Bank of America N.T. & S. A., 1966.

Benoit, Emile, and Boulding, Kenneth E., eds. *Disarmament and the Economy.* New York: Harper and Row, 1963.

Blodgett, Ralph H. *Analysis of the Population of the Cape Kennedy Area.* Gainesville, Fla.: Bureau of Economic and Business Research, University of Florida, 1965.

———. *Analysis of Retail Trade in the Cape Kennedy Area.* Gainesville, Fla.: Bureau of Economic and Business Research, University of Florida, 1965.

———. *Analysis of Selected Service Trades in the Cape Kennedy Area.* Gainesville, Fla.: Bureau of Economic and Business Research, University of Florida, 1965.

———. *The NASA Impact on the Economic Growth and Development of the Cape Kennedy Region.* Gainesville, Fla.: Bureau of Economic and Business Research, University of Florida, 1964.

Bolton, Roger E. *Defense Purchases and Regional Growth.* Washington, D.C.: Brookings Institution, 1966.

Boyer, P. F., and Burford, Roger L. *The Impact of the NASA Programs on the New Orleans Area Economy.* Baton Rouge, La.: College of Business Administration, Louisiana State University, 1964.

Bresse, Gerald, Klingenmeier, Russell, Cahill, Harold, Whelan, James, Church, Archer, and Whiteman, Dorothy. *The Impact of Large Installations on Nearby Areas: Accelerated Urban Growth.* Beverly Hills, Calif.: Sage Publications, 1965.

"California's Search for New Sources of Economic Growth." *California's Future Economic Growth.* Edited by Werner Z. Hirsch and Richard N. Baisden. San Francisco, Calif.: Diablo Press, 1965.

Coil, Joan B. *Personnel Management in an R & D Organization: An Analysis of the Phasedown of the Gemini Program Office at the Manned Spacecraft Center.* Houston, Tex.: Manned Spacecraft Center, NASA, 1967.

Federal Reserve Bank of Atlanta. "3,2,1—Blast Off! NASA's Impact on the District States." *Monthly Review* (April, 1965).

———. "Slowdown in Space Programs: Its Impact on the Southeast." *Monthly Review* (May, 1969).

Florida State University, Institute for Social Research, Urban Research Center. *Urban Data Abstract: Brevard, 1968.* Tallahassee, Fla.: Florida State University, 1968.

Grigg, Charles M., and Dillingham, William P. *A Study of the Impact of the NASA Space Program on the Finances of Local Government in the Cape Kennedy Area.* Tallahassee, Fla.: Institute of Governmental Research and Institute for Social Research, Florida State University, 1965.

Holman, Mary A., and Konkel, Ronald M. "Manned Space Flight and Employment in the South." *Monthly Labor Review* (March, 1968).

———, and Konkel, Ronald M. "Space Employment in Los Angeles: A Declining Role in the Aerospace Industry?" *California Management Review* (Fall, 1968).

Joiner, Helen. *The Redstone Arsenal Complex in the Pre-Missile Era.* Hunts-

ville, Alabama: Historical Division, Army Missile Command, Redstone Arsenal, June 22, 1966.
Konkel, Ronald M. "Space Employment and Economic Growth in Houston and New Orleans, 1961–1966." Thesis submitted to the Department of Economics of the Graduate School of Tulane University, 1968.
Myers, Robert A. *Planning for Impact: A Case Study of the Impact of the Space Program on Huntsville.* New York: New York University, 1967.
Nelson, John P. *East Central Florida Profile: A Compilation of Selected Facts and Figures.* Titusville, Fla.: Joint Community Impact Coordination Committee, 1964.
Peden, Guy T., and Eatherly, Billy J. *Economic Growth in the Initial Impact Area of the National Aeronautics and Space Administration's Mississippi Test Operations in Hancock County.* State College, Miss.: Bureau of Business and Economic Research, Mississippi State University, 1965.
Puffer, Frank W., and Williams, Robert M. "An Econometric Model for Southern California and Forecasts for 1967." *California Management Review* (Spring, 1967).
Stanford Research Institute. *The Structure and Dynamics of the R & D Industry, with Special Reference to NASA Programs in the Los Angeles Area.* Technical Memorandum Report. Menlo Park, Calif.: Stanford Research Institute, 1965. Research under NASA Contract NASr-49(21).
Swenson, Loyd S. "The Fertile Crescent: The South's Role in the National Space Program." *Southwestern Historical Quarterly* (January, 1968).

FEDERAL SPENDING IN RELATION TO ECONOMIC STABILITY

Burns, Arthur F. "Progress Toward Economic Stability." *American Economic Review* (March, 1960).
Fulton, Joseph F. "Employment Impact of Changing Defense Programs." *Monthly Labor Review* (May, 1964).
Hickman, Bert G. *Growth and Stability of the Postwar Economy.* Washington, D.C.: Brookings Institution, 1960.
Holman, Mary A. "The Dependence of Business Firms on Government Employment." *Quarterly Review of Economics and Business,* 8, 4 (Winter, 1968).
Lee, Maurice W. *Macroeconomics: Fluctuations, Growth, and Stability.* 4th ed. Homewood, Ill.: Richard D. Irwin, 1967.
Lewis, Wilfred. *Federal Fiscal Policy in the Postwar Recessions.* Washington, D.C.: Brookings Institution, 1962.
Miller, Thomas G., Jr. *Strategies for Survival in the Aerospace Industry.* Cambridge, Mass.: Arthur D. Little, 1964.
National Bureau of Economic Research. *Public Finances: Needs, Sources, and Utilization.* A Conference of the Universities-National Bureau Committee of Economic Research. Princeton, N.J.: Princeton University Press, 1961.
"North American Rockwell: The First Space-Age Merger." *Forbes* (July, 1968).
Oliver, Richard P. "The Employment Effect of Defense Expenditures." *Monthly Labor Review* (September, 1967).
Weidenbaum, Murray L. "Concentration and Competition in the Military Market." *Quarterly Review of Economics and Business,* 8, 1 (Spring, 1968).

FEDERAL PROCUREMENT IN RELATION TO ECONOMIC CONCENTRATION AND MONOPOLISTIC CONTROL

Adams, Walter, and Gray, Horace M. *Monopoly in America: The Government as Promoter.* New York: Macmillan, 1955.

American Bar Association. *Antitrust Developments: 1955–1968, A Supplement to the Report of the Attorney General's National Committee to Study the Antitrust Laws.* Chicago: American Bar Association, 1968.

Bachman, Otto J., et al. *Patents and the Corporation.* 2d ed. Bedford, Mass.: Patents and the Corporation, 1959.

Bain, Joe S. *Barriers to New Competition: Their Character and Consequences in Manufacturing Industries.* Cambridge, Mass.: Harvard University Press, 1956.

Baldwin, William L. *The Structure of the Defense Market 1955–1964.* Durham, N.C.: Duke University Press, 1967.

Barber, Richard J. *The Politics of Research.* Washington, D.C.: Public Affairs Press, 1966.

Bohm, Robert A. *Empirical Evidence of the Geographic and Industrial Distribution of Aerospace Expenditures.* St. Louis, Mo.: Washington University, 1966.

Boyle, Stanley E. "Competition in the Synthetic Rubber Industry." *Economic Journal* (April, 1961).

Brazzel, John M. *Regional Equity and the Geographic Distribution of Federal R & D Funds.* Columbia, Mo.: School of Business and Public Administration, University of Missouri, 1967.

———. *Federal R & D Expenditures and the Geographic Distribution of R & D Activity.* Columbia, Mo.: School of Business and Public Administration, University of Missouri, 1967.

Buehler, Vernon M. "Economic Impact of Defense Programs." U.S. Bureau of the Budget, *Statistical Reporter* (July, 1967).

Burgess, John T. *An Economic Analysis of Research and Development Procurement: The NASA Subcontract Program.* St. Louis, Mo.: Department of Economics, Washington University, 1967.

C-E-I-R. *Economic Impact Study of 1963 Defense-Related Procurement, By Industry and Geographic Area.* Arlington, Va.: C-E-I-R, 1965.

Collins, N. R., and Preston, L. E. "The Size Structure of the Largest Industrial Firms, 1909–1958." *American Economic Review,* LI (December, 1961).

Eaton, William W. "Patent Problem, Who Owns the Rights?" *Harvard Business Review* (July/August, 1967).

Hamburg, Daniel. *Essays on the Economics of Research and Development.* New York: Random House, 1966.

Harbridge House, Inc. *Government Patent Policy Study.* Vol. I: *Final Report.* Boston, Mass.: Harbridge House, 1968. Undertaken under contract with the Federal Council for Science and Technology.

———. *Government Patent Policy Study.* Vol. II: *Effect of Patent Policy on Government R & D Programs.* Boston, Mass.: Harbridge House, 1968. Undertaken under contract with the Federal Council for Science and Technology.

———. *Government Patent Policy Study.* Vol. III: *Government Promotion of Federally-Sponsored Inventions.* Boston, Mass.: Harbridge House, 1968. Undertaken under contract with the Federal Council for Science and Technology.

———. *Government Patent Policy Study.* Vol. IV: *Effect of Government Patent Policy on Commercial Utilization and Business Competition.* Boston, Mass.: Harbridge House, 1968. Undertaken under contract with the Federal Council for Science and Technology.

Hoffenberg, Marvin. *Analysis of NASA Postcard Subcontract Data.* Los Angeles: Institute of Government and Public Affairs, University of California, 1966.

Holman, Mary A. "The Economics of Government Ownership of Patented Inventions." In *Nurturing New Ideas: Legal Rights and Economic Roles.* Edited by L. J. Harris. Washington, D. C.: Bureau of National Affairs Press, 1969.

———. "Inventions from Government-Financed R & D, A New Resource?" *Land Economics* (August, 1965).

———. "Utilization of Government-Owned Patented Inventions." *Patent Trademark, and Copyright Journal,* 7, 2, 3 (Summer and Fall, 1963).

———, and Konkel, Ronald M. "Manned Space Flight Employment in the South." *Monthly Labor Review* (March, 1968).

Johnson, Harry G. "Federal Support of Basic Research: Some Economic Issues." *Basic Research and National Goals.* A Report to the Committee on Science and Astronautics by the National Academy of Sciences. Washington, D.C., March, 1965.

Kahn, Alfred E. "The Role of Patents." In *Competition, Cartels and Their Regulations.* Edited by J. P. Miller. Amsterdam, Holland: North Holland Publishing Company, 1962.

Kaysen, Carl, and Turner, Donald F. *Antitrust Policy: An Economic and Legal Analysis.* Cambridge, Mass.: Harvard University Press, 1959.

Kreps, Theodore J. *An Evaluation of Antitrust Policy: Its Relation to Economic Growth, Full Employment, and Prices.* Study Paper Number 22 of the Joint Economic Committee, 86th Cong., 2d Sess., 1960.

Kurz, Mordecai. *Research and Development, Technical Change, and the Competitive Mechanism.* Stanford, Calif.: Institute for Mathematical Studies in the Social Sciences, Stanford University, 1962. Technical Report No. 118.

Machlup, Fritz. *An Economic Review of the Patent System.* Study No. 15 of the Subcommittee on Patents, Trademarks, and Copyrights of the Senate Committee on the Judiciary. Washington, D.C.: U.S. Government Printing Office, 1968.

Markham, Jesse W. "Inventive Activity: Government Controls and the Legal Environment." *Rate and Direction of Inventive Activity: Economic and Social Factors.* Princeton, N.J.: Princeton University Press, 1962.

———, Worley, James S., and Brothers, Dwight S. "The Value of the American Patent System: An Inquiry into Possible Approaches to Its Measurements." *Patent, Trademark and Copyright Journal,* I, 1 (June, 1957).

Massel, Mark S. "Government Purchasing and Research and Development." In *Competition and Monopoly: Legal and Economic Issues*. Garden City, N.Y.: Doubleday, 1964.

Melman, Seymour, *The Impact of the Patent System on Research*. Study No. 11, U.S. Senate, Subcommittee on Patents, Trademarks, and Copyrights of the Committee on the Judiciary, 85th Cong., 2d Sess., 1958.

National Industrial Conference Board. *Antitrust in an Expanding Economy*. Transcript of a special one-day conference, May 16, 1962. New York: National Industrial Conference Board, 1962.

———. *Basic Antitrust Questions in the Mid-Sixties*. Transcript of the Fifth Conference on Antitrust Issues in Today's Economy, March 3, 1966. New York: National Industrial Conference Board, 1966.

Peck, Merton J., and Scherer, Frederic M. *The Weapons Acquisition Process: An Economic Analysis*. Cambridge, Mass.: Graduate School of Business Administration, Harvard University, 1962.

Riefler, Roger E., and Downing, Paul B. "Regional Effect of Defense Effort on Employment." *Monthly Labor Review* (July, 1968).

Rose, Sanford. "Bigness Is a Numbers Game: Is Economic Power in the United States Getting More Concentrated?" *Fortune* (November, 1969).

Rossman, Joseph, and Sanders, Barkev S. "The Patent Utilization Study." *Patent, Trademark, and Copyright Journal*, I, 1 (June, 1957).

Scherer, Frederic M. *The Weapons Acquisition Process: Economic Incentives*. Cambridge, Mass.: Graduate School of Business Administration, Harvard University, 1964.

Shapero, Albert, Howell, Richard P., and Tombaugh, James R. *An Exploratory Study of the Structure and Dynamics of the R & D Industry: Denver, Tucson, and Orlando*. Menlo Park, Calif.: Stanford Research Institute, 1964.

———, Howell, Richard P., and Tombaugh, James R. *The Structure and Dynamics of the Defense R & D Industry: The Los Angeles and Boston Complexes*. Menlo Park, Calif.: Stanford Research Institute, 1965.

Shephard, William G. "Trends of Concentration in American Manufacturing Industries, 1947–1958." *Review of Economics and Statistics* (May, 1964).

Singer, Eugene M. *Antitrust Economics: Selected Legal Cases and Economic Models*. Englewood Cliffs, N.J.: Prentice-Hall, 1968.

Small, Sylvia S. "Concentration of Industrial Employment." *Monthly Labor Review*, 87, 9 (September, 1964).

Solo, Robert A. "Patent Policy for Government-Sponsored R & D." *Patent, Trademark, and Copyright Journal*, 10, 2 (Summer, 1966).

Stekler, Herman O. *The Structure and Performance of the Aerospace Industry*. Berkeley, Calif.: University of California Press, 1965.

Stewart, Charles T. "Regional Allocation of Public Funds for Economic Development." *Land Economics*, XLIII, 4 (November, 1967).

———. *A Study of Changing Regional Patterns of Research and Development and Science-Based Technology in Influencing Regional Development*. Washington, D.C.: Department of Economics, The George Washington University, 1964.

Swenson, Norman P. *Factors Influencing the Geographic Distribution of*

Procurement Awards. St. Louis, Mo.: Washington University, 1966. Working Paper 6530.

U.S. Attorney General. *Report of the Attorney General's National Committee to Study the Antitrust Laws*. Stanley N. Barnes and S. Chesterfield Oppenheim, Co-chairmen. Washington, D.C.: U.S. Government Printing Office, 1967.

U.S. Congress. House. Committee on Government Research. *Investigation of Expenditures for Federally Conducted Research and Development Programs*. 3 Parts. 88th Cong., 1st and 2d Sess., 1963 and 1964.

U.S. Congress. Senate. Committee on Government Operations. Subcommittee on Government Research. *Equitable Distribution of R & D Funds by Government Agencies*. 3 Parts. 90th Cong., 1st Sess., May, 1967.

——. Committee on the Judiciary. Subcommittee on Antitrust and Monopoly. *Concentration Ratios in Manufacturing Industries, 1963*. Report prepared by the U.S. Bureau of the Census. 89th Cong., 2d Sess., 1966.

——. Committee on the Judiciary. Subcommittee on Antitrust and Monopoly. *Economic Concentration*. Part I: *Overall and Conglomerate Aspects*. Hearings before the Subcommittee on Antitrust and Monopoly of the Committee on the Judiciary, 88th Cong., 2d Sess., 1964.

——. *Economic Concentration*. Part II: *Mergers and Other Factors Affecting Industry Concentration*. Hearings before the Subcommittee on Antitrust and Monopoly of the Committee on the Judiciary, 89th Cong., 1st Sess., 1965.

——. *Economic Concentration*. Part III: *Concentration, Invention, and Innovation*. Hearings before the Subcommittee on Antitrust and Monopoly of the Committee on the Judiciary, 89th Cong., 1st Sess., 1965.

——. *Economic Concentration*. Part IV: *Concentration and Efficiency*. Hearings before the Subcommittee on Antitrust and Monopoly of the Committee on the Judiciary, 89th Cong., 1st Sess., 1965.

——. *Economic Concentration*. Part V: *Concentration and Divisional Reporting*. Hearings before the Subcommittee on Antitrust and Monopoly of the Committee on the Judiciary, 90th Cong., 1st Sess., 1967.

——. *Economic Concentration*. Part V(a): *The Fortune Directory of the 500 Largest United States Industrial Corporations, 1954–1965*. Hearings before the Subcommittee on Antitrust and Monopoly of the Committee on the Judiciary, 90th Cong., 1st Sess., 1967.

——. Committee on the Judiciary. Subcommittee on Patents, Trademarks, and Copyrights. *Government Patent Policy*. Hearings before the Subcommittee on Patents, Trademarks and Copyrights of the Committee on the Judiciary, Part II, 87th Cong., 2d Sess., 1961.

——. *Patent Practices of the Federal Aviation Agency*, 86th Cong., 2d Sess., 1961.

——. *National Patent Policy*. Hearings before the Subcommittee on Patents, Trademarks, and Copyrights of the Committee on the Judiciary, 87th Cong., 1st Sess., 1961.

——. Select Committee on Small Business. *Economic Aspects of Government Patent Policies*. Hearings before the Select Committee on Small Business, 88th Cong., 1st Sess., March 7–14, 1963.

——. *The Role and Effect of Technology in the Nation's Economy: A Review of the Effect of Government Research and Development on Economic Growth.* Hearings before a subcommittee of the Select Committee on Small Business, 6 Parts, 88th Cong., 1st Sess., 1963.

U.S. Federal Trade Commission. *Report on Industrial Concentration and Product Diversification in the 1,000 Largest Manufacturing Companies: 1950.* Washington, D.C.: U.S. Government Printing Office, 1957.

Watson, Donald S., Bright, Harold F., and Burns, Arthur E. "Federal Patent Policies in Contracts for Research and Development." *Patent, Trademark, and Copyright Journal,* 4, 4 (Winter, 1960).

______, and Holman, Mary A. "Concentration of Patents from Government-Financed Research in Industry." *Review of Economics and Statistics,* XLIX, 3 (August, 1967).

______, and Holman, Mary A. "The Concentration of Patent Ownership in Corporations." *Journal of Industrial Economics* (Spring, 1970).

______, and Holman, Mary A. "The Federal Government's Propensity to Patent." *Patent, Trademark, and Copyright Journal,* 10, 1 (Spring, 1966).

Weidenbaum, Murray L. *Competition in High Technology Government Markets.* St. Louis, Mo.: Washington University, 1967.

______. "Concentration and Competition in the Military Market." *Quarterly Review of Economics and Business,* 8, 1 (Spring, 1967).

______. *Federal Financing of Research and Development and the Regional Distribution of Income.* St. Louis, Mo.: Washington University, 1966. Working Paper 6526.

______. *Shifting the Composition of Government Spending: Implications for the Regional Distribution of Income.* St. Louis, Mo.: Washington University, 1965. Working Paper 6520.

Woodfill, Douglas. *Competition, Efficiency, and Military R & D Contracts.* Berkeley, Calif.: University of California, 1965. Paper No. 26.

FEDERAL PROCUREMENT FOR RESEARCH AND DEVELOPMENT IN RELATION TO ECONOMIC GROWTH

Adelman, Irma. *Theories of Economic Growth and Development.* Stanford, Calif.: Stanford University Press, 1961.

Ames, Edward. "Research, Invention, Development, and Innovation." *American Economic Review,* LI, 3 (June, 1961).

Baird, John, and Holley, Virginia F. *Some Major Impacts of the National Space Program: Impacts upon Aviation and Aeronautics.* Vol. II. Menlo Park, Calif.: Stanford Research Institute, 1968.

Bauer, Raymond A. *A Methodological Essay on the Impact of Technology: Second-Order Consequences.* Cambridge, Mass.: M.I.T. Press, 1969.

Bayce, Arthur E. *Some Major Impacts of the National Space Program: Impacts of New Materials Technology.* Vol. IV. Menlo Park, Calif.: Stanford Research Institute, 1968.

Beckmann, M. J., and Sato, Ryuzo. "Aggregate Production Functions and Types of Technical Progress: A Statistical Analysis." *American Economic Review,* LIX, 1 (March, 1969).

Boettinger, Henry M. "Big Gap in Economic Theory: Is Technical Innovation an Embarrassment or Opportunity for Today's Economists?" *Harvard Business Review*, 45, 4 (July/August, 1967).

Bright, James R., ed. *Technological Forecasting for Industry and Government: Methods and Applications*. Englewood Cliffs, N.J.: Prentice-Hall, 1968.

Brooks, Harvey. "National Science Policy and Technology Transfer." *Proceedings of a Conference on Technology Transfer and Innovation*. Washington, D.C.: U.S. Government Printing Office, 1967.

Brown, Murray. *On the Theory and Measurement of Technological Change*. London: Cambridge University Press, 1966.

Browne, Theodore D., et al. *Project for the Analysis of Technology Transfer—The Initial Year*. Denver, Colo.: Industrial Economics Division, Denver Research Institute, 1968.

Cochran, David. *The Benefits of the National Space Program and Their Appreciation and Understanding by the American Public*. 2 vols. Menlo Park, Calif.: Stanford Research Institute, 1967.

Coddington, Dean C., et al. *Project for the Analysis of Technology Transfer: First Annual Report*. Denver, Colo.: Industrial Economics Division, Denver Research Institute, University of Denver, November, 1968.

Denison, Edward F. *The Sources of Economic Growth in the United States and the Alternatives Before Us*. Supplementary Paper No. 13 of the Committee for Economic Development. New York: Committee for Economic Development, 1962.

______. *Why Growth Rates Differ: Postwar Experience of Nine Western Countries*. Washington, D.C.: Brookings Institution, 1967.

Denver Research Institute, Industrial Economics Division. *Technology Transfer—A Selected Bibliography*. Denver, Colo.: Denver Research Institute, 1968.

Doctors, Samuel I. *The Role of Federal Agencies in Technology Transfer*. Cambridge, Mass.: M.I.T. Press, 1969.

Eisner, Robert. "Investment and the Frustrations of Econometricians." *American Economic Review*, LIX, 2 (May, 1969).

Enos, John L. "Inventions and Innovation in the Petroleum Refining Industry." *The Rate and Distribution of Incentive Activity*. Princeton, N.J.: Princeton University Press, 1962.

Gamarra, Nancy T. *Selection of Statements on Erroneous Predictions and Negative Comments Concerning Exploration, Territorial Expansion, and Scientific and Technological Development*. Prepared at the request of the Senate Committee on Aeronautical and Space Sciences. Washington, D.C.: Library of Congress, Legislative Reference Service, 1967.

Gilmore, John S., et al. *The Channels of Technology Acquisition in Commercial Firms and the NASA Dissemination Program*. Denver, Colo.: Denver Research Institute, 1967.

______, and Coddington, Dean C. *Defense Industry Diversification*. U.S. Arms Control and Disarmament Agency Publication No. 30. Washington, D.C.: U.S. Government Printing Office, 1966.

______, and Coddington, Dean C. "Diversification Guides for Defense Firms." *Harvard Business Review* (May/June, 1966).

Glasser, Peter E., et al. *Space Technology Transfer and Developing Nations.* Cambridge, Mass.: Arthur D. Little, 1968.

Hough, Roger W., et al. *Some Major Impacts of the National Space Program: Identification of New Occupations,* Vol. I. Menlo Park, Calif.: Stanford Research Institute, 1968.

Hutchings, Edward, and Hutchings, Elizabeth, eds. *Scientific Progress and Human Values.* Proceedings of the Conference Celebrating the 75th Anniversary of the California Institute of Technology. New York: Elsevier Publishing Co., 1967.

Isenson, Raymond S. "Technological Forecasting: Lessons from Project Hindsight." In *Technological Forecasting for Industry and Government.* Edited by James R. Bright. Englewood Cliffs, N.J.: Prentice-Hall, 1968.

Kendrick, John W. *Productivity Trends in the United States.* Princeton, N.J.: Princeton University Press for the National Bureau of Economic Research, 1961.

Levitt, I. M. *Some Major Impacts of the National Space Program: Astronomy as an Example of Scientific Impacts.* Vol. III. Menlo Park, Calif.: Stanford Research Institute, 1968.

Machlup, Fritz. *The Production and Distribution of Knowledge in the United States.* Princeton, N.J.: Princeton University Press, 1962.

Mansfield, Edwin. *The Economics of Technological Change.* New York: W. W. Norton, 1968.

———. *Industrial Research and Technological Innovation: An Econometric Analysis.* New York: W. W. Norton, 1968.

———. "Technological Change: Measurement, Determinants, and Diffusion." *The Employment Impact of Technological Change.* Appendix Vol. II. Report of the National Commission on Technology, Automation, and Economic Progress. Washington, D.C.: U.S. Government Printing Office, 1966.

McLaughlin, Curtis P., Rosenbloom, Richard S., and Wolek, Francis W. *Technology Transfer and the Flow of Technical Information in a Large Industrial Corporation.* Cambridge, Mass.: Harvard University Press, 1965.

Midwest Research Institute. *Medical Applications of NASA-Developed Science and Technology—Final Report.* Kansas City, Mo.: Midwest Research Institute, 1966.

Miller, Thomas G., Jr. *Strategies for Survival in the Aerospace Industry.* Cambridge, Mass.: Arthur D. Little, 1964.

Milton, Helen S., and Johnson, Ellis A. *Sponsorship of Research: A Survey of the Scientific Literature, 1920–1960.* Bethesda, Md.: Operations Research Office, Johns Hopkins University, September, 1961.

Mishan, Ezra J. *The Costs of Economic Growth.* New York: Frederick A. Praeger, 1967.

Mullis, Charles W., ed. *Report of the National Conference on Technology Utilization and Economic Growth.* Conference sponsored by the Aerospace Research Applications Center, Indiana University; the National Aeronautics and Space Administration; and the Office of State Technical Services, U.S. Department of Commerce. Bloomington, Ind.: Aerospace Research Applications Center, 1967.

Myers, Sumner. *Technology Transfer and Industrial Innovation*. Washington, D.C.: National Planning Association, 1967.

National Bureau of Economic Research and Social Science Research Council. *The Rate and Direction of Inventive Activity: Economic and Social Factors*. Princeton, N.J.: Princeton University Press, 1962.

National Planning Association and National Science Foundation. *Proceedings of a Conference on Technology Transfer and Innovation, May 15–17, 1966*. NSF Report 67–5. Washington, D.C.: U.S. Government Printing Office, 1967.

Nelson, Richard R., Peck, Merton J., and Kalachek, Edward D. *Technology, Economic Growth and Public Policy*. Washington, D.C.: Brookings Institution, 1967.

Organization for Economic Co-operation and Development. *Social Change and Economic Growth*. Paris: OECD, 1966.

Prehoda, R. W. *Some Major Impacts of the National Space Program: Public Health, Medicine, and Biological Research*. Vol. VI. Menlo Park, Calif.: Stanford Research Institute, 1968.

Price, Derek J. de Solla. *Science Since Babylon*. New Haven, Conn.: Yale University Press, 1961.

Rosenbloom, Richard S. *Technology Transfer—Process and Policy: An Analysis of the Utilization of Technological By-products of Military and Space R & D*, Special Report No. 62. Washington, D.C.: National Planning Association, 1965.

Schmookler, Jacob. *Invention and Economic Growth*. Cambridge, Mass.: Harvard University Press, 1966.

______. "Technological Change and Economic Theory." *American Economic Review*, LV, 2 (May, 1965).

Shockley, William. "Individual Variations of Productivity in Research Laboratories." *Science*, CXXII (November 4, 1955).

Siegel, Irving H. "Technological Change and Long-Run Forecasting." *Journal of Business*, XXVI (July, 1953).

Suits, C. Guy. "Organization of Technical Activities at the General Electric Company." *Organizing the Research Function for Profit*. Research and Development Series No. 1. New York: American Management Association, 1956.

Sutton, Sterling Scott. "Scientific Output: Methods of Measurement." Unpublished thesis submitted to The George Washington University, Washington, D.C., June, 1964.

U.S. Congress. House. Committee on Science and Astronautics. Subcommittee on NASA Oversight. *Staff Study: The National Space Program—Its Values and Benefits*, 90th Cong., 1st Sess., 1967.

U.S. Congress. Senate. Select Committee on Small Business. Subcommittee on Science and Technology. *Report: Policy Planning for Technology Transfer*, 90th Cong., 1st Sess., 1967.

______. Select Committee on Small Business. Subcommittee on Science and Technology. *Report: The Prospects for Technology Transfer*. 90th Cong., 2d Sess., 1968.

U.S. Department of Commerce. Bureau of the Census. *Long-Term Economic*

Growth: 1860–1965. Washington, D.C.: U.S. Government Printing Office, 1966.

———. Office of Business Economics. "Some Major Issues in Productivity Analysis." *Survey of Current Business,* 49, 5, Part II (May, 1969).

U.S. Department of Labor. *Technological Trends in 36 Major American Industries*. Washington, D.C.: Bureau of Labor Statistics, Office of Productivity and Technological Development, 1965.

U.S. National Aeronautics and Space Administration. Office of Technology Utilization. *Summary Descriptive Information on a Random Selection of Transfer Examples*. Washington, D.C.: National Aeronautics and Space Administration, 1969.

U.S. National Commission on Technology, Automation, and Economic Progress. *Technology and the American Economy*. Washington, D.C.: U.S. Government Printing Office, 1966.

U.S. National Science Foundation. *Successful Industrial Innovations*. Washington, D.C.: U.S. Government Printing Office, 1969.

Wall Street Journal. "Space By-Products: Moon Shot Technology Spawns Ideas for Home, Business, and Medicine," July 9, 1969.

Watson, Donald S., and Holman, Mary A. *An Evaluation of the Patent Policies of the National Aeronautics and Space Administration*. Report published by the U.S. House Committee on Science and Astronautics, 89th Cong., 2d Sess., 1966.

Wells, J. G., and Waterman, R. H. "Space Technology: Pay-Off from Spin-Off." *Harvard Business Review* (July/August, 1964).

Index